T0189639

Use R!

More information about this series at http://www.springer.com/series/6991

Daniel Borcard • François Gillet • Pierre Legendre

Numerical Ecology with R

Second Edition

Daniel Borcard
Université de Montréal
Département de sciences biologiques
Montréal, Québec, Canada H3C 3J7

François Gillet
Université Bourgogne Franche-Comté
UMR Chrono-environnement
Besançon, France

Pierre Legendre
Université de Montréal
Département de sciences biologiques
Montréal, Québec, Canada H3C 3J7

ISSN 2197-5736 ISSN 2197-5744 (electronic)
Use R!
ISBN 978-3-319-71403-5 ISBN 978-3-319-71404-2 (eBook)
https://doi.org/10.1007/978-3-319-71404-2

Library of Congress Control Number: 2017961342

Printed on acid-free paper

This Springer imprint is published by the registered company Springer International Publishing AG part of Springer Nature.
The registered company address is: Gewerbestrasse 11, 6330 Cham, Switzerland

Preface

Ecology is sexy. Teaching ecology is therefore the art of presenting a fascinating topic to well-predisposed audiences. It is not easy: the complexities of modern ecological science go well beyond the introductory chapters taught in high schools or the marvellous movies about ecosystems presented on TV. But well-predisposed audiences are ready to make the effort. *Numerical* ecology is another story. For some unclear reasons, a majority of ecology-oriented people are strangely reluctant when it comes to quantifying nature and using mathematical tools to help understand it. As if nature was inherently non-mathematical, which it is certainly not: mathematics is the common language of all sciences. Teachers of biostatistics and numerical ecology thus have to overcome this reluctance: before even beginning to teach the subject itself, they must convince their audience of the interest and necessity of it.

During many decades ecologists, be they students or researchers (in the academic, private or government spheres), used to plan their research and collect data with few, if any, statistical consideration, and then entrusted the "statistical" analyses of their results to a person hired especially for that purpose. That person may well have been a competent statistician, and indeed in many cases the progressive integration of statistics into the whole process of ecological research was triggered by such people. In other cases, however, the end product was a large amount of data summarized using a handful of basic statistics and tests of significance that were far from revealing all the richness of the structures hidden in the data tables. The separation of the ecological and statistical worlds presented many problems. The most important were that the ecologists were unaware of the array of methods available at the time, and the statisticians were unaware of the ecological hypotheses to be tested and the specific requirements of ecological data (the double-zero problem is a good example). Apart from preventing the data to be exploited properly, this double unawareness prevented the development of methods specifically tailored to ecological problems.

The answer to this situation is to form mathematically inclined ecologists. Fortunately, more and more such people have appeared during the recent decades. The result of their work is a huge development of statistical ecology, the availability of several excellent textbooks, and the increasing awareness of the responsibility of ecologists with regard to the proper design and analysis of their research. This awareness makes the task easier for teachers as well.

Until the first years of this millennium, however, a critical ingredient was still missing for the teaching to be efficient and for the practice of statistics to become generalized among ecologists: a set of standard packages available to everyone, everywhere. A biostatistics or numerical ecology course means nothing without practical exercises. A course linked to commercial software is much better, but it is bound to restrict future applications if the researcher moves and loses access to the software that he or she knows. Furthermore, commercial packages are in most cases written for larger audiences than the community of ecologists and they may not include all the functions required for analysing ecological data. The **R** language resolved that issue, thanks to the dedication of the many researchers who created and freely contributed extensive, well-designed, and well-documented packages. Now the teacher no longer has to say: "this is the way PCA works... on paper;" she or he can say instead: "this is the way PCA works, now I will show you on-screen how to run one, and in a few minutes you will be able to run your own, and do it anywhere in the world on your own data!"

Another fundamental property of the **R** language is that it is meant as a self-learning environment. A book on **R** is therefore bound to follow that philosophy, and must provide the support necessary for anyone wishing to explore the subject by himself or herself. This book has been written to provide a bridge between the theory and practice of numerical ecology, that anyone can cross. Our dearest hope is that it will make many happy teachers and happy ecologists.

Since they are living entities, both the field of numerical ecology and the **R** language evolve. As a result, much has happened in both fields since the publication of the first edition of *Numerical Ecology with R* in 2011. Therefore, it was time not only to update the code provided in the first edition, but also to present new methods, provide more insight into existing ones, offer more examples and a wider array of applications of the major methods. We also took the opportunity to present the code in a more attractive way, generated by R Markdown in RStudio®, with different colours for functions, objects, arguments and comments.

Our dearest hope is that all this will make many more happy teachers and happy ecologists.

Montréal, QC, Canada — Daniel Borcard
Besançon, France — François Gillet
Montréal, QC, Canada — Pierre Legendre

Contents

About the Authors

Daniel Borcard is lecturer of Biostatistics and Ecology and researcher in Numerical Ecology at Université de Montréal, Québec, Canada. His research interests include Numerical Ecology, Ecology of communities, and Soil Ecology/Zoology.

François Gillet is professor of Community Ecology and Ecological Modelling at Université Bourgogne Franche-Comté, Besançon, France, and visiting professor at École Polytechnique Fédérale de Lausanne, Switzerland. His research deals with the structure, diversity, ecology and dynamics of plant communities.

Pierre Legendre is professor of Quantitative Biology and Ecology at Université de Montréal, fellow of the Royal Society of Canada, and Web of Science Highly Cited Researcher in Environment/Ecology. He is the founder of the field of numerical ecology.

Supplementary Material

All the necessary data files, the scripts used in the chapters, as well as the **R** functions and packages that are not available through the CRAN web site, can be downloaded from our web page http://adn.biol.umontreal.ca/~numericalecology/numecolR/.

Chapter 1
Introduction

1.1 Why Numerical Ecology?

Although multivariate analysis of ecological data already existed and was being actively developed in the 1960's, it really flourished in the years 1970 and later. Many textbooks were published during these years, among them the seminal *Écologie numérique* (Legendre and Legendre 1979), and its English translation *Numerical Ecology* (Legendre and Legendre 1983). The authors of these books unified under one single roof a very wide array of statistical and other numerical techniques and presented them in a comprehensive way, not only to help researchers understand the available methods of statistical analysis, but also to explain how to choose and apply them in an ordered, logical way to reach their research goals. Mathematical explanations were not absent from these books, and provided a precious insider look into the various techniques, which was appealing to readers wishing to go beyond the simple user level.

Since then, numerical ecology has become ubiquitous. Every serious researcher or practitioner has become aware of the tremendous interest of exploiting painfully acquired data as efficiently as possible. Other manuals have been published (e.g. Orlóci and Kenkel 1985; Jongman et al. 1995; McCune and Grace 2002; McGarigal et al. 2000; Zuur et al. 2007; Greenacre and Primicerio 2013; Wildi 2013). A second English edition of *Numerical Ecology* was published in 1998, followed by a third in 2012, broadening the perspective and introducing numerous methods that were unavailable at the times of the previous editions. The progress continues. In this book we present some of the developments that we consider most important, albeit in a more user-oriented way than in the abovementioned manuals, using the **R** language. For the most recent methods, we provide explanations at a more fundamental level when we consider it appropriate and helpful.

Not all existing methods of data analysis are addressed in this book, of course. Apart from the most widely used and fruitful methods, our choices are based on our own experience as quantitative community ecologists. However, small sections

D. Borcard et al., *Numerical Ecology with R*, Use R!,
https://doi.org/10.1007/978-3-319-71404-2_1

have sometimes been added to briefly describe other avenues than the main ones, without going into details.

1.2 Why R?

The **R** language has experienced such a tremendous development and reached such a wide array of users during the recent years that a justification of its application to numerical ecology is not required. Development also means that more and more domains of numerical ecology are now covered, up to the point where, computationally speaking, some of the most recent methods are actually only available through **R** packages.

This book is not intended as a primer in **R**, however. To find that kind of support, readers should consult the CRAN web page (http://www.R-project.org). The link to *Manuals* provides many free electronic documents, and the link to *Books* many references. Readers are expected to have a minimal working knowledge of the basics of the language, e.g. formatting data and importing them into **R**, awareness of the main classes of objects handled in this environment (vectors, matrices, data frames and factors), as well as the basic syntax necessary to manipulate, create and otherwise use objects within **R**. Nevertheless, Chap. 2 starts at an elementary level as far as multivariate objects are concerned, since these are the main targets of most analyses addressed throughout the book, while not necessarily being most familiar to many users.

The book is by far not exhaustive as to the array of functions devoted to any of the methods. Usually we present one or several variants, but often other functions serving similar purposes are available in **R**. Centring the book on a small number of well-integrated packages and adding some functions of our own when necessary helps users up the learning curve while keeping the amount of package-level idiosyncrasies at a reasonable level. Our choices should not suggest that other existing packages are inferior to the ones used in the book.

1.3 Readership and Structure of the Book

The intended audience of this book is the researchers, practitioners, graduate students and teachers who already have a background in general and multivariate statistics and wish to apply their knowledge to their data using the **R** language, as well as people willing to accompany their learning of the discipline with practical applications. Although an important part of this book follows the organization and symbolism of Legendre and Legendre (2012) and many references to that book are made herein, readers may draw their training from other sources without problem.

Combining an application-oriented book such as this one with a detailed exposé of the methods used in numerical ecology would have led to an impossibly long and cumbersome opus. However, all chapters start with a short introduction summarizing its subject matter, to ensure that readers are aware of the scope of the chapter and can appreciate the point of view from which the methods are addressed. Depending on the amount of documentation already existing in statistical textbooks, some introductions are longer than others.

Overall, the book guides readers through an applied exploration of the major methods of multivariate data analysis, as seen through the eye of an ecologist. Starting with some exploratory approaches (Chap. 2), it proceeds logically with the construction of the key building blocks of most techniques, i.e. association measures and matrices (Chap. 3), and then submits example data to three families of approaches: clustering (Chap. 4), ordination and canonical ordination (Chaps. 5 and 6), spatial analysis (Chap. 7), and finally community diversity (Chap. 8). The methods' aims thus range from descriptive to explanatory and to predictive and encompass a wide variety of approaches that should provide readers with an extensive toolbox that can address a wide palette of questions arising in contemporary multivariate ecological analysis.

1.4 How to Use This Book

The book is meant as a companion when working at the computer. The authors pictured a reader studying a chapter by reading the text and simultaneously executing the code. To fully understand the various methods, it is preferable to go through the chapters sequentially, since each builds upon the previous ones. At the beginning of each chapter, an empty **R** console is assumed to be open. All the necessary data files, the scripts used in the chapters, as well as the **R** functions and packages that are not available through the CRAN web site, can be downloaded from our web page (http://adn.biol.umontreal.ca/~numericalecology/numecolR/). Some of the home-made functions duplicate existing ones, providing alternative solutions (for instance different or expanded graphical outputs), while others have been written to streamline complex sequences of operations.

Although the code provided can be run in one single copy-and-paste shot within each chapter (with some rare exceptions for interactive functions), the best procedure is to proceed through the code slowly and explore each set of commands carefully. Although the use and meaning of some arguments is explained within the code or in the text, readers are warmly invited to use and abuse of the **R** documentation files (function name following a question mark) to learn about and explore the various options available. Our aim is not to describe all options of all functions, which would be an impossible and useless task. We are confident that an avid user, willing to go beyond the provided examples, will be kept busy for months exploring the options that he or she deems the most interesting.

Within each chapter, after the introduction, readers are invited to import the data as well as the **R** packages necessary for the exercises of the whole chapter. The **R** code used in each chapter is self-contained, i.e., it can usually be run in one step even if some analyses are based on results produced in previous chapters. If such objects are needed, they are recomputed at the beginning of the chapter.

In everyday use, one generally does not produce an **R** object for every single operation, nor does one create and name a new graphical window for every plot. We do that in the book to provide readers with all the entities necessary to backtrack the procedures, compare results and explore variants. Therefore, after having run most of the code in a chapter, if one decides to explore another path using some intermediate result, the corresponding object will be available without need to re-compute it. This is particularly handy for results of computer-intensive methods (like some based on large numbers of random permutations).

In the code sections of the book, all calls to graphical windows have been deleted for brevity. They are found in the electronic code scripts, however. Furthermore, the book shows several, but not all, graphical outputs for reference.

Sometimes, readers are made aware of some special features of the code or of tricks used to obtain particular results, by means of hint boxes located at the bottom of code sections.

Although many methods are applied to the example data, ecological interpretation is not provided in all cases. Sometimes questions are left open to readers, as an incentive to verify if she or he has correctly understood the method, and hence its application and the numerical or graphical outputs.

Lastly, for some methods, programming-oriented readers are invited to write their own code. These incentives are placed in boxes called "code-it-yourself corners". When examples are provided, they are meant for pedagogical purposes and do not pretend at computational efficiency. The aim of these boxes is to help interested readers code in **R** the matrix algebra equations presented in Legendre and Legendre (2012) and obtain the main outputs that ready-made packages provide. The whole idea is of course to reach the deepest possible understanding of the mathematical working of some key methods.

1.5 The Data Sets

Apart from rare cases where *ad hoc* fictitious data are built for special purposes, the applications rely on two main data sets that are readily available in **R**. However, data provided in R packages can be modified over the years. Therefore we prefer to provide them also in the electronic material accompanying this book, because this ensures that the results obtained by the readers will be exactly the same as those presented in the book. The two data sets are briefly presented here. The first (Doubs) data set is explored in more detail in Chap. 2, and readers are encouraged to apply the same exploratory methods to the second one.

1.5.1 The Doubs Fish Data

In an important doctoral thesis, Verneaux (1973; see also Verneaux et al. 2003) proposed to use fish species to characterize ecological zones along European rivers and streams. He showed that fish communities were good biological indicators of these water bodies. Starting from the river source, Verneaux proposed a typology in four zones, and he named each one after a characteristic species: the trout zone (from the brown trout *Salmo trutta fario*), the grayling zone (from *Thymallus thymallus*), the barbel zone (from *Barbus barbus*) and the bream zone (from the common bream *Abramis brama*). The two upper zones are considered as the "Salmonid region" and the two lowermost ones form the "Cyprinid region". The corresponding ecological conditions, with much variation among rivers, range from relatively pristine, well oxygenated and oligotrophic to eutrophic and oxygen-deprived waters.

The Doubs data set that is used in the present book (**`Doubs.RData`**) consists of five data frames, three of them containing a portion of the data used by Verneaux for his studies. These data have been collected at 30 sites along the Doubs River, which runs near the France-Switzerland border in the Jura Mountains. The first matrix contains coded abundances of 27 fish species, the second matrix contains 11 environmental variables related to the hydrology, geomorphology and chemistry of the river, and the third matrix contains the geographical coordinates (Cartesian, X and Y in km) of the sites. The Cartesian coordinates have been obtained as follows. One of us (FG) returned to Verneaux's thesis to obtain more accurate positions of the sampling sites than available in existing databases. These new locations were coded in GPS angular coordinates (WGS84), and transformed into Cartesian coordinates by using function **`geoXY()`** of package **`SoDA`** in **R**. Earlier versions of these data have already served as test cases in the development of numerical techniques (Chessel et al. 1987). Two additional data frames are provided in the present book's material: **`latlong`** contains the latitudes and longitudes of the sampling sites, and **`fishtraits`** contains four quantitative variables and six binary variables describing the diet. Values are taken from various sources, mainly fishbase.org (Froese and Pauly 2017), checked and adapted to the regional context by François Degiorgi.[1]

Working with the original environmental data available in Verneaux's thesis, one of us (FG) made some corrections to the data available in **R** and restored the variables to their original units, which are presented in Table 1.1.

Since the fish species of this data set have well-defined ecological requirements that have been often exploited in ecological and applied environmental studies, it is useful to provide their full Latin and English names. This is done here in Table 1.2.

---

[1]Many thanks to Dr. Degiorgi for this precious work.

Table 1.1 Environmental variables of the Doubs data set used in this book and their units

Variable	Code	Units
Distance from the source	`dfs`	km
Elevation	`ele`	m a.s.l.
Slope	`slo`	‰
Mean minimum discharge	`dis`	$m^3 \cdot s^{-1}$
pH of water	`pH`	–
Hardness (Ca concentration)	`har`	$mg \cdot L^{-1}$
Phosphate concentration	`pho`	$mg \cdot L^{-1}$
Nitrate concentration	`nit`	$mg \cdot L^{-1}$
Ammonium concentration	`amm`	$mg \cdot L^{-1}$
Dissolved oxygen	`oxy`	$mg \cdot L^{-1}$
Biological oxygen demand	`bod`	$mg \cdot L^{-1}$

Table 1.2 Labels, Latin names, family and English names of the fish species of the Doubs dataset

Label	Latin name	Family	English name
Cogo	*Cottus gobio*	*Cottidae*	Bullhead
Satr	*Salmo trutta fario*	*Salmonidae*	Brown trout
Phph	*Phoxinus phoxinus*	*Cyprinidae*	Eurasian minnow
Babl	*Barbatula barbatula*	*Nemacheilidae*	Stone loach
Thth	*Thymallus thymallus*	*Salmonidae*	Grayling
Teso	*Telestes souffia*	*Cyprinidae*	Vairone
Chna	*Chondrostoma nasus*	*Cyprinidae*	Common nase
Pato	*Parachondrostoma toxostoma*	*Cyprinidae*	South-west European nase
Lele	*Leuciscus leuciscus*	*Cyprinidae*	Common dace
Sqce	*Squalius cephalus*	*Cyprinidae*	European chub
Baba	*Barbus barbus*	*Cyprinidae*	Barbel
Albi	*Alburnoides bipunctatus*	*Cyprinidae*	Schneider
Gogo	*Gobio gobio*	*Cyprinidae*	Gudgeon
Eslu	*Esox lucius*	*Esocidae*	Northern pike
Pefl	*Perca fluviatilis*	*Percidae*	European perch
Rham	*Rhodeus amarus*	*Cyprinidae*	European bitterling
Legi	*Lepomis gibbosus*	*Centrarchidae*	Pumpkinseed
Scer	*Scardinius erythrophtalmus*	*Cyprinidae*	Rudd
Cyca	*Cyprinus carpio*	*Cyprinidae*	Common carp
Titi	*Tinca tinca*	*Cyprinidae*	Tench
Abbr	*Abramis brama*	*Cyprinidae*	Freshwater bream
Icme	*Ameiurus melas*	*Ictaluridae*	Black bullhead
Gyce	*Gymnocephalus cernua*	*Percidae*	Ruffe
Ruru	*Rutilus rutilus*	*Cyprinidae*	Roach
Blbj	*Blicca bjoerkna*	*Cyprinidae*	White bream
Alal	*Alburnus alburnus*	*Cyprinidae*	Bleak
Anan	*Anguilla anguilla*	*Anguillidae*	European eel

Latin names after fishbase.org (Froese and Pauly 2017)

Table 1.3 Environmental variables of the oribatid mite data set used in this book and their units

Variable	Code	Units
Substrate density (dry matter)	`SubsDens`	$g \cdot dm^{-3}$
Water content	`WatrCont`	$g \cdot dm^{-3}$
Substrate	`Substrate`	7 unordered classes
Shrubs	`Shrub`	3 ordered classes
Microtopography	`Topo`	Blanket – Hummock

1.5.2 The Oribatid Mite Data

Oribatid mites (Acari: Oribatida) are a very diversified group of small (0.2 to 1.2 mm) soil-dwelling, mostly microphytophagous and detritivorous arthropods. A well-aerated soil or a complex substrate like *Sphagnum* mosses present in bogs and wet forests can harbour up to several hundred thousand (10^5) individuals per square metre. Local assemblages are sometimes composed of over a hundred species, including many rare ones. This diversity makes oribatid mites an interesting target group to study community-environment relationships at very local scales.

The example data set is composed of 70 cores of mostly *Sphagnum* mosses collected in a peat moss mat bordering a small lake (Lac Geai) on the territory of the *Station de biologie des Laurentides* of Université de Montréal, Québec, Canada in June 1989. The data were collected in order to test various ecological hypotheses about the relationships between living communities and their environment when the latter is spatially structured, and develop statistical techniques for the analysis of the spatial structure of living communities. It has since become a classical test data set, used in several publications (e.g. Borcard et al. 1992; Borcard and Legendre 1994; Borcard et al. 2004; Wagner 2004; Legendre 2005; Dray et al. 2006; Griffith and Peres-Neto 2006). These data are available in packages **`vegan`** and **`ade4`**.

The data set ("mite.RData") comprises three files that contain the abundances of 35 morphospecies, 5 substrate and microtopographic variables, and the X-Y Cartesian coordinates of the 70 cores (in cm). The environmental variables are the following (Table 1.3):

The cores have been sampled on a 10.0 m × 2.6 m strip of various substrates forming a transect between a mixed forest and the lake's free water on the shore of an acidic lake. Figure 1.1 shows the 70 soil cores and the types of substrate.

1.6 A Quick Reminder About Help Sources

The **R** language was designed to be a self-learning tool. So you can use and abuse of the various ways to ask questions, display code, run examples that are imbedded in the framework. Some important help tools are presented here (Table 1.4).

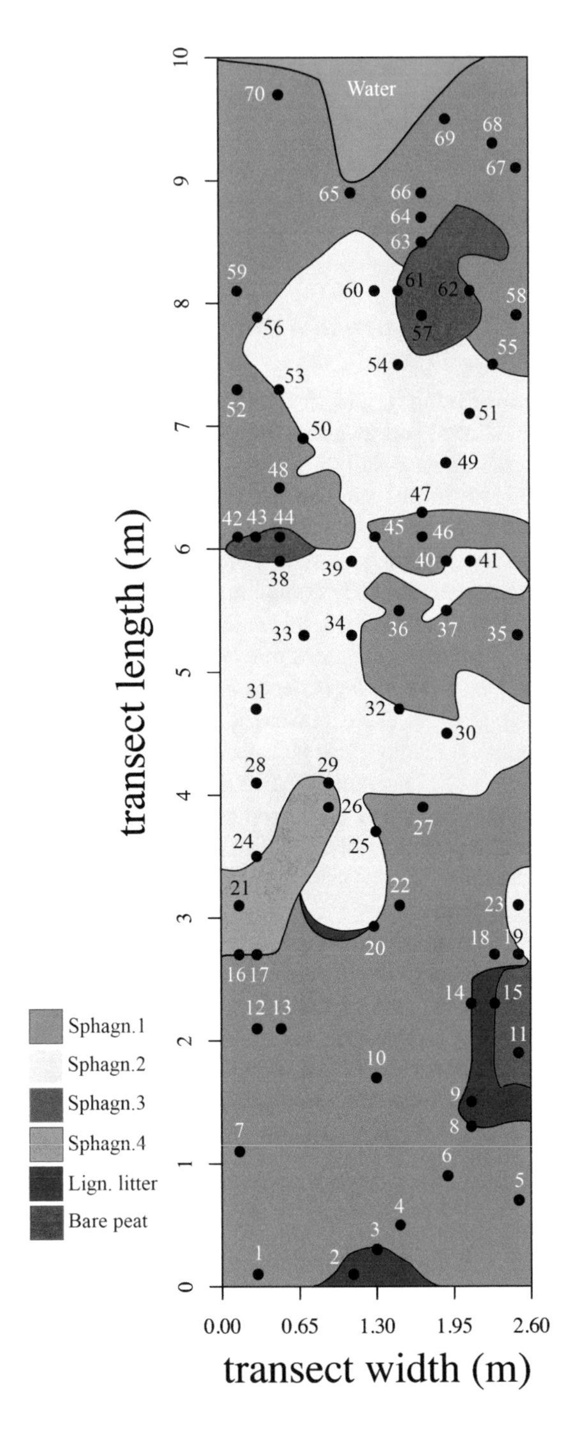

Fig. 1.1 Map of the mite data sampling area, showing the location of the 70 cores and the type of substrate (Details: see Borcard and Legendre 1994)

Table 1.4 Several help resources in **R**

Action	Use	Example	Remarks
?(question mark)	Obtain information about a function	`?decostand`	The package to which the function belongs must be active
`??` (double question mark)	Obtain information on the basis of a keyword	`??diversity`	The search is done in all packages installed in the computer
Type the name of a function	Display the code of the function on-screen	`diversity`	Not all functions can be displayed fully; some contain compiled code
`help (package="....")`	Display information on the package, including a list of all functions and data.	`help (package="ade4")`	
`data (package="...")`	List the datasets contained in a package	`data (package="vegan")`	
http://cran.r-project.org/	Broader search than above; access to discussion lists	Search on the CRAN web site: click on the "Search" link and choose one of the links	Outside the **R** master webserver.
http://www.rseek.org/	Seek any **R** function	Search "line plot"	Outside the **R** console

1.7 Now It Is Time…

… to get your hands full of code, numerical outputs and plots. Revise the basics of the methods, explore the code, analyse it, change it, try to apply it to your data and interpret your results. Above all, we hope to show that doing numerical ecology in **R** is fun!

Chapter 2
Exploratory Data Analysis

2.1 Objectives

Nowadays, most ecological research is done with hypothesis testing and modelling in mind. However, Exploratory Data Analysis (**EDA**), with its visualization tools and simple statistics, is still required at the beginning of the statistical analysis of multidimensional data, in order to:

- get an overview of the data;
- transform or recode some variables;
- orient further analyses.

As a worked example, we will explore the classical Doubs River dataset to introduce some techniques of EDA using **R** functions found in standard packages. In this chapter you will:

- learn or revise some bases of the **R** language;
- learn some EDA techniques applied to multidimensional ecological data;
- explore the Doubs dataset in hydrobiology as a first worked example.

2.2 Data Exploration

2.2.1 Data Extraction

The Doubs data used here are available in a .RData file found among the files provided with the book; see Chap. 1.

D. Borcard et al., *Numerical Ecology with R*, Use R!,
https://doi.org/10.1007/978-3-319-71404-2_2

```
# Load required packages
library(vegan)
library(RgoogleMaps)
library(googleVis)
library(labdsv)

# Source additional functions that will be used later in this
# Chapter. Our scripts assume that files to be read are in
# the working directory.
source("panelutils.R")

# Load the data. File Doubs.Rdata is assumed to be
# in the working directory
load("Doubs.RData")

# The file Doubs.RData contains the following objects:
#    spe: species (community) data frame (fish abundances)
#    env: environmental data frame
#    spa: spatial data frame – cartesian coordinates
#    fishtraits: functional traits of fish species
#    latlong: spatial data frame – latitude and longitude
```

Hints *At the beginning of a session, make sure to place all necessary data files and scripts in a single folder and define this folder as your working directory, either through the menu or by using function* **setwd()**.

Although it is not necessary, we strongly recommend that you use RStudio as script manager, which adds many interesting features to standard text editors. The R code in the companion materials of this book is optimized for RStudio and complies with the R Core Team's guidelines for good practices in R programming. Once all necessary files are placed in the same folder and RStudio is configured to run R scripts, just double-click on an R script file and the corresponding folder will be automatically defined as the current working directory.

*Users of the standard **R** console can use the **R** built-in text editor to write **R** code and run any selected portion using easy keyboard commands (<Control+Return> or <Command+Return> depending on the machine you are using). To open a new file, click on the File menu, then click on New script. Dragging an **R** script, for example our file "chap2.R", onto the R icon, will automatically open it in a new file managed by the **R** text editor.*

If you are uncertain of the class of an object, type **class**(object_name).

2.2.2 Species Data: First Contact

We can start data exploration, which will first focus on the community data (object **spe** loaded as an element of the Doubs.RData file above). Verneaux used a semi-quantitative, species-specific, abundance scale (0–5), so that comparisons between

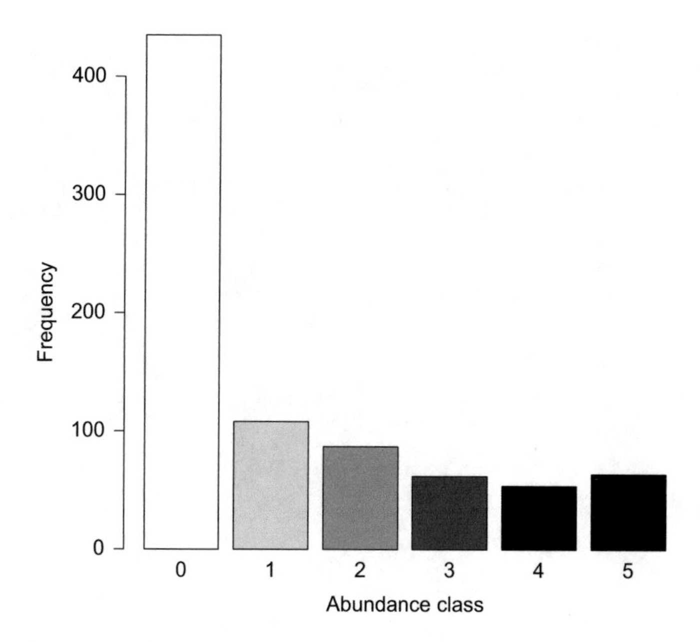

Fig. 2.1 Barplot of abundance classes

species abundances will make sense. The maximum value, 5, corresponds to the class with the maximum number of individuals captured by electrical fishing in the Doubs River and its tributaries (i.e. not only in this data set) by Verneaux. Therefore, species-specific codes cannot be understood as unbiased estimates of the true abundances (number or density of individuals) or biomasses at the sites.

We will first apply some basic **R** functions and draw a barplot (Fig. 2.1):

```
## Exploration of a data frame using basic R functions
spe                       # Display the whole data frame in the
                          # console
                          # Not recommended for large datasets!
spe[1:5, 1:10]            # Display only 5 lines and 10 columns
head(spe)                 # Display only the first 6 lines
tail(spe)                 # Display only the last 6 rows
nrow(spe)                 # Number of rows (sites)
ncol(spe)                 # Number of columns (species)
dim(spe)                  # Dimensions of the data frame (rows,
                          # columns)
```

```
colnames(spe)              # Column labels (descriptors = species)
rownames(spe)              # Row labels (objects = sites)
summary(spe)               # Descriptive statistics for columns

## Overall distribution of abundances (dominance codes)
# Minimum and maximum of abundance values in the whole data set
range(spe)
# Minimum and maximum value for each species
apply(spe, 2, range)
# Count the cases for each abundance class
(ab <- table(unlist(spe)))
# Barplot of the distribution, all species confounded
barplot(ab,
  las = 1,
  xlab = "Abundance class",
  ylab = "Frequency",
  col = gray(5 : 0 / 5)
)
# Number of absences
sum(spe == 0)
# Proportion of zeros in the community data set
sum(spe == 0) / (nrow(spe) * ncol(spe))
```

Hint *Observe how the shades of grey of the bars have been defined in the function* **barplot()**. *The argument* `col = gray(5 : 0 / 5)` *means "I want five shades of grey with levels ranging from 5/5 (i.e., white) to 0/5 (black)".*

Look at the barplot of abundance classes. How do you interpret the high frequency of zeros (absences) in the data frame?

2.2.3 Species Data: A Closer Look

The commands above give an idea of the data structure. But codes and numbers are not very attractive or inspiring, so let us illustrate some features. We will first create a map of the sites (Fig. 2.2):

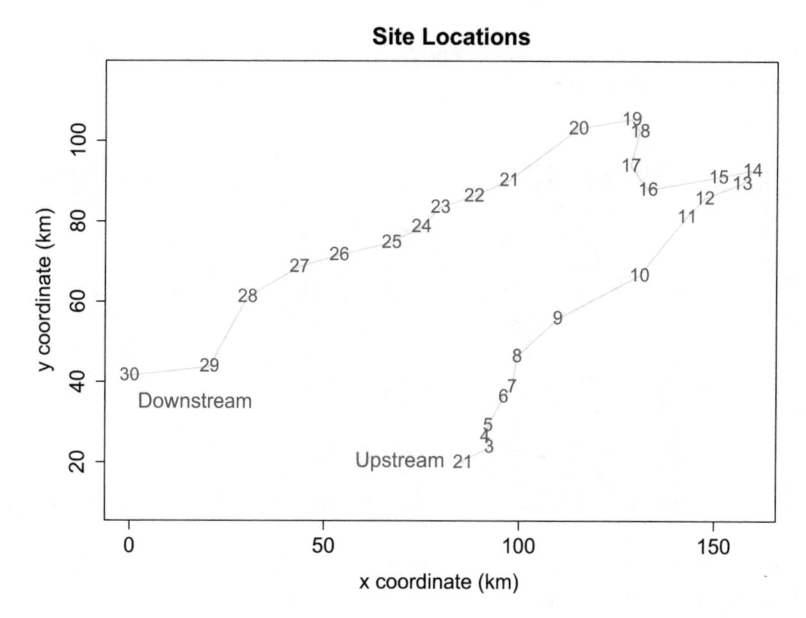

Fig. 2.2 Map of the 30 sampling sites along the Doubs River. Sites 1 and 2 are very close to each other

```
## Map of the locations of the sites
# Geographic coordinates x and y from the spa data frame
plot(spa,
  asp = 1,
  type = "n",
  main = "Site Locations",
  xlab = "x coordinate (km)",
  ylab = "y coordinate (km)"
)
# Add a blue line connecting the sites along the Doubs River
lines(spa, col = "light blue")
# Add the site labels
text(spa, row.names(spa), cex = 0.8, col = "red")
# Add text blocks
text(68, 20, "Upstream", cex = 1.2, col = "red")
text(15, 35, "Downstream", cex = 1.2, col = "red")
```

When the data set covers a sufficiently large area, it is possible to project the sites onto a Google Maps® map:

```
## Sites projected onto a Google Maps® background
# By default the plot method of the googleVis package uses
# the standard browser to display its output.
nom <- latlong$Site
latlong2 <- paste(latlong$LatitudeN, latlong$LongitudeE, sep = ":")
df <- data.frame(latlong2, nom, stringsAsFactors = FALSE)

mymap1 <- gvisMap(df,
  locationvar = "latlong2",
  tipvar = "nom",
  options = list(showTip = TRUE)
)
plot(mymap1)
```

Now the river looks more real, but where are the fish? To show the distributions and abundances of the four species used to characterize ecological zones in European rivers (Fig. 2.3), one can type:

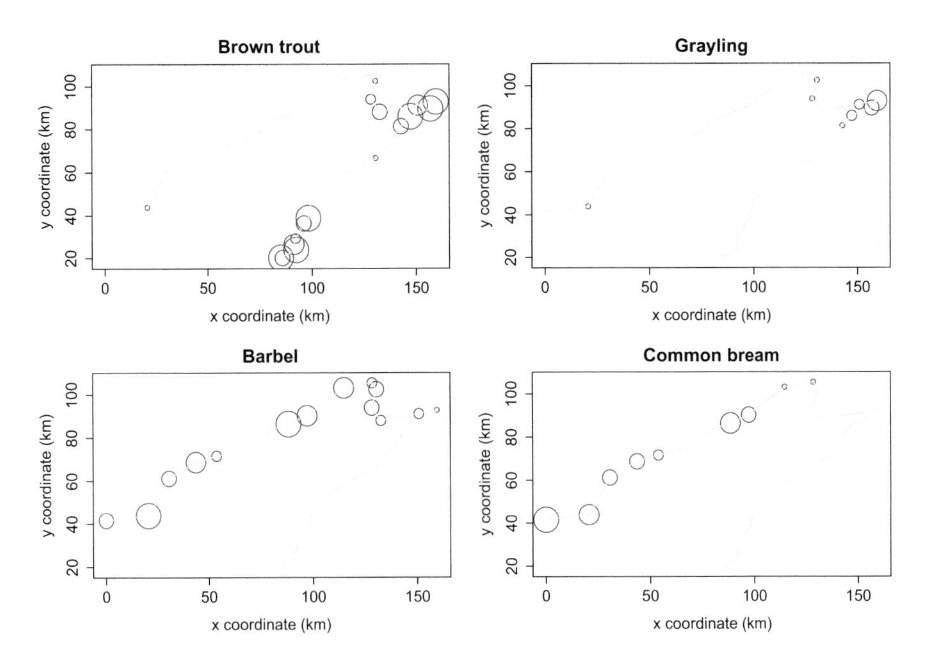

Fig. 2.3 Bubble maps of the abundances of four fish species

```
## Maps of some fish species
# Divide the plot window into 4 frames, 2 per row
par(mfrow = c(2,2))
# Plot four species
plot(spa,
  asp = 1,
  cex.axis = 0.8,
  col = "brown",
  cex = spe$Satr,
  main = "Brown trout",
  xlab = "x coordinate (km)",
  ylab = "y coordinate (km)"
)
lines(spa, col = "light blue")
plot(spa,
  asp = 1,
  cex.axis = 0.8,
  col = "brown",
  cex = spe$Thth,
  main = "Grayling",
  xlab = "x coordinate (km)",
  ylab = "y coordinate (km)"
)
lines(spa, col = "light blue")
plot(spa,
  asp = 1,
  cex.axis = 0.8,
  col = "brown",
  cex = spe$Baba,
  main = "Barbel",
  xlab = "x coordinate (km)",
  ylab = "y coordinate (km)"
)
lines(spa, col = "light blue")
plot(spa,
  asp = 1,
  cex.axis = 0.8,
  col = "brown",
  cex = spe$Abbr,
  main = "Common bream",
  xlab = "x coordinate (km)",
  ylab = "y coordinate (km)"
)
lines(spa, col = "light blue")
```

Hint *Note the use of the* `cex` *argument in the* **`plot()`** *function:* `cex` *is used to define the size of an item in a graph. Here its value is a vector belonging to the* `spe` *data frame, i.e., the abundances of a given species (e.g.,* `cex = spe$Satr`*). The result is a series of bubbles whose diameter at each site is proportional to the species abundance. Also, since the object spa contains only two variables x and y, the list of arguments has been simplified by replacing the first two arguments for horizontal and vertical axes by the name of the spa data frame.*

From these maps you can understand why Verneaux chose these four species as ecological indicators. More about the environmental conditions later.

At how many sites does each species occur? Calculate the relative occurrences of the species (proportions of the number of sites) and plot histograms (Fig. 2.4):

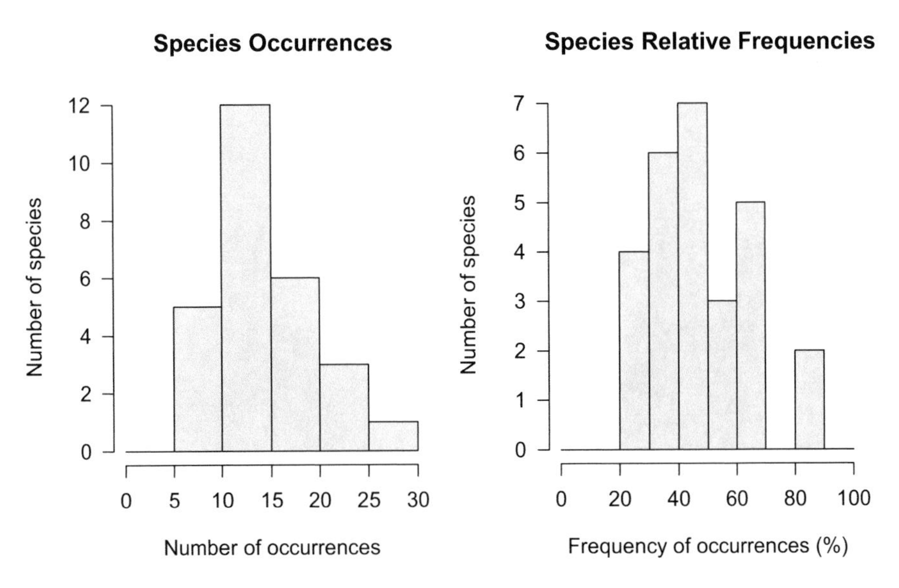

Fig. 2.4 Frequency histograms: species occurrences and relative frequencies at the 30 sites

```
## Compare species: number of occurrences
# Compute the number of sites where each species is present
# To sum by columns, the second argument of apply(), MARGIN,
# is set to 2
spe.pres <- apply(spe > 0, 2, sum)
# Sort the results in increasing order
sort(spe.pres)
# Compute percentage frequencies
spe.relf <- 100 * spe.pres/nrow(spe)
# Round the sorted output to 1 digit
round(sort(spe.relf), 1)
# Plot the histograms
par(mfrow = c(1,2))
hist(spe.pres,
  main = "Species Occurrences",
  right = FALSE,
  las = 1,
  xlab = "Number of occurrences",
  ylab = "Number of species",
  breaks = seq(0, 30, by = 5),
  col = "bisque"
)
hist(spe.relf,
  main = "Species Relative Frequencies",
  right = FALSE,
  las = 1,
  xlab = "Frequency of occurrences (%)",
  ylab = "Number of species",
  breaks = seq(0, 100, by = 10),
  col = "bisque"
)
```

Hint *Examine the use of the* **apply()** *function, applied here to the columns of the data frame* spe. *Note that the first part of the function call* (spe > 0) *evaluates the values in the data frame to* TRUE/FALSE, *and the number of* TRUE *cases per column is counted by summing.*

Now that we have seen at how many sites each species is present, we may want to know how many species are present at each site (species richness, Fig. 2.5):

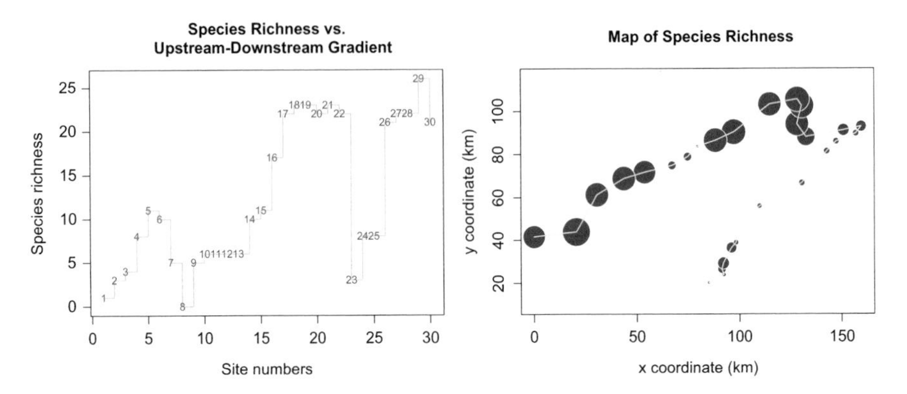

Fig. 2.5 Species richness along the river

```
## Compare sites: species richness
# Compute the number of species at each site
# To sum by rows, the second argument of apply(), MARGIN, is
# set to 1
sit.pres <- apply(spe > 0, 1, sum)
# Sort the results in increasing order
sort(sit.pres)
par(mfrow = c(1, 2))
# Plot species richness vs. position of the sites along the river
plot(sit.pres,type = "s",
  las = 1,
  col = "gray",
  main = "Species Richness vs. \n Upstream-Downstream Gradient",
  xlab = "Site numbers",
  ylab = "Species richness"
)
text(sit.pres, row.names(spe), cex = .8, col = "red")
# Use geographic coordinates to plot a bubble map
plot(spa,
  asp = 1,
  main = "Map of Species Richness",
  pch = 21,
  col = "white",
  bg = "brown",
  cex = 5 * sit.pres / max(sit.pres),
  xlab = "x coordinate (km)",
  ylab = "y coordinate (km)"
)
lines(spa, col = "light blue")
```

Hint *Observe the use of the* `type = "s"` *argument of the* **`plot()`** *function to draw steps between values.*

Can you identify the richness hotspots along the river?

More elaborate measures of diversity will be presented in Chap. 8.

2.2.4 Ecological Data Transformation

There are instances where one needs to transform the data prior to analysis. The main reasons are given below with examples of transformations:

- Make descriptors that have been measured in different units comparable. Standardization to z-scores (i.e., centring and reduction) and ranging (to a [0,1] interval) make variables dimensionless. Following that, their variances can be added, e.g. in principal component analysis (see Chap. 5);
- Transform the variables to have a normal (or at least a symmetric) distribution and stabilize their variances (through square root, fourth root, log transformations, etc.);
- Make the relationships among variables linear (e.g., log-transform the response variable if the relationship is exponential);
- Modify the weights of the variables or objects prior to a multivariate analysis, e.g., give the same variance to all variables, or the same length, (or norm) to all object vectors;
- Code categorical variables into dummy binary variables or Helmert contrasts.

Species abundances are dimensionally homogenous (expressed in the same physical units), quantitative (count, density, cover, biovolume, biomass, frequency, etc.) or semi-quantitative (two or more ordered classes) variables, and restricted to positive or null values (zero meaning absence). For these, **simple transformations** can be used to reduce the importance of observations with very high values; **`sqrt()`** (square root), **`^0.25`** (fourth root), or **`log1p()`** (log(y + 1) to keep absences as zeros) are commonly applied **R** functions (see also Chap. 3). In extreme cases, to give the same weight to all positive abundances irrespective of their values, the data can be transformed to binary 1–0 form (presence-absence).

The **`decostand()`** function of the **`vegan`** package provides many options for common standardization of ecological data. In this function, **standardization** refers to transformations that have the objective to make the rows or columns of the data table comparable to one another because they will have acquired some property. In contrast to simple transformations such as square root, log or presence-absence, the values are not transformed individually but relative to other values in the data table. Standardizations can be done relative to sites (e.g. relative abundances per site) or species (abundances scaled with respect to the species maximum abundance or its total abundance), or simultaneously to both site and species totals (with the chi-square transformation), depending on the focus of the analysis.

Note that **`decostand()`** has a `log` argument for logarithmic transformation. But here the transformation is $log\_b(y) + 1$ for $y > 0$, where b is the base of the logarithm. Zeros remain untouched. The base of the logarithm is provided by the argument `logbase`. This transformation, proposed by Anderson et al. (2006), is not equivalent to $\log(y + 1)$. Increasing the value of the base increases the severity of the downscaling of large values.

Here are some examples of data standardization illustrated by boxplots (Fig. 2.6).

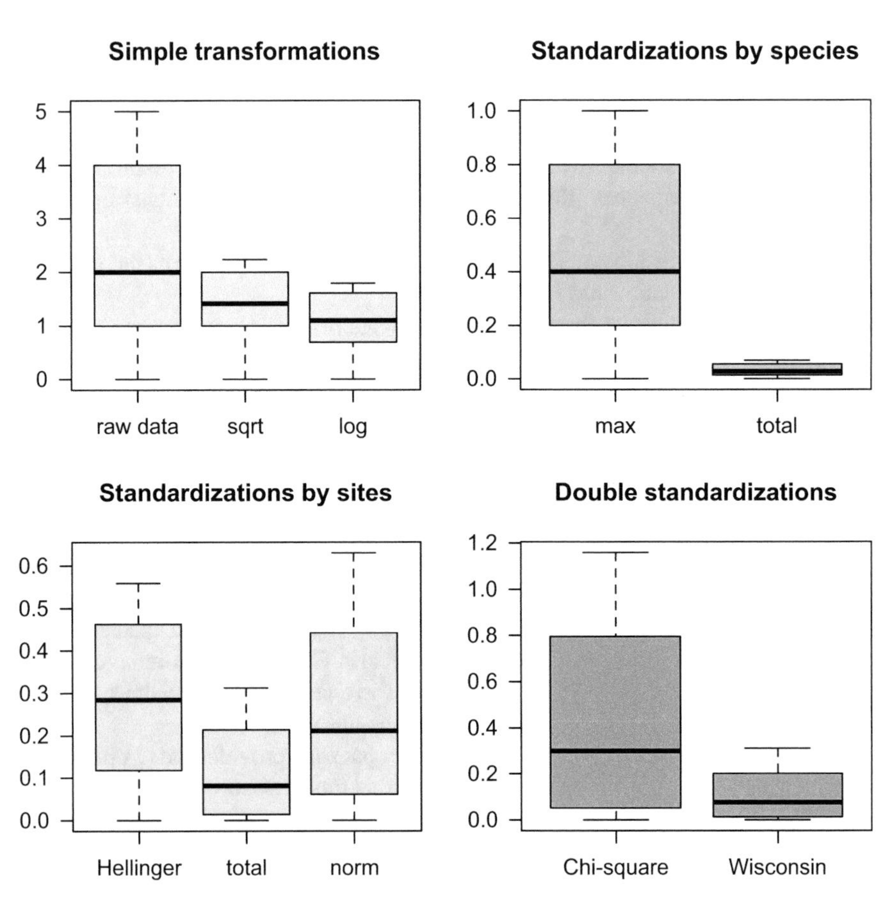

Fig. 2.6 Boxplots of transformed abundances of a common species, *Barbatula barbatula* (stone loach)

```
# Get help on the decostand() function
?decostand

## Simple transformations
# Partial view of the raw data (abundance codes)
spe[1:5, 2:4]
# Transform abundances to presence-absence (1-0)
spe.pa <- decostand(spe, method = "pa")
spe.pa[1:5, 2:4]

## Standardization by columns (species)
# Scale abundances by dividing them by the maximum value of
# each species
# Note: MARGIN = 2 (column, default value) for argument "max"
spe.scal <- decostand(spe, "max")
spe.scal[1:5, 2:4]
# Display the maximum in each transformed column
apply(spe.scal, 2, max)
```

Did the scaling work properly? It is good to keep an eye on your results by producing a plot or computing summary statistics.

```
# Scale abundances by dividing them by the species totals
# (relative abundance by species)
# Note: here, override the default MARGIN = 1 argument of "total"
spe.relsp <- decostand(spe, "total", MARGIN = 2)
spe.relsp[1:5, 2:4]
# Display the sum by column
# Classical: apply(spe.relsp, 2, sum)
colSums(spe.relsp)

## Standardization by rows (sites)
# Scale abundances by dividing them by the site totals
# (profiles of relative abundance by site)
spe.rel <- decostand(spe, "total") # default MARGIN = 1
spe.rel[1:5, 2:4]
# Display the sum of row vectors to determine if the scaling
# worked properly
rowSums(spe.rel)

# Give a length (norm) of 1 to each row vector
# This is called the chord transformation
spe.norm <- decostand(spe, "normalize") # default MARGIN = 1
spe.norm[1:5, 2:4]
```

```
# Verify the norm of the transformed row vectors
# Write a 1-line function that computes the norm of vector x
vec.norm <- function(x) sqrt(sum(x ^ 2))
# Then, apply that function to the rows of matrix spe.norm
apply(spe.norm, 1, vec.norm)
```

The scaling above is called the 'chord transformation': the Euclidean distance function applied to chord-transformed data produces a chord distance matrix (Chap. 3). The chord transformation is useful prior to PCA and RDA (Chap. 5 and 6) and k-means partitioning (Chap. 4). The chord transformation can also be applied to log-transformed data (see Chap. 3).

```
# Compute square root of relative abundances per site
spe.hel <- decostand(spe, "hellinger")
spe.hel[1:5, 2:4]
# Check the norm of row vectors
apply(spe.hel, 1, vec.norm)
```

This is called the Hellinger transformation. The Euclidean distance function applied to Hellinger-transformed data produces a Hellinger distance matrix (Chap. 3). The Hellinger transformation is useful prior to PCA and RDA (Chap. 5 and 6) and k-means partitioning (Chap. 4).

Note: the Hellinger transformation can also be obtained by applying the chord transformation to square-root-transformed species data.

```
## Double standardization by columns and rows
# Chi-square transformation
spe.chi <- decostand(spe, "chi.square")
spe.chi[1:5, 2:4]
# Check what happened to site 8 where no species was found
spe.chi[7:9, ]
# Note: decostand produced values of 0 for 0/0 instead of NaN
```

The Euclidean distance function applied to chi-square-transformed data produces a chi-square distance matrix (Chap. 3).

```
# Wisconsin standardization
# Abundances are first ranged by species maxima and then
# by site totals
spe.wis <- wisconsin(spe)
spe.wis[1:5,2:4]

## Boxplots of transformed abundances of a common species
# (the stone loach, species #4)
par(mfrow = c(2,2))
boxplot(spe$Babl,
  sqrt(spe$Babl),
  log1p(spe$Babl),
  las = 1,
  main = "Simple transformations",
  names = c("raw data", "sqrt", "log"),
  col = "bisque"
)
boxplot(spe.scal$Babl,
  spe.relsp$Babl,
  las = 1,
  main = "Standardizations by species",
  names = c("max", "total"),
  col = "lightgreen"
)
boxplot(spe.hel$Babl,
  spe.rel$Babl,
  spe.norm$Babl,
  las = 1,
  main = "Standardizations by sites",
  names = c("Hellinger", "total", "norm"),
  col = "lightblue"
)
boxplot(spe.chi$Babl,
  spe.wis$Babl,
  las = 1,
  main = "Double standardizations",
  names = c("Chi-square", "Wisconsin"),
  col = "orange"
)
```

Hint *Take a look at the line:* `vec.norm <- function(x) sqrt(sum(x^2))`. *It is an example of a small function built on the fly to fill a gap in the functions available in standard* **R** *packages: this function computes the norm (length) of a vector using a matrix algebraic form of Pythagora's theorem. For more matrix algebra, visit the* ***Code It Yourself*** *corners.*

Compare the effects of these transformations and standardizations on the ranges and distributions of the scaled abundances.

Another way of comparing the effects of transformations on species abundances is to plot them along the river course:

```
## Plot raw and transformed abundances along the upstream-
## downstream river gradient
par(mfrow = c(2, 2))
plot(env$dfs,
  spe$Satr,
  type = "l",
  col = 4,
  main = "Raw data",
  xlab = "Distance from the source [km]",
  ylab = "Raw abundance code"
)
lines(env$dfs, spe$Thth, col = 3)
lines(env$dfs, spe$Baba, col = "orange")
lines(env$dfs, spe$Abbr, col = 2)
lines(env$dfs, spe$Babl, col = 1, lty = "dotted")

plot(env$dfs,
  spe.scal$Satr,
  type = "l",
  col = 4,
  main = "Species abundances ranged
  by maximum",
  xlab = "Distance from the source [km]",
  ylab = "Ranged abundance"
)
lines(env$dfs, spe.scal$Thth, col = 3)
lines(env$dfs, spe.scal$Baba, col = "orange")
lines(env$dfs, spe.scal$Abbr, col = 2)
lines(env$dfs, spe.scal$Babl, col = 1, lty = "dotted")

plot(env$dfs,
  spe.hel$Satr,
  type = "l",
  col = 4,
  main =  "Hellinger-transformed abundances",
  xlab = "Distance from the source [km]",
  ylab = "Standardized abundance"
)
lines(env$dfs, spe.hel$Thth, col = 3)
lines(env$dfs, spe.hel$Baba, col = "orange")
lines(env$dfs, spe.hel$Abbr, col = 2)
lines(env$dfs, spe.hel$Babl, col = 1, lty = "dotted")

plot(env$dfs,
```

```
    spe.chi$Satr,
    type = "l",
    col = 4,
    main = "Chi-square-transformed abundances",
    xlab = "Distance from the source [km]",
    ylab = "Standardized abundance"
)
lines(env$dfs, spe.chi$Thth, col = 3)
lines(env$dfs, spe.chi$Baba, col = "orange")
lines(env$dfs, spe.chi$Abbr, col = 2)
lines(env$dfs, spe.chi$Babl, col = 1, lty = "dotted")
legend("topright",
    c("Brown trout", "Grayling", "Barbel", "Common bream",
    "Stone loach"),
    col = c(4, 3, "orange", 2, 1),
    lty = c(rep(1, 4), 3)
)
```

Compare the graphs and explain the differences.

In some cases (often vegetation studies), data are collected using abundance scales that are meant to represent specific properties: number of individuals (abundance classes), cover (dominance classes), or both (e.g. Braun-Blanquet abundance-dominance scale). The scales being ordinal and somewhat arbitrary, the resulting data do not easily lend themselves to a simple transformation. In such cases one may have to convert scales by attributing values according to the data at hand. For discrete scales it can be done by function **vegtrans()** of package **labdsv**.

For example, assuming we knew how to convert the fish abundance codes (ranging from 0 to 5 in our **spe** dataset) to average numbers of individuals, we could do it by providing two vectors, one with the current scale and one with the converted scale. **Beware**: this would not make sense for this fish data set, whose abundances are species-specific (see Sect. 2.2).

```
## Conversion of the fish abundance using an arbitrary scale
current <- c(0, 1, 2, 3, 4, 5)
converted <- c(0, 1 ,5, 10, 20, 50)
spe.conv <- vegtrans(spe, current, converted)
```

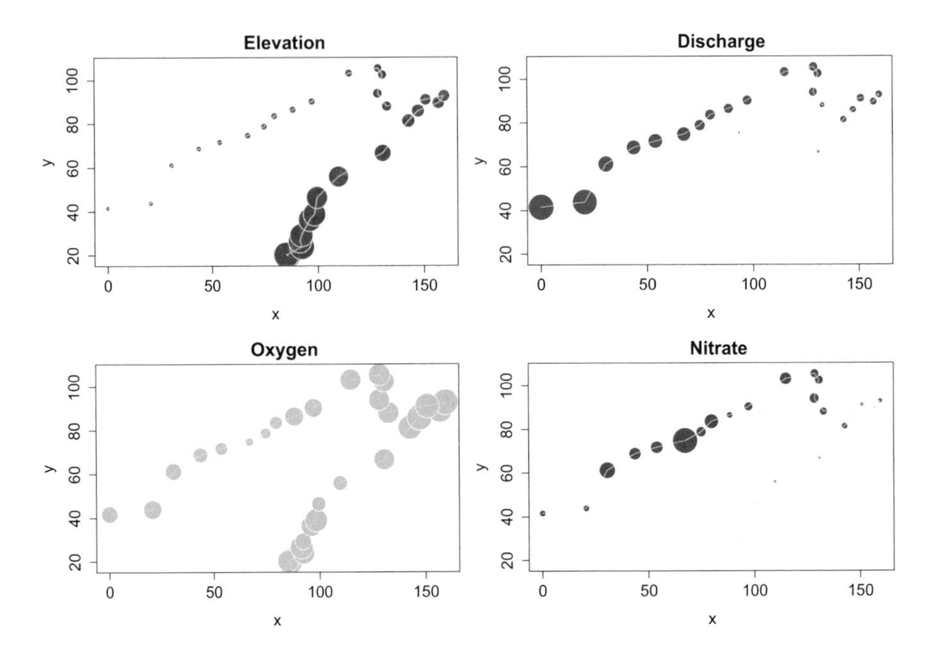

Fig. 2.7 Bubble maps of four environmental variables

2.2.5 Environmental Data

Now that we are acquainted with the species data, let us turn to the environmental data file (object **env**).

First, go back to Sect. 2.2 and apply the basic functions presented there to the object **env**. While examining the **summary()**, note how the variables differ from the species data in values and spatial distributions.

Draw maps of some of the environmental variables, first in the form of bubble maps (Fig. 2.7):

```
par(mfrow = c(2, 2))
plot(spa,
  asp = 1,
  cex.axis = 0.8,
  main = "Elevation",
  pch = 21,
  col = "white",
  bg = "red",
  cex = 5 * env$ele / max(env$ele),
  xlab = "x",
  ylab = "y"
)
```

```
lines(spa, col = "light blue")
plot(spa,
  asp = 1,
  cex.axis = 0.8,
  main = "Discharge",
  pch = 21,
  col = "white",
  bg = "blue",
  cex = 5 * env$dis / max(env$dis),
  xlab = "x",
  ylab = "y"
)
lines(spa, col = "light blue")
plot(spa,
  asp = 1,
  cex.axis = 0.8,
  main = "Oxygen",
  pch = 21,
  col = "white",
  bg = "green3",
  cex = 5 * env$oxy / max(env$oxy),
  xlab =  "x",
  ylab = "y"
)
lines(spa, col = "light blue")
plot(spa,
  asp = 1,
  cex.axis = 0.8,
  main = "Nitrate",
  pch = 21,
  col = "white",
  bg = "brown",
  cex = 5 * env$nit / max(env$nit),
  xlab = "x",
  ylab = "y"
)
lines(spa, col = "light blue")
```

Hint *See how the* `cex` *argument is used to make the size of the bubbles comparable among plots. Play with these values to see the changes in the graphical output.*

Which ones of these maps display an upstream-downstream gradient? How could you explain the spatial patterns of the other variables?

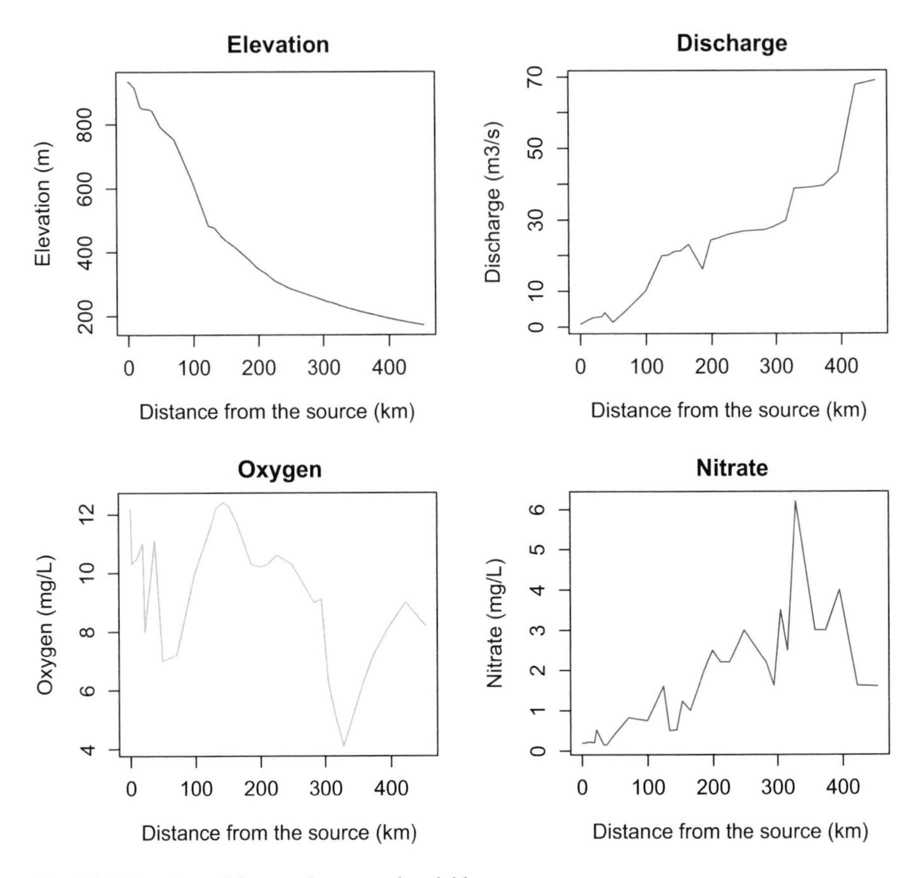

Fig. 2.8 Line plots of four environmental variables

Now, examine the variation of some descriptors along the river by means of line plots (Fig. 2.8):

```
## Line plots
par(mfrow = c(2, 2))
plot(env$dfs, env$ele,
  type = "l",
  xlab = "Distance from the source (km)",
  ylab = "Elevation (m)",
  col = "red", main = "Elevation"
)
```

```
plot(env$dfs, env$dis,
  type = "l",
  xlab = "Distance from the source (km)",
  ylab = "Discharge (m3/s)",
  col = "blue",
  main = "Discharge"
)
plot(env$dfs, env$oxy,
  type = "l",
  xlab = "Distance from the source (km)",
  ylab = "Oxygen (mg/L)",
  col = "green3",
  main = "Oxygen"
)
plot(env$dfs, env$nit,
  type = "l",
  xlab = "Distance from the source (km)",
  ylab = "Nitrate (mg/L)",
  col = "brown",
  main = "Nitrate"
)
```

To explore graphically the bivariate relationships among the environmental variables, we can use the powerful **`pairs()`** graphical function, which draws a matrix of scatter plots (Fig. 2.9).

Moreover, we can add a LOWESS smoother to each bivariate plot and draw histograms in the diagonal plots, showing the frequency distribution of each variable, using external functions found in the **`panelutils.R`** script.

```
## Scatter plots for all pairs of environmental variables
# Bivariate plots with histograms on the diagonal and smooth
# fitted curves
pairs(env,
  panel = panel.smooth,
  diag.panel = panel.hist,
  main = "Bivariate Plots with Histograms and Smooth Curves"
)
```

Hint *Each scatterplot shows the relationship between two variables identified on the main diagonal. The abscissa of the scatterplot is the variable above or under it, and the ordinate is the variable to its left or right.*

Fig. 2.9 Scatter plots between all pairs of environmental variables with LOWESS smoothers

From the histograms, do many variables seem normally distributed?

Note that normality is not required for explanatory variables in regression analysis and in canonical ordination.

Do many scatter plots show linear or at least monotonic relationships?

Simple transformations, such as the log transformation, can be used to improve the distributions of some variables (make them more symmetrical and closer to normal distributions). Furthermore, because environmental variables are dimensionally heterogeneous (expressed in different units and scales), many statistical analyses require their standardization to zero mean and unit variance. These centred and scaled variables are called *z*-scores. We can now illustrate transformations and standardization with our example data (Fig. 2.10).

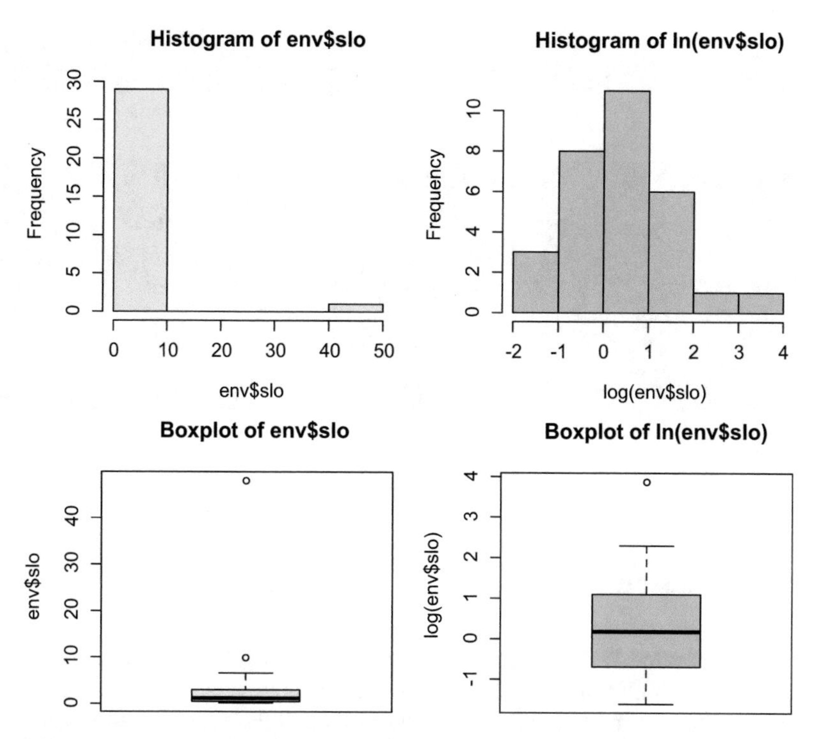

Fig. 2.10 Histograms and boxplots of the untransformed (left) and log-transformed (right) "slope" variable

```
## Simple transformation of an environmental variable
range(env$slo)
# Log-transformation of the slope variable (y = ln(x))
# Compare histograms and boxplots of the raw and transformed values
par(mfrow = c(2, 2))
hist(env$slo,
  col = "bisque",
  right = FALSE
)
hist(log(env$slo),
  col = "light green",
  right = FALSE,
  main = "Histogram of ln(env$slo)"
)
boxplot(env$slo,
  col = "bisque",
  main = "Boxplot of env$slo",
  ylab = "env$slo"
)
boxplot(log(env$slo),
  col = "light green",
  main = "Boxplot of ln(env$slo)",
  ylab = "log(env$slo)"
)
```

Hint *Normality of a vector can be tested by using the Shapiro-Wilk test, available through function* `shapiro.test()`.

```
## Standardization of all environmental variables
# Center and scale = standardize the variables (z-scores)
env.z <- decostand(env, "standardize")
apply(env.z, 2, mean)   # means = 0
apply(env.z, 2, sd) # standard deviations = 1

# Same standardization using the scale() function (which returns
# a matrix)
env.z <- as.data.frame(scale(env))
```

2.3 Conclusion

The tools presented in this chapter allow researchers to get a general impression of their data. Although you will see much more elaborate analyses in the following chapters, keep in mind that a first exploratory look at the data can tell much about them. Information about simple parameters and distributions of variables is important to consider and will help in the correct selection of more advanced analyses. Graphical representations like bubble maps are useful to reveal how the variables are spatially organized; they may help generate hypotheses about the processes acting behind the scene. Boxplots and simple statistics may be necessary to reveal unusual or aberrant values.

EDA is often neglected by scientists who are eager to jump to more sophisticated analyses. We hope we have convinced you that it should have an important place in the toolbox of ecologists.

Chapter 3
Association Measures and Matrices

3.1 Objectives

Most methods of multivariate analysis, in particular ordination and clustering techniques, are explicitly or implicitly[1] based on the comparison of all possible pairs of objects or descriptors. The comparisons take the form of association measures (often called coefficients or indices), which are assembled in a square and symmetric[2] association matrix, of dimensions $n \times n$ when objects are compared, or $p \times p$ when variables are compared. Since the subsequent analyses are done on association matrices, the choice of an appropriate measure is crucial. In this Chapter you will:

- quickly revise the main categories of association coefficients;
- learn how to compute, examine and visually compare dissimilarity matrices (Q mode) and dependence matrices (R mode);
- apply these techniques to a classical dataset;
- learn or revise some basics of programming functions with the **R** language.

3.2 The Main Categories of Association Measures (Short Overview)

It is beyond the scope of this book to explain the various indices in detail, but it is useful to provide a wrap-up of the main categories of measures. This will facilitate the choice of an appropriate index in many situations, and improve the

---

[1]The association measure among objects may be implicit. It is the Euclidean distance in principal component analysis (PCA, Chap. 5) and *k*-means partitioning (Chap. 4), for example, and the chi-square distance in correspondence analysis (CA, Chap. 5).

[2]See footnote 4 in Sect. 3.2.2.

D. Borcard et al., *Numerical Ecology with R*, Use R!,
https://doi.org/10.1007/978-3-319-71404-2_3

understanding of the applications proposed below. Note that we will use the expressions "measure", "index" and "coefficient" as synonyms to refer to the quantities used to compare pairs of objects or variables[3].

3.2.1 *Q Mode and R Mode*

When pairs of objects are compared, the analysis is said to be in the **Q mode**. When pairs of descriptors are compared the analysis is said to be in the **R mode**. This distinction is important because the association measures in Q and R-mode analyses are not the same.

In **Q mode**, the association measure is the **dissimilarity** or the **similarity** between pairs of objects, e.g., Euclidean distance, Jaccard similarity. In **R mode**, one uses a measure of **dependence** among variables, such as the covariance or correlation coefficient.

3.2.2 *Symmetrical or Asymmetrical Coefficients in Q Mode: The Double-Zero Problem*

Virtually all dissimilarity or similarity measures used in ecology are *symmetric* in one sense: the value of the coefficient between objects n_1 and n_2 is the same as the value of the coefficient between objects n_2 and n_1. The same holds for dependence measures in the R mode. The problem addressed here is different. It concerns the treatment of double-zeros in the comparison of pairs of objects.

In certain cases, the zero value has the same meaning as any other value along the scale of a descriptor. For instance, the absence (0 mg/L) of dissolved oxygen in the deep layers of a lake is an ecologically meaningful information: the concentration is below the measurement threshold and this poses severe constraints on aerobic life forms, whatever the reason for this condition.

On the contrary, the zero value in a matrix of species abundances (or presence-absence) is much more tricky to interpret. The presence of a species at two sites generally implies that these sites provide a similar set of minimal conditions allowing the species to survive; these conditions are the dimensions of the species' ecological niche. The absence of a species from a relevé or site, however, may be due to a variety of causes: the species' niche may be occupied by a replacement species, or the species may not have reached that site even though the conditions are

[3]Although the term "coefficient" is sometimes narrowly defined as a multiplicative factor of a variable in a mathematical expression, it has been applied for decades in multivariate data analysis to the broader sense used in this book.

favourable, or the absence of the species is due to different non-optimal conditions on *any* of the important dimensions of its ecological niche, or the species is present but has not been observed or captured by the researcher, or the species does not show a regular distribution among the sites under study. The key points here are that (1) in most situations, the absence of a species from two sites cannot readily be counted as an indication of resemblance between these sites, because this double absence may be due to completely different reasons, and (2) the number of un-interpretable double zeros in a species matrix depends on the number of species and thus increases strongly with the number of rare species in the matrix.

The information "presence" thus has a clearer interpretation than the information "absence". One can distinguish two classes of association measures based on this problem: the coefficients that consider the double zeros (sometimes also called "negative matches") as indications of resemblance (like any other value) are said to be **symmetrical**, the others, **asymmetrical**[4]. In most cases, it is preferable to use asymmetrical coefficients when analysing species data, unless one has compelling reasons to consider the absence of a species from two sites as being due to the same cause, and this for all cases of double-zeros. Possible examples of such exceptions are controlled experiments with known community compositions or ecologically homogeneous areas with disturbed zones.

3.2.3 *Association Measures for Qualitative or Quantitative Data*

Some variables are qualitative (nominal or categorical, either binary or multiclass), others are semi-quantitative (ordinal) or quantitative (discrete or continuous). Association coefficients exist for all types of variables, but most of them fall into two classes: coefficients for binary variables (hereunder called *binary coefficients* for short, although it is the variables that are binary, not the values of the association measures) and coefficients for quantitative variables (called *quantitative coefficients* hereafter).

3.2.4 *To Summarize…*

Keep track on what kind of association measure you need. Before any analysis, ask the following questions:

[4]The use of the words *symmetrical/asymmetrical* for this distinction, as opposed to *symmetric/asymmetric* (same value of the coefficient between n_1 and n_2 as between n_2 and n_1), follows Legendre and Legendre (2012). Legendre and De Cáceres (2013) use the expressions *double-zero symmetrical* and *double-zero asymmetrical*.

- Are you comparing objects (Q-mode) or variables (R-mode analysis)?
- Are you dealing with species data (which usually leads to asymmetrical coefficients) or other types of variables (symmetrical coefficients)?
- Are your data binary (binary coefficients), quantitative (quantitative coefficients) or mixed or of other types (e.g. ordinal; special coefficients)?

In the following sections, you will explore various possibilities. In most cases, more than one association measure is available to study a given problem.

3.3 Q Mode: Computing Dissimilarity Matrices Among Objects

In the Q mode, we will use six packages: **`stats`** (included in the standard installation of **R**), **`vegan`**, **`ade4`**, **`adespatial`**, **`cluster`** and **`FD`**. Note that this list of packages is by far not exhaustive, but it should satisfy the needs of most ecologists.

Although the literature provides similarity as well as dissimilarity measures, in **R** all similarity measures are converted to dissimilarities to compute a square matrix of class **`"dist"`** in which the diagonal (distance between each object and itself) is 0 and can be ignored. The conversion formula varies with the package used, and this is not without consequences:

- in **`stats`**, **`FD`** and **`vegan`**, the conversion from similarities S to dissimilarities D is done by computing $D = 1 - S$;
- in **`ade4`**, it is done as $\sqrt{1-S}$. This allows several indices to become Euclidean[5], a geometric property that will be useful in some analyses, e.g. in principal coordinate analysis (see Chap. 5). We will come to it when it becomes relevant. Dissimilarity matrices computed by other packages using coefficients that are not Euclidean can often be made Euclidean by computing **`D2 <- sqrt(D)`**;
- in **`adespatial`**, most similarity coefficients are converted as $D = 1 - S$, the exceptions being three classical indices for presence-absence data, namely Jaccard, Sørensen and Ochiai (see below), which are converted as $\sqrt{1-S}$. For all computed coefficients, function **`dist.ldc()`** of that package produces messages indicating if the selected coefficient is Euclidean or non-Euclidean, or if **`sqrt(D)`** would be Euclidean;
- in **`cluster`**, all available measures are dissimilarities, so no conversion has to be made;

[5]Metric dissimilarity measures, which are also called *distances*, share four properties: minimum 0, positiveness, symmetry and triangle inequality. Furthermore, the points can be represented in a Euclidean space. It may happen, though, that some dissimilarity matrices are metric (triangle inequality respected) but non Euclidean (all points cannot be represented in a Euclidean space). See Legendre and Legendre (2012) p. 500.

- Therefore, although we will stick to the conventional names of the most classical coefficients, as they can be found in textbooks, it will be implicit from now on that all similarity measures have been converted to dissimilarities when computed by **R** functions. For instance, the Jaccard (1901) community index is originally a similarity, but the output of the computation of that coefficient in **stats**, **vegan**, **adespatial** and **ade4** is a dissimilarity matrix; it is produced as $D = (1 - S)$ in **stats** and **vegan** and as $D = \text{sqrt}(1 - S)$ in **adespatial** and **ade4** .

3.3.1 Q Mode: Quantitative Species Data

Let us use the fish species dataset **spe** again. We will consider the data as quantitative although, strictly speaking, the values do not represent raw fish abundances.

Quantitative species data generally require asymmetrical dissimilarity measures. In this category, frequently used coefficients are the percentage difference, often (incorrectly) referred to as the Bray-Curtis dissimilarity D_{14} [6] (also known as the reciprocal of the Steinhaus similarity index, S_{17}), the chord distance D_3, the chi-square distance D_{21}, and the Hellinger distance D_{17}. Let us compute dissimilarity matrices using some of these indices. In the process, we shall use the package **gclus** for visualization.

- **Percentage difference** (aka **Bray-Curtis**) dissimilarity matrices can be computed directly from raw data, although true abundances are often log-transformed, because D_{14} gives the same importance to absolute differences in abundance irrespective of the order of magnitude of the abundances. In this coefficient, a difference of 5 individuals has the same weight when the abundances are 3 and 8 as when the abundances are 6203 and 6208.
- The **chord** distance is a Euclidean distance computed on site vectors normalized to length 1; this normalization is called the chord transformation. The normalization is done by **vegan**'s function **decostand()**, argument `normalize` [7]. The chord distance can also be computed in a single step by using function **dist.ldc()** [8] of package **adespatial** with the argument `chord`.

[6]In this book, symbols and numbering of similarity and dissimilarity measures are taken from Legendre and Legendre (2012).

[7]This way of presenting several distance measures (as pre-transformations followed by computation of the Euclidean distance) was described by Legendre and Gallagher (2001). More about this topic in Sect. 3.5.

[8]The name dist.**ldc** refers to the authors of the article presenting the properties and use of 16 coefficients for the study of beta diversity: **L**egendre and **D**e **C**áceres (2013). We will revisit this function and some of its indices in Chap. 8.

- The **Hellinger** distance is a Euclidean distance between site vectors where the abundance values are first divided by the site total abundance, and the result is square-root transformed; this is called the Hellinger transformation. The Hellinger transformation is also the chord transformation of square-root transformed abundance data. The Hellinger transformation is obtained in one step by **decostand** with the argument `hellinger`. The Hellinger distance can also be computed in a single step by using function **dist.ldc()** of package **adespatial**, whose argument `hellinger` is the default.
- The **log-chord distance** is a chord distance applied to log-transformed abundance data. It can be obtained by first transforming the raw abundance data by $\ln(y + 1)$, followed by the computation of the chord transformation or the chord distance, as explained above. The log-chord distance can also be computed in a single step by using function **dist.ldc()** of package **adespatial** with the argument `log.chord`.

Legendre and Borcard (2018) have shown than the chord, Hellinger and log-chord distances can be construed as chord distances computed on data transformed by functions that are members of a series of Box-Cox normalizing transformations (Eq. 3.1):

$$f(y) = \left(y^{\lambda} - 1\right)/\lambda \qquad (3.1)$$

where $\lambda = 1$ for the plain chord transformation, 0.5 for the Hellinger transformation and 0 for the log-chord transformation. Note that $\lambda = 0$ actually refers to the limit of $f(y)$ when λ approaches 0, which is $\ln(y)$, or $\ln(y + 1)$ for community composition data (see Chap. 2). This sequence of transformations allows the normalization of increasingly asymmetric frequency distributions. Any other exponent between 1 and 0, for example exponent 0.25 (double square root), could be used to pre-transform the data before the chord transformation is applied.

```
# Load the required packages
library(ade4)
library(adespatial)
library(vegan)
library(gclus)
library(cluster)
library(FD)

# Source additional functions that will be used later in this
# Chapter. Our scripts assume that files to be read are in
# the working directory.
source("coldiss.R")
source("panelutils.R")
```

```
# Load the data
# File Doubs.Rdata is assumed to be in the working directory
load("Doubs.Rdata")
# Remove empty site 8
spe <- spe[-8, ]
env <- env[-8, ]
spa <- spa[-8, ]

## Q-mode dissimilarity and distance measures for
## (semi-)quantitative data
# Percentage difference (aka Bray-Curtis) dissimilarity matrix
# on raw species data
spe.db <- vegdist(spe)  # method = "bray" (default)
head(spe.db)
# Percentage difference (aka Bray-Curtis) dissimilarity matrix
# on log-transformed abundances
spe.dbln <- vegdist(log1p(spe))
head(spe.dbln)
# Chord distance matrix
spe.dc <- dist.ldc(spe, "chord")
   # Alternate, two-step computation in vegan:
   spe.norm <- decostand(spe, "nor")
   spe.dc <- dist(spe.norm)
head(spe.dc)
# Hellinger distance matrix
spe.dh <- dist.ldc(spe) # Hellinger is the default distance
   # Alternate, two-step computation in vegan:
   spe.hel <- decostand(spe, "hel")
   spe.dh <- dist(spe.hel)
head(spe.dh)
# Log-chord distance matrix
spe.logchord <- dist.ldc(spe, "log.chord")
   # Alternate, three-step computation in vegan:
   spe.ln <- log1p(spe)
   spe.ln.norm <- decostand(spe.ln, "nor")
   spe.logchord <- dist(spe.ln.norm)
head(spe.logchord)
```

Hint *Type* **`?log1p`** *to see what you have done to the data prior to building the second percentage difference dissimilarity matrix. Why use* **`log1p`** *rather than* **`log`**?

3.3.2 Q Mode: Binary (Presence-Absence) Species Data

When the only data available are binary, or when the abundances are irrelevant, or, sometimes, when the data table contains quantitative values of uncertain or unequal quality, the analyses are done on presence-absence (1–0) data.

The Doubs fish dataset is quantitative, but for the sake of the example we shall compute binary coefficients from these data (using the appropriate arguments when needed): all values larger than 0 will be given a value equal to 1. The exercise consists in computing several dissimilarity matrices based on appropriate similarity coefficients: the Jaccard (S_7) and Sørensen (S_8) similarities. For each pair of sites, the Jaccard similarity is the ratio between the number of double 1's and the number of species, excluding the species represented by double zeros in the pair of objects considered. Therefore, a Jaccard similarity of 0.25 means that 25% of the total number of species observed at two sites were present in both sites and 75% in one site only. The Jaccard dissimilarity computed in **R** is either (1–0.25) or $\sqrt{1-0.25}$ depending on the package used. The Sørensen similarity (S_8) gives double weight to the number of double 1's; its reciprocal (complement to 1) is equivalent to the percentage difference (aka Bray-Curtis) dissimilarity computed on species presence-absence data.

A further interesting relationship is that the Ochiai similarity (S_{14}), which is also appropriate for species presence-absence data, is related to the chord, Hellinger and log-chord distances. Computing either one of these distances on presence-absence data, followed by division by $\sqrt{2}$, produces $\sqrt{1-\mathit{Ochiai\ similarity}}$. This reasoning shows that the chord, Hellinger and log-chord transformations are meaningful for species presence-absence data. This is also the case for the chi-square transformation: the Euclidean distance computed on data transformed in that way produces the chi-square distance, which is appropriate for both quantitative and presence-absence data.

```
## Q-mode dissimilarity measures for binary data

# Jaccard dissimilarity matrix using function vegdist()
spe.dj <- vegdist(spe, "jac", binary = TRUE)
head(spe.dj)
head(sqrt(spe.dj))
# Jaccard dissimilarity matrix using function dist()
spe.dj2 <- dist(spe, "binary")
head(spe.dj2)
# Jaccard dissimilarity matrix using function dist.binary()
spe.dj3 <- dist.binary(spe, method = 1)
head(spe.dj3)
# Sorensen dissimilarity matrix using function dist.ldc()
spe.ds <- dist.ldc(spe, "sorensen")
```

```
# Sorensen dissimilarity matrix using function vegdist()
spe.ds2 <- vegdist(spe, method = "bray", binary = TRUE)
# Sorensen dissimilarity matrix using function dist.binary()
spe.ds3 <- dist.binary(spe, method = 5)
head(spe.ds)
head(spe.ds2)
head(sqrt(spe.ds2))
head(spe.ds3)
# Ochiai dissimilarity matrix
spe.och <- dist.ldc(spe, "ochiai")   # or
spe.och <- dist.binary(spe, method = 7)
head(spe.och)
```

Note: no preliminary binary transformation of the data (`decostand(.., "pa")`*) is necessary since all functions that compute binary dissimilarity indices make the data binary before computing the coefficients.* **`dist.binary()`** *does that automatically whereas* **`vegdist()`** *requires argument* `binary = TRUE`*. In* **`dist.ldc()`***, the default is* `binary = FALSE` *except for the Jaccard, Sørensen and Ochiai indices.*

The display of several values of the alternate versions of the Jaccard and Sørensen dissimilarity matrices show differences. Go back to the introduction of Sect. 3.3 to understand why.

Hint *Explore the arguments of the functions* **`vegdist()`**, **`dist.binary()`** *and* **`dist()`** *to see which coefficients are available. Some of them are available in more than one function. In* **`dist()`**, *argument* `binary` *produces (1 – Jaccard). In the help file of* **`dist.binary()`**, *be careful: the numbering of the coefficients does* ***not*** *follow Legendre and Legendre (2012), but Gower and Legendre (1986).*

Association matrices are generally intermediate entities in data analyses. They are rarely examined directly. However, when not too many objects are involved, it may be useful to display them in a way that emphasizes their main features. We suggest that you plot several dissimilarity matrices using our additional function **`coldiss()`**. A reordering feature is included in **`coldiss()`**, which uses the function **`order.single()`** of the **gclus** package to reorder each dissimilarity matrix, so that similar sites are displayed close together along the main diagonal. Therefore, you can compare the results obtained before and after reordering each matrix.

The package **gclus** is called within the **`coldiss()`** function, so that it must have been installed prior to running the code that follows. Figure 3.1 shows an example.

Fig. 3.1 Heat maps of a percentage difference (aka Bray-Curtis) dissimilarity matrix computed on the raw fish data

```
## Graphical display of association matrices
# Colour plots (also called heat maps, or trellis diagrams in the
# data analysis literature) using the coldiss() function

# Usage:
# coldiss(D = dissimilarity.matrix,
#         nc = 4,
#         byrank = TRUE,
#         diag = FALSE)
# If D is not a dissimilarity matrix (max(D) > 1), then D is
# divided by max(D)
# nc number of colours (classes)
# byrank =  TRUE    equal-sized classes
# byrank =  FALSE   equal-length intervals
# diag = TRUE   print object labels also on the diagonal

## Compare dissimilarity and distance matrices obtained from the
## species data. Four colours are used with equal-length intervals
# Percentage difference (aka Bray-Curtis) dissimilarity matrix on
# raw species abundance data
coldiss(spe.db, byrank = FALSE, diag = TRUE)
# Same but on log-transformed data
coldiss(spe.dbln, byrank = FALSE, diag = TRUE)
```

Compare the two percentage difference plots (raw and log-transformed data; the latter is not presented here). The differences are due to the log transformation. In the untransformed dissimilarity matrix, small differences in abundant species have the same importance as small differences in species with few individuals.

```
# Chord distance matrix
coldiss(spe.dc, byrank = FALSE, diag = TRUE)
# Hellinger distance matrix
coldiss(spe.dh, byrank = FALSE, diag = TRUE)
# Log-chord distance matrix
coldiss(spe.logchord, byrank = FALSE, diag = TRUE)
```

Compare the four colour plots. They all represent distance or dissimilarity matrices built upon quantitative abundance data. Are they similar?

```
# Jaccard dissimilarity matrix
coldiss(spe.dj, byrank = FALSE, diag = TRUE)
```

Compare the Jaccard plot with the previous ones. The Jaccard plot is based on binary data. Does this influence the result? Is the difference more important than between the various plots based on quantitative coefficients?

Although these examples deal with species data, the Doubs transect is characterized by strong ecological gradients (e.g. oxygen, nitrate content; see Chap. 2). In such a well-defined context, it may be interesting to assume for discussion that species are absent for similar reasons from a given section of the stream, and compute an association matrix based on a symmetrical coefficient for comparison purposes. Here is an example using the simple matching coefficient S_1 (presented in Sect. 3.3.4).

```
# Simple matching dissimilarity
# (called the Sokal and Michener index in ade4)
spe.s1 <- dist.binary(spe, method = 2)
coldiss(spe.s1^2, byrank = FALSE, diag = TRUE)
```

Compare this symmetrical dissimilarity matrix with the Jaccard matrix. Which dissimilarities are the most affected by taking, or not, double zeros into account?

The Code It Yourself corner #1

Write several lines of code to compute Jaccard's "coefficient of community" (S_7) between sites #15 and 16 of the **spe** *data frame. Sites 15 and 16 now have row* *numbers 14 and 15 respectively because we deleted row #8 which is devoid of species.*

Jaccard's similarity index has the following formula for two objects x_1 *and* x_2*:*

$S_{(x1,x2)} = a/(a + b + c)$

where a is the number of double 1's, and b and c are the numbers of 0–1 and 1–0 combinations.

After the computation, convert the similarity value to a dissimilarity using the formulas used in **vegan** *and* **ade4** *(not the same formula).*

For the aficionados: display the code used in the **ade4** *function* **dist.binary** *(just type* `dist.binary`*). Look how the quantities a, b, c and d are computed in just four lines for whole data frames. A fine example of programming elegance by Daniel Chessel and Stéphane Dray.*

3.3.3 Q Mode: Quantitative Data (Excluding Species Abundances)

For quantitative variables with a clear interpretation of double zeros, the queen of the symmetrical distance measures is the **Euclidean distance** D_1. "*It is computed using Pythagora's formula, from site-points positioned in a* p*-dimensional space called a* metric *or* Euclidean space" (Legendre and Legendre 2012, p. 299).

The Euclidean distance has no upper limit and its value is strongly influenced by the scale of each descriptor. For instance, changing the scale of a set of measurements from g/L to mg/L multiplies the contribution of that descriptor to the Euclidean distance by 1000. Therefore, the use of the Euclidean distance on raw data is restricted to datasets that are dimensionally homogeneous, e.g. geographic coordinates expressed in km. Otherwise, D_1 is computed on standardized variables (*z*-scores). Standardization is also applied to situations where one wishes to give the same weight to all variables in a set of dimensionally homogeneous descriptors.

Here you could compute a matrix of Euclidean distances on the (standardized) environmental variables of our **env** dataset. We shall remove one variable, **dfs** (distance from the source), since it is a spatial rather than an environmental descriptor. The results will be displayed using **coldiss()** (Fig. 3.2).

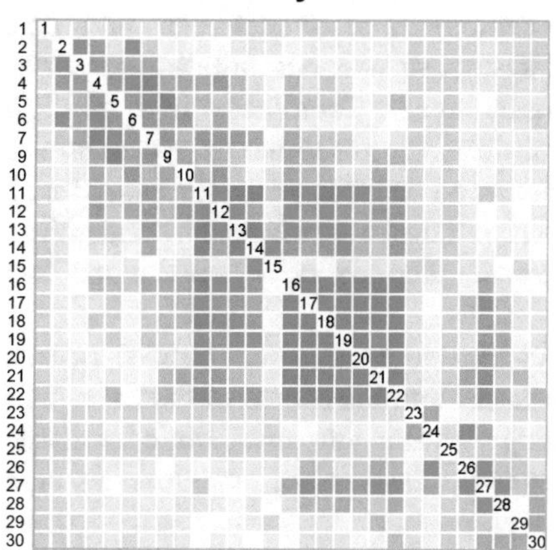

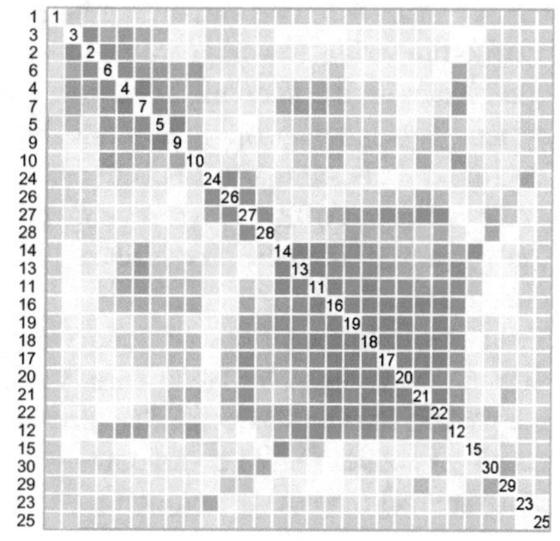

Fig. 3.2 Heat maps of a matrix of Euclidean distances on the standardized environmental variables

```
# Remove the 'dfs' variable from the env dataset
env2 <- env[,-1]

# Euclidean distance matrix of the standardized env2 data frame
env.de <- dist(scale(env2))
coldiss(env.de, nc = 16, diag = TRUE)
```

Hint *See how the environmental variables have been standardized "on the fly" using the function* **scale()**.

Such plots of dissimilarity matrices can be used for a quick comparison. For instance, you could plot the Hellinger species-based distance matrix and the environmental distance matrix, both using equal-sized categories (`byrank` = `TRUE`, the default), in order to compare them visually:

```
# Hellinger distance matrix of the species data
# Use nc = 16 equal-sized classes
coldiss(spe.dh, nc = 16, diag = TRUE)
```

Compare the left-hand plots of this and the previous pair of heat maps, since they present the sites in the same order. Do you observe common features?

The Euclidean distance is of course appropriate to compute matrices of geographical distances based on variables giving the coordinates of the sites in any units on a 1 or 2-dimensional orthogonal (Cartesian) system such as cm, m or km (or UTM coordinates if the sites all belong to the same zone). Coordinates in a spherical system (e.g. latitude-longitude) must be transformed prior to the computation of a Euclidean distance matrix. This transformation can be done by function **geoXY()** of the package **SoDA**. Note that geographical coordinates should *not* be standardized (only centred if necessary), since this would alter the ratio between the two dimensions.

In the following lines of code you will also compute a matrix of Euclidean distance on a single variable: **dfs**, the distance from the source of the river. This matrix will thus represent the distances among sites along the river, while the matrix based on spatial (X-Y) coordinates will represent the distance among points on a geographical map (as the crow flies, so to say).

```
# Euclidean distance matrix on spatial coordinates (2D)
spa.de <- dist(spa)
coldiss(spa.de, nc = 16, diag = TRUE)

# Euclidean distance matrix on distance from the source (1D)
dfs.df <- as.data.frame(env$dfs, row.names = rownames(env))
riv.de <- dist(dfs.df)
coldiss(riv.de, nc = 16, diag = TRUE)
```

Why are the X-Y plot and the Euclidean distance from the source plot so different?

3.3.4 Q Mode: Binary Data (Excluding Species Presence-Absence Data)

The simplest symmetrical similarity measure for binary data is the "simple matching coefficient" S_1. For each pair of sites, it is the ratio between the number of double 1's plus double 0's and the total number of variables.

The fish environment dataset is exclusively made of quantitative variables, so we shall create fictitious data to demonstrate the computation of S_1. We shall resort to this method from time to time, just to show how it is convenient to create data sets of known characteristics in **R**, for instance for simulation purposes.

```
# Compute five binary variables with 30 objects each.
# Each variable has a predefined number of 0 and 1
# Variable 1: 10 x 1 and 20 x 0; the order is randomized
var1 <- sample(c(rep(1, 10), rep(0, 20)))
# Variable 2: 15 x 0 and 15 x 1, one block each
var2 <- c(rep(0, 15), rep(1, 15))
# Variable 3: alternation of 3 x 1 and 3 x 0 up to 30 objects
var3 <- rep(c(1, 1, 1, 0, 0, 0), 5)
# Variable 4: alternation of 5 x 1 and 10 x 0 up to 30 objects
var4 <- rep(c(rep(1, 5),  rep(0, 10)),  2)
# Variable 5: 16 objects with randomized distribution of 7 x 1
# and 9 x 0,  followed by 4 x 0 and 10 x 1
var5.1 <- sample(c(rep(1, 7),  rep(0, 9)))
var5.2 <- c(rep(0, 4),  rep(1, 10))
var5 <- c(var5.1,  var5.2)

# Variables 1 to 5 are put into a data frame
(dat <- data.frame(var1, var2, var3, var4, var5))
dim(dat)

# Computation of a matrix of simple matching coefficients
# (called the Sokal and Michener index in ade4)
dat.s1 <- dist.binary(dat, method = 2)
coldiss(dat.s1, diag = TRUE)
```

3.3.5 Q Mode: Mixed Types Including Categorical (Qualitative Multiclass) Variables

Among the association measures that can handle nominal data correctly, one is readily available in **R**: Gower's similarity S_{15}. This coefficient has been devised to handle data containing variables of various mathematical types, each variable receiving a treatment corresponding to its category. The final (dis)similarity between two objects is obtained by averaging the partial (dis)similarities computed for all variables separately. We shall use Gower's similarity as a symmetrical index; when a variable is declared as a *factor* in a data frame, the simple matching rule is applied, i.e., for each pair of objects the similarity is 1 for that variable if the factor has the same level in the two objects and 0 if the level is different. One function to compute Gower's dissimilarity is **`daisy()`** of package **`cluster`**. Avoid the use of **`vegdist`**() (`method = "gower"`), which is appropriate for quantitative and presence-absence data, but not for multiclass categorical variables.

`daisy()` can handle data frames made of mixed-type variables, provided that each variable is correctly defined. Optionally, the user can provide an argument (in the form of a list) specifying the types of some or all variables in the dataset. When there are missing values (coded NA) in the data, the function excludes from the comparison of two sites a variable where one or the other object (or both) has a missing value.

gowdis() of package **FD** is the most complete function to compute Gower's coefficient. It computes the distance for mixed variables including asymmetrical binary variables. It provides three ways of handling ordinal variables, including the method of Podani (1999). It offers the same treatment of missing values as in **daisy()** and allows users to attribute different weights to the variables.

Let us again create an artificial dataset containing four variables: two random quantitative variables and two factors:

```
# Fictitious data for Gower (S15) index
# Random normal deviates with zero mean and unit standard deviation
var.g1 <- rnorm(30, 0, 1)
# Random uniform deviates from 0 to 5
var.g2 <- runif(30, 0, 5)
# Factor with 3 levels (10 objects each)
var.g3 <- gl(3, 10,  labels = c("A", "B", "C"))
# Factor with 2 levels,  orthogonal to var.g3
var.g4 <- gl(2, 5, 30,  labels = c("D", "E"))
```

Together, var.g3 and var.g4 represent a 2-way crossed balanced design.

```
dat2 <- data.frame(var.g1, var.g2, var.g3, var.g4)
summary(dat2)
```

Hints *Function* **gl()** *is quite handy to generate factors, but by default it uses numerals as labels for the levels. Use argument* `labels` *to provide alphanumeric characters instead of numbers.*

Note the use of **data.frame()** *to assemble the four variables. Unlike* **cbind()**, **data.frame()** *preserves the classes of the variables. Variables 3 and 4 thus retain their class* `"factor"`.

Let us first compute and view the complete S_{15} matrix. Then repeat the computation using the two factors (`var.g3` and `var.g4`) only:

```
# Computation of a matrix of Gower dissimilarity using
# function daisy()

# Complete data matrix (4 variables)
dat2.S15 <- daisy(dat2, "gower")
range(dat2.S15)
coldiss(dat2.S15, diag = TRUE)

# Data matrix with the two orthogonal factors only
dat2partial.S15 <- daisy(dat2[, 3:4], "gower")
coldiss(dat2partial.S15, diag = TRUE)
head(as.matrix(dat2partial.S15))

# What are the dissimilarity values in the dat2partial.S15 matrix?
levels(factor(dat2partial.S15))
```

The values correspond to pairs of objects that share the same levels for 2, 1 or no factor. Pairs with the highest dissimilarity values share no common levels.

```
# Computation of a matrix of Gower dissimilarity using
# function gowdis() of package FD
?gowdis
dat2.S15.2 <- gowdis(dat2)
range(dat2.S15.2)
coldiss(dat2.S15.2, diag = TRUE)

# Data matrix with the two orthogonal factors only
dat2partial.S15.2 <- gowdis(dat2[ , 3:4])
coldiss(dat2partial.S15.2, diag = TRUE)
head(as.matrix(dat2partial.S15.2))

# What are the dissimilarity values in the dat2partial.S15.2
# matrix?
levels(factor(dat2partial.S15.2))
```

3.4 R Mode: Computing Dependence Matrices Among Variables

Correlation-type coefficients must be used to compare variables in the R mode. These include the Pearson as well as the non-parametric correlation coefficients (Spearman, Kendall) for quantitative or ordinal data, and contingency statistics for the comparison of categorical variables (chi-square statistic and derived forms). For presence-absence data, binary coefficients such as the Jaccard, Sørensen and Ochiai coefficients can be used in the R mode to compare species.

3.4.1 R Mode: Species Abundance Data

Covariances as well as parametric and non-parametric correlation coefficients are often used to compare species distributions through space or time. Note that double-zeros as well as the joint variations in abundance contribute to increase the correlations. In the search for species associations, Legendre (2005) applied one of the transformations described in Sect. 3.5 in order to remove the effect of the total abundance per site prior to the calculation of parametric and non-parametric correlations. Some concepts of species association use only the positive covariances or correlations to recognize associations of co-varying species.

Besides correlations, the chi-square distance, which was used in the Q mode, can also be computed on transposed matrices (R mode).

The example below shows how to compute and display an R-mode chi-square dissimilarity matrix of the 27 fish species:

```
# Transpose matrix of species abundances
spe.t <- t(spe)
# Chi-square pre-transformation followed by Euclidean distance
spe.t.chi <- decostand(spe.t, "chi.square")
spe.t.D16 <- dist(spe.t.chi)
coldiss(spe.t.D16, diag = TRUE)
```

Can you identify groups of species in the right-hand display?

3.4.2 R Mode: Species Presence-Absence Data

For binary species data, the Jaccard (S_7), Sørensen (S_8) and Ochiai (S_{14}) coefficients can also be used in the R mode. Apply S_7 to the fish presence-absence data after transposition of the matrix (object `spe.t`):

```
# Jaccard index on fish presence-absence
spe.t.S7 <- vegdist(spe.t, "jaccard", binary = TRUE)
coldiss(spe.t.S7, diag = TRUE)
```

Compare the right-hand display with the one obtained with the chi-square distance. Are the species groups consistent?

3.4.3 R Mode: Quantitative and Ordinal Data (Other than Species Abundances)

To compare dimensionally homogeneous quantitative variables, one can use either the covariance or Pearson's r correlation coefficient. Note, however, that these indices are *linear*, so that they may perform poorly to detect monotonic but nonlinear relationships among variables. If the variables are not dimensionally homogeneous, Pearson's r must be preferred to the covariance, since the correlation r is actually the covariance computed on standardized variables.

Comparison among ordinal variables, or among quantitative variables that may be monotonically but not linearly related, can be achieved using rank correlation coefficients like Spearman's ρ (rho) or Kendall's τ (tau).

Here are some examples based on the fish environmental data **env**. The function **cor()** (**stats** package) computes correlations among the columns of the *untransposed* matrix, i.e., the original matrix where the variables are in columns. First example: Pearson r (Fig. 3.3):

```
# Pearson r linear correlation among environmental variables
env.pearson <- cor(env) # default method = "pearson"
round(env.pearson, 2)
# Reorder the variables prior to plotting
env.o <- order.single(env.pearson)

# pairs() is a function to plot a matrix of bivariate scatter
# plots. panelutils.R is a set of functions that add useful
# features to pairs().
pairs(env[ ,env.o],
      lower.panel = panel.smooth,
      upper.panel = panel.cor,
      diag.panel = panel.hist,
      main = "Pearson Correlation Matrix")
```

Identify the variables correlated with variable 'dfs' (the distance from the source). What story does that tell?

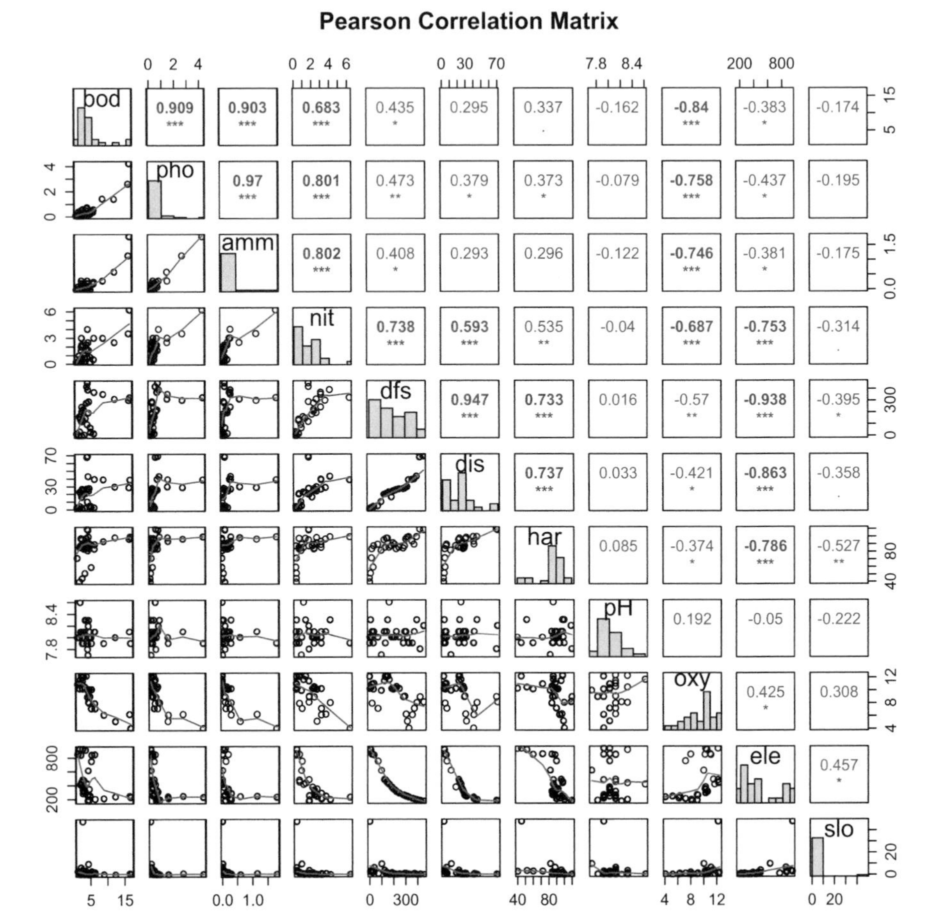

Fig. 3.3 Multipanel display of pairwise relationships between environmental variables with Pearson r correlations

The same variables are now compared using Kendall's τ:

```
# Kendall tau rank correlation among environmental variables,
# no colours
env.ken <- cor(env, method = "kendall")
env.o <- order.single(env.ken)
pairs(env[ ,env.o], lower.panel = panel.smoothb,
    upper.panel = panel.cor, no.col = TRUE,
    method = "kendall", diag.panel = panel.hist,
    main = "Kendall Correlation Matrix")
```

Judging by these plots of bivariate relationships, would you favour the use of Kendall's tau or Pearson's r*?*

3.4.4 R Mode: Binary Data (Other than Species Abundance Data)

The simplest way of comparing pairs of binary variables is to compute a matrix of Pearson's *r*. In that case Pearson's *r* is called the *point correlation coefficient* or Pearson's φ. That coefficient is closely related to the chi-square statistic for 2 × 2 tables without correction for continuity: $\chi^2 = n\varphi^2$ where *n* is the number of objects.

3.5 Pre-transformations for Species Data

In Sect. 3.2.2, we explained why species abundance data should be treated in a special way, avoiding the use of double zeros as indications of resemblance among sites; linear methods, which explicitly or implicitly use the Euclidean distance among sites or the covariance or correlation among variables, are therefore not appropriate for such data. Unfortunately, many of the most powerful statistical tools available to ecologists, like ANOVA, *k*-means partitioning (see Chap. 4), principal component analysis (PCA, see Chap. 5) and redundancy analysis (RDA, see Chap. 6) *are* methods of linear analysis. Consequently, these methods were more or less "forbidden" to species data until Legendre and Gallagher (2001) showed that several asymmetrical association measures (i.e., measures that are appropriate to species data) can be obtained in two computation steps: a transformation of the raw data followed by the calculation of the Euclidean distance. These two steps reconstruct the asymmetrical distance among sites, therefore allowing the use of all linear methods of analysis with species data.

As will be seen in the following chapters, in many cases one simply has to apply the pre-transformation to the species data, and then feed these to the linear methods of data analysis: PCA, RDA, *k*-means, and so on.

Legendre and Gallagher proposed five pre-transformations of the species data[9]. Four of them are available in **vegan** as arguments of the function **decostand()**: profiles of relative abundances by site (`"total"`)[10], site normalization, also called the chord transformation (`"normalize"`), Hellinger transformation (`"hellinger"`), and chi-square double standardization (`"chi.square"`). We can add the log-chord transformation to this list. See Sects. 2.2.4 and 3.3.1 for examples. All these transformations express the data as relative abundances per sites

[9]These authors proposed two forms of chi-square transformation. These forms are closely related, so that implementing only one is sufficient for data analysis.

[10]Note that Legendre and De Cáceres (2013) have shown that, contrary to the other transformations, the distance between species profiles lacks important properties to study beta diversity and should therefore be avoided in this wide context.

(site profiles) in some way; this removes from the data the total abundance per site, which is the response of the species to the total productivity of the sites. In the Hellinger and log-chord transformations, the relative abundance values are respectively square-rooted or their log is taken, which further reduces the importance of the highest abundance values.

Table 3.1 Some distance and dissimilarity functions in Q mode that are useful to ecologists and are available in widely used R packages. The symbol ⇒ means that applying the function designed for quantitative data to presence-absence data produces the same result as computing the corresponding function designed for presence-absence data.

Quantitative data		Presence-absence data
Community composition data		
Ružička dissimilarity **vegdist(.,"jac")** **dist.ldc(., "ruzicka")**	⇒	Jaccard dissimilarity **vegdist(., "jac", binary=TRUE)** **dist.ldc(., "jaccard")** **dist.binary(., method=1)**
Chord distance **dist.ldc(., "chord")** **decostand(., "norm")** followed by **vegdist(., "euc")**	⇒	Ochiai dissimilarity **dist.ldc(., "ochiai")** **dist.binary(., method=7)**
Hellinger distance **dist.ldc(., "hellinger")** **decostand(.,"hel")** followed by **vegdist(.,"euc")**	⇒	Ochiai dissimilarity **dist.ldc(., "ochiai")** **dist.binary(., method=7)**
Log-chord distance **dist.ldc(., "log.chord")** **decostand(log1p(.), "norm")** followed by **vegdist(., "euc")**	⇒	Ochiai dissimilarity **dist.ldc(., "ochiai")** **dist.binary(., method=7)**
Percentage difference dissimilarity **dist.ldc(., "percentdiff")** **vegdist(., "bray")**	⇒	Sørensen dissimilarity **dist.ldc(., "sorensen")** **dist.binary(., method=5)**
Chi-square distance **dist.ldc(., "chisquare")** **decostand(., "chi.square")** followed by **vegdist(.,"euc")**		Chi-square distance **dist.ldc(., "chisquare")** **decostand(.,"chi.square")** followed by **vegdist(.,"euc")**
Canberra distance **dist.ldc(., "canberra")** **vegdist(., "canberra")**		
Other variables, mixed physical units		
Standardized variables: Euclidean distance **vegdist(., "euc")**		Standardized variables: Simple matching coefficient **dist.binary(., method=2)**
Non-standardized variables: Gower dissimilarity **daisy(., "gower")**		

3.6 Conclusion

Although association matrices are in most cases intermediate entities in a data analysis, this chapter has shown that their computation deserves close attention. Many choices are available, and crucial decisions must be made at this step of the analytical procedure. The graphical tools presented in this chapter are precious helps to make these decisions, but great care must be taken to remain on solid theoretical ground. The success of the analytical steps following the computation of an association matrix largely depends on the choice of an appropriate association measure. Commonly used distance functions in the Q mode, available in **R** packages, are presented in Table 3.1.

The similarity coefficients for presence-absence data and the Gower similarity may be transformed into dissimilarities using $\sqrt{1-S}$ to avoid the production of negative eigenvalues and complex eigenvectors in principal coordinate analysis. This transformation is made automatically by **ade4**, but not by **vegan**. **dist.ldc()** of **adespatial** does this transformation for Jaccard, Sørensen and Ochiai.

In Table 3.1, functions **decostand()** and **vegdist()** belong to package **vegan**; function **dist.binary()** belongs to package **ade4**; function **daisy()** belongs to package **cluster**. Other **R** functions propose dissimilarity coefficients. Some are mentioned in the course of this chapter. The Euclidean distance can be computed either by **vegdist(.,"euc")** as shown in the table, by **daisy(.,"euc")**, by **dist.ldc(.,"euc")** or by **dist(.)**.

Chapter 4
Cluster Analysis

4.1 Objectives

In most cases, data exploration (Chap. 2) and the computation of association matrices (Chap. 3) are preliminary steps towards deeper analyses. In this chapter you will go further by experimenting one of the large groups of analytical methods used in ecology: clustering. Practically, you will:

- learn how to choose among various clustering methods and compute them;
- apply these techniques to the Doubs River data to identify groups of sites and fish species.
- explore two methods of constrained clustering, a powerful modelling approach where the clustering process is constrained by an external data set.

4.2 Clustering Overview

The objective of clustering is to recognize discontinuous subsets in an environment that is sometimes discrete (as in taxonomy), but most often perceived as continuous in ecology. This requires some degree of abstraction, but ecologists may want to get a simplified, structured view of their data; generating a typology is one way to achieve that goal. In some instances, typologies are compared to independent classifications (based on theory or on other typologies obtained from independent data). What we present here is a collection of methods used to decide whether objects are similar enough to be allocated to a group, and identify the distinctions or separations between groups.

Clustering consists in partitioning the collection of objects (or descriptors in R-mode) under study. A hard *partition* is a division of a set (collection) into subsets, such that each object or descriptor belongs to one and only one subset for that partition (Legendre and Rogers 1972). For instance, a species cannot be

D. Borcard et al., *Numerical Ecology with R*, Use R!,
https://doi.org/10.1007/978-3-319-71404-2_4

simultaneously the member of two genera: membership is binary (0 or 1). Some methods, less commonly used, consider fuzzy partitions, in which membership is continuous (between 0 and 1). Depending on the clustering model, the result can be a single partition or a series of hierarchically nested partitions. Except for its constrained version, clustering is *not* a statistical method in the strict sense since it does not test any a priori hypothesis. Post-hoc cluster robustness tests are available, however (see Sect. 4.7.3.3). Clustering helps bring out some features hidden in the data; it is the user who decides if these structures are interesting and worth interpreting in ecological terms.

Note that most clustering methods are computed from association matrices, which stresses the importance of the choice of an appropriate association coefficient.

Clustering methods differ by their algorithms, which are the effective methods for solving a problem using a finite sequence of instructions. One can recognize the following families of clustering methods:

1. *Sequential or simultaneous algorithms*. Most clustering algorithms are sequential and consist in the repetition of a given procedure until all objects have found their place. The less frequent simultaneous algorithms find the solution in a single step.
2. *Agglomerative or divisive*. Among the sequential algorithms, agglomerative procedures begin with the discontinuous collection of objects, which are successively grouped into larger and larger clusters until a single, all-encompassing cluster is obtained. Divisive methods, on the contrary, start with the collection of objects considered as one single group, and divide it into subgroups, and so on until the objects are completely separated. In either case it is left to the user to decide which of the intermediate partitions should be retained, given the problem under study.
3. *Monothetic versus polythetic*. Divisive methods may be monothetic or polythetic. Monothetic methods use a single descriptor (the one that is considered the best for that level) at each step for partitioning, whereas polythetic methods use all descriptors; in most polythetic cases, the descriptors are combined into an association matrix.
4. *Hierarchical versus non-hierarchical methods*. In hierarchical methods, the members of inferior-ranking clusters become members of larger, higher-ranking clusters. Most of the time, hierarchical methods produce non-overlapping clusters. Non-hierarchical methods (e.g. *k*-means partitioning) produce a single partition, without any hierarchy among the groups.
5. *Probabilistic versus non-probabilistic methods*. Probabilistic methods define groups in such a way that the within-group association matrices have a given probability of being homogeneous. Probabilistic methods are sometimes used to define species associations.
6. *Unconstrained or constrained methods*. Unconstrained clustering relies upon the information of a single data set, whereas constrained clustering uses two matrices: the one being clustered, and a second one containing a set of explanatory variables which provides a constraint (or guidance) as to where to group or divide the data of the first matrix.

These categories are not represented equally in the ecologist's toolbox. Most methods presented below are sequential, agglomerative and hierarchical (Sects. 4.3, 4.3.1, 4.3.2, 4.4, 4.5 and 4.6), but others, like *k*-means partitioning, are divisive and non-hierarchical (Sect. 4.8). Two methods are of special interest: Ward's hierarchical clustering and *k*-means partitioning are both least-squares methods. That characteristic relates them to the linear model. In addition, we will explore two methods of constrained clustering, one with sequential constraint (Sect. 4.14) and the other, more general, called multivariate regression tree analysis (MRT, Sect. 4.12).

Hierarchical clustering results are generally represented as dendrograms or similar tree-like graphs. Non-hierarchical procedures produce groups of objects (or variables), which may either be used in further analyses, presented as end-results (for instance species associations) or, when the project has a spatial component, mapped on the area under study.

A clustering of the sites of a species data set is informative by itself, but ecologists often want to interpret it by means of external, environmental variables. We will explore two ways of achieving this goal (Sect. 4.9).

Although most methods in this chapter will be applied to sites, clustering can also be applied to species in order to define species assemblages. Section 4.10 addresses this topic.

The search for indicator or characteristic species in groups of sites is a particularly important question in fundamental and applied ecology. It is presented in Sect. 4.11.

Finally, a brief section (Sect. 4.15) will be devoted to two methods of fuzzy clustering, a non-hierarchical approach that considers partial memberships of objects to clusters.

Before entering the subject, let us prepare our **R** session by loading the necessary packages and preparing the data tables.

```
# Load the required packages
library(ade4)
library(adespatial)
library(vegan)
library(gclus)
library(cluster)
library(pvclust)
library(RColorBrewer)
library(labdsv)
library(rioja)
library(indicspecies)
library(mvpart)
library(MVPARTwrap)
library(dendextend)
library(vegclust)
library(colorspace)
library(agricolae)
library(picante)
```

```
# Source additional functions that will be used later in this
# chapter. Our scripts assume that files to be read are in
# the working directory.
source("drawmap.R")
source("drawmap3.R")
source("hcoplot.R")
source("test.a.R")
source("coldiss.R")
source("bartlett.perm.R")
source("boxplerk.R")
source("boxplert.R")

# Function to compute a binary dissimilarity matrix from clusters
grpdist <- function(X) {
  require(cluster)
  gr <- as.data.frame(as.factor(X))
  distgr <- daisy(gr, "gower")
  distgr
}

# Load the data
# File Doubs.Rdata is assumed to be in the working directory
load("Doubs.Rdata")
# Remove empty site 8
spe <- spe[-8, ]
env <- env[-8, ]
spa <- spa[-8, ]
latlong <- latlong[-8, ]
```

4.3 Hierarchical Clustering Based on Links

4.3.1 Single Linkage Agglomerative Clustering

Also called *nearest neighbour sorting*, this method agglomerates objects on the basis of their shortest pairwise dissimilarities (or greatest similarities): the fusion of an object (or a group) with a group at a given similarity (or dissimilarity) level only requires that one object of each of the two groups about to agglomerate be linked to one another at that level. Two groups agglomerate at the dissimilarity separating the closest pair of their members. This makes agglomeration easy. Consequently, the dendrogram resulting of a single linkage clustering often shows chaining of objects: a pair is linked to a third object, which in turn is linked with another one, and so on. The result may therefore be difficult to interpret in terms of partitions, but gradients are revealed quite clearly. The list of the first connections making an object member of a cluster, or allowing two clusters to fuse, is called the *chain of primary connections*; this chain forms the *minimum spanning tree* (MST). This entity is presented here and will be used in later analyses (Chap. 7).

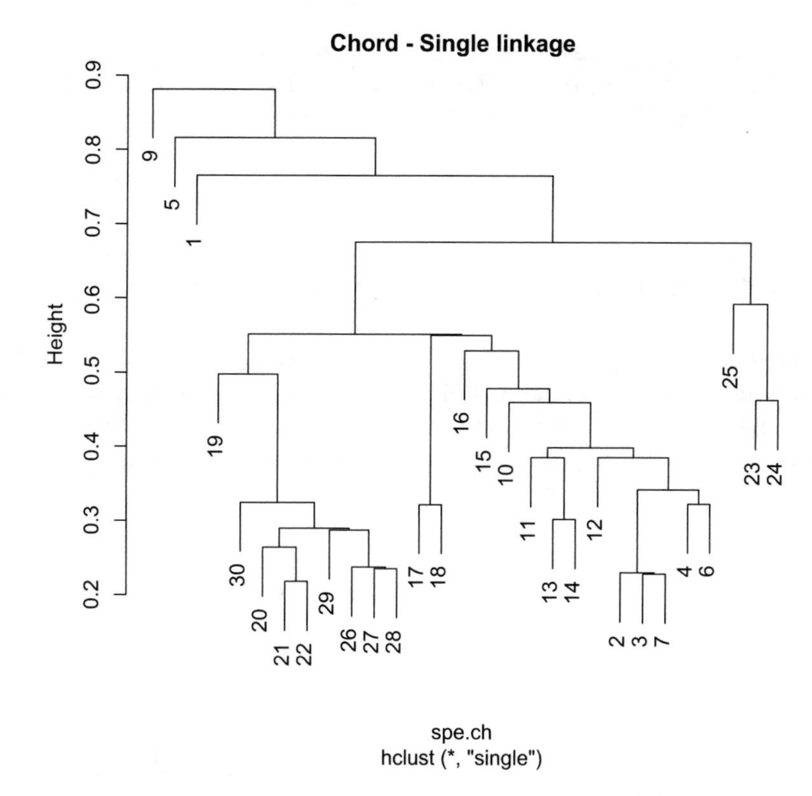

Fig. 4.1 Single linkage agglomerative clustering of a matrix of chord distances among sites (species data)

Common hierarchical clustering methods are available through the function **hclust()** of the **stats** package. You will now compute and illustrate (Fig. 4.1) your first cluster analysis on the basis of an association matrix computed in Chap. 3 and recomputed here for convenience:

```
# Compute matrix of chord distance among sites
spe.norm <- decostand(spe, "normalize")
spe.ch <- vegdist(spe.norm, "euc")

# Attach site names to object of class 'dist'
attr(spe.ch, "labels") <- rownames(spe)

# Compute single linkage agglomerative clustering
spe.ch.single <- hclust(spe.ch, method = "single")
# Plot a dendrogram using the default options
plot(spe.ch.single,
   labels = rownames(spe),
   main = "Chord - Single linkage")
```

Based on this first result, how would you describe the data set? Do you see a single simple gradient, or else distinguishable groups of sites? Can you identify some chaining of the sites? What about sites 1, 5 and 9?

4.3.2 Complete Linkage Agglomerative Clustering

Contrary to single linkage clustering, complete linkage clustering (also called *furthest neighbour sorting*) allows an object (or a group) to agglomerate with another group only at the dissimilarity corresponding to that of the most distant pair of objects; thus, *a fortiori*, all members of the two groups are linked (Fig. 4.2):

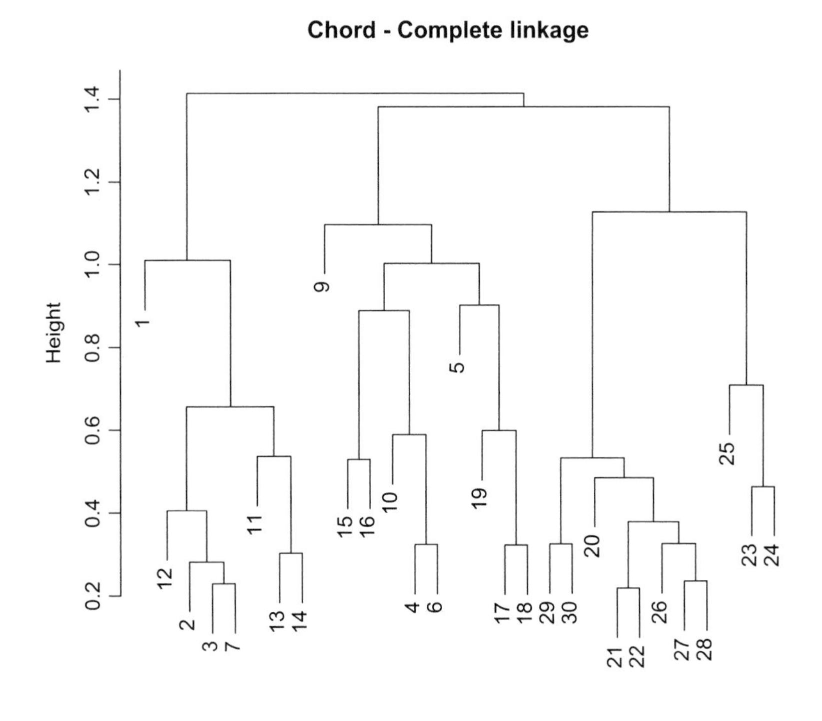

Fig. 4.2 Complete linkage agglomerative clustering of a matrix of chord distance among sites (species data)

```
# Compute complete-linkage agglomerative clustering
spe.ch.complete <- hclust(spe.ch, method = "complete")
plot(spe.ch.complete,
   labels = rownames(spe),
   main = "Chord - Complete linkage")
```

Given that these are sites along a river (with the numbers following the stream), does this result tend to place sites that are neighbours along the river in the same groups?
How can two perfectly valid clustering methods produce such different results when applied to the same data?

The comparison between the two dendrograms (Figs. 4.1 and 4.2) shows the difference in the philosophy and the results of the two methods: single linkage allows an object to agglomerate easily to a group, since a link to a single object of the group suffices to induce fusion. This is a "closest friend" procedure, so to say. The resulting dendrogram does not always show clearly separated groups, but can be used to identify gradients in the data. At the opposite, complete linkage clustering is much more contrasting. A group admits a new member only at a dissimilarity corresponding to the furthest object of the group: one could say that the admission requires unanimity of the members of the group. It follows that the larger a group is, the more difficult it is to agglomerate with it. Complete linkage, therefore, tends to produce many small separate groups, which tend to be rather spherical in multivariate space and agglomerate at large distances. Therefore, this method is interesting to search for and identify discontinuities in data.

4.4 Average Agglomerative Clustering

This family comprises four methods that are based on average dissimilarities among objects or on centroids of clusters. The differences among them are in the way of computing the positions of the groups (arithmetic average versus centroids) and in the weighting or non-weighting of the groups according to the number of objects they contain when computing fusion levels. Table 4.1 summarizes their names and properties.

The best-known method of this family, UPGMA, allows an object to join a group at the mean of the dissimilarities between this object and all members of the group. When two groups join, they do it at the mean of the dissimilarities between all members of one group and all members of the other. Let us apply it to our data (Fig. 4.3):

Table 4.1 The four methods of average clustering. The names in quotes are the corresponding arguments of function **hclust()**

	Arithmetic average	Centroid clustering
Equal weights	Unweighted pair-group method using arithmetic averages (UPGMA) "`average`"	Unweighted pair-group method using centroids (UPGMC) "`centroid`"
Unequal weights	Weighted pair-group method using arithmetic averages (WPGMA) "`mcquitty`"	Weighted pair-group method using centroids (WPGMC) "`median`"

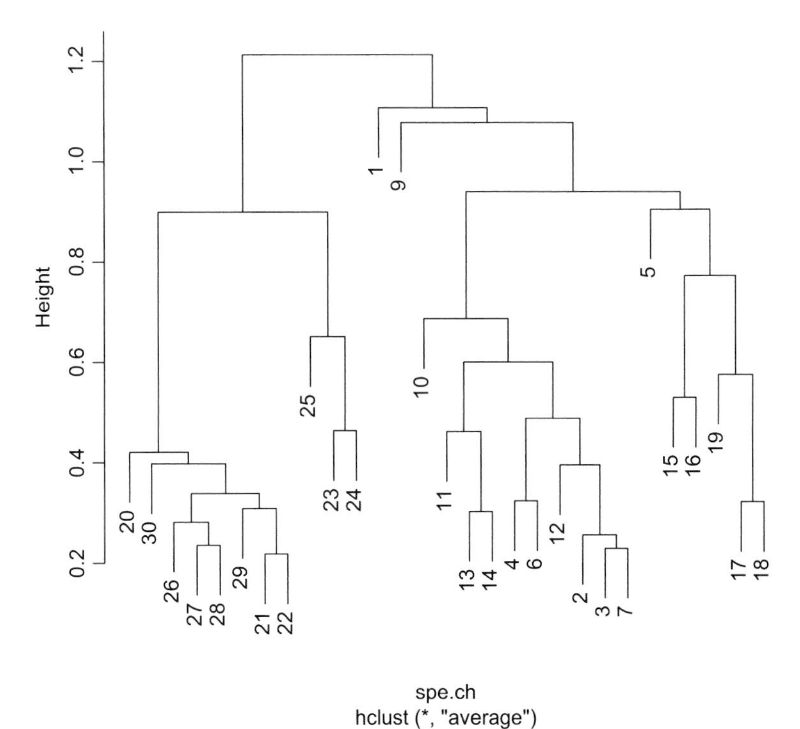

Fig. 4.3 UPGMA clustering of a matrix of chord distance among sites (species data)

> *The result looks somewhat intermediate between a single and a complete linkage clustering. This is often the case.*

```
# Compute UPGMA agglomerative clustering
spe.ch.UPGMA <- hclust(spe.ch, method = "average")
plot(spe.ch.UPGMA,
   labels = rownames(spe),
   main = "Chord - UPGMA")
```

```
# Compute centroid clustering
spe.ch.centroid <- hclust(spe.ch, method = "centroid")
plot(spe.ch.centroid,
   labels = rownames(spe),
   main = "Chord - Centroid")
```

The resulting dendrogram is an ecologist's nightmare. Legendre and Legendre (2012, p. 376) explain how reversals are produced and suggest interpreting them as polychotomies rather than dichotomies.

Note that UPGMC and WPGMC can sometimes lead to reversals in the dendrograms. The result no longer forms a series of nested partitions and may be difficult to interpret. An example is obtained as follows (Fig. 4.4):

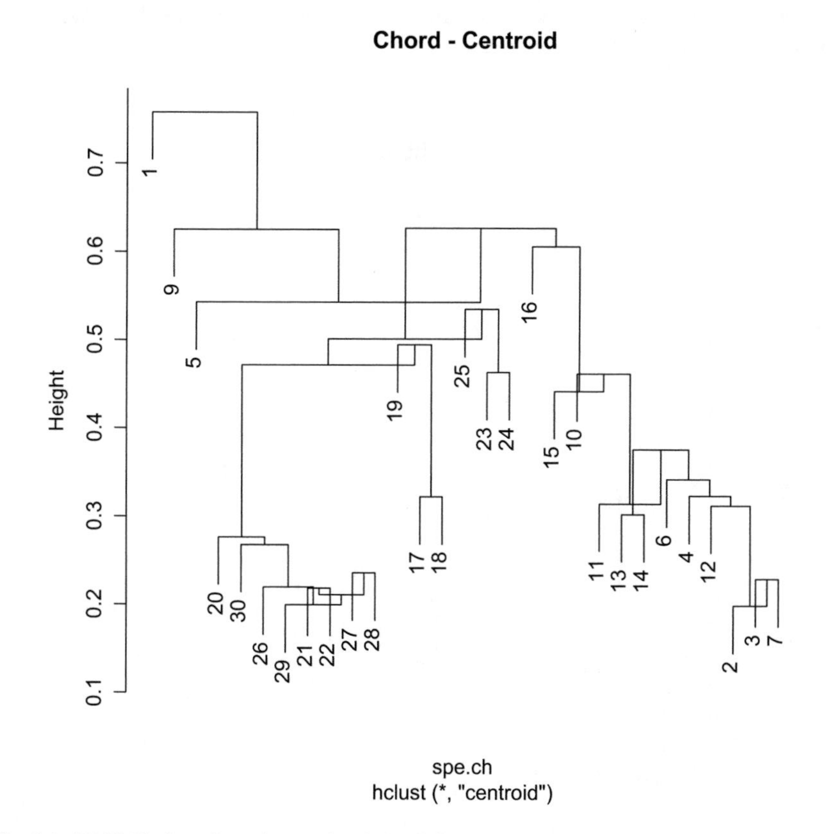

Fig. 4.4 UPGMC clustering of a matrix of chord distance among sites (species data)

4.5 Ward's Minimum Variance Clustering

This method is based on the linear model criterion of least squares. The objective is to define groups in such a way that the within-group sum of squares (i.e., the squared error of ANOVA) is minimized. The within-cluster sum of squared errors can be computed as the sum of the squared distances among members of a cluster divided by the number of objects. Note also that although the computation of within-group sums-of-squares is based on a Euclidean model, the Ward method will produce meaningful results from dissimilarities that are Euclidean or not.

In the literature, two different algorithms are found for Ward clustering; one implements Ward's (1963) minimum variance clustering criterion, the other does not (Murtagh and Legendre 2014). Function **`hclust()`** was modified in R 3.1.1; `method = "ward.D2"` now implements the Ward (1963) criterion where dissimilarities are squared before cluster updating, whereas `method = "ward.D"` does not implement that criterion. The latter was implemented by **`hclust()`** with `method = "ward"` in **R** versions up to 3.0.3. Fig. 4.5 shows the dendrogram resulting from a `ward.D2` clustering.

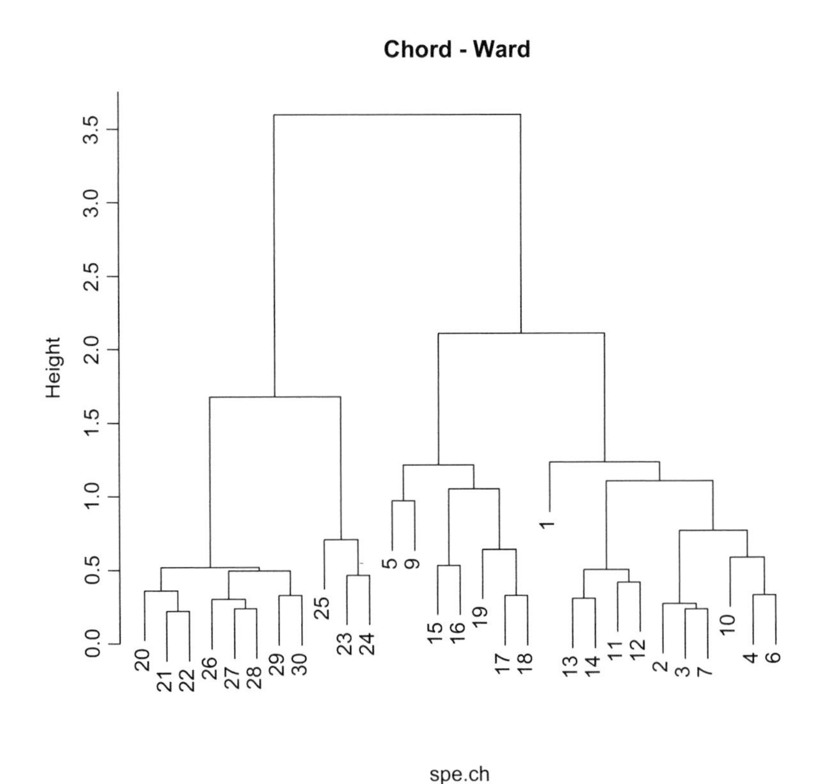

Fig. 4.5 Ward clustering of a matrix of chord distance among sites (species data)

```
# Compute Ward's minimum variance clustering
spe.ch.ward <- hclust(spe.ch, method = "ward.D2")
plot(spe.ch.ward,
   main = "Chord - Ward")
```

Hint *Below you will see more options of the* **plot()** *function which produces dendrograms of objects of class* **hclust**. *Another path is to change the class of such an object using function* **as.dendrogram()**, *which opens yet more possibilities. Type* **?dendrogram** *for details.*

4.6 Flexible Clustering

Lance and Williams (1966, 1967) proposed a model encompassing all the clustering methods described above, which are obtained by changing the values of four parameters α_h, α_i, β and γ. See Legendre and Legendre (2012, p. 370). **hclust()** is implemented using the Lance and Williams algorithm. As an alternative to the examples above, flexible clustering is available in the **R** package **cluster**, function **agnes()**, using arguments `method` and `par.method`. In **agnes()**, flexible clustering is called by argument `method` = `"flexible"` and the parameters are given to the argument `par.method` as a numeric vector of length 1 (α_h only), 3 (α_h, α_i and β, with $\gamma = 0$) or 4. In the simplest application of this rather complex clustering method, flexibility is commanded by the value of parameter β, hence the name "beta-flexible clustering". Let us illustrate this by the computation of a flexible clustering with $\beta = -0.25$. If one provides only one value to `par.method`, **agnes()** considers it as the value of α_h in a context where $\alpha_h = \alpha_i = (1 - \beta)/2$ and $\gamma = 0$ (Legendre and Legendre 2012 p. 370). Therefore, to obtain, as in the following example, $\beta = -0.25$, the value to give is `par.method` = `0.625` because $\alpha_h = (1 - \beta)/2 = (1 - (-0.25))/2 = 0.625$. See the documentation file of **agnes()** for more details.

Beta-flexible clustering with beta = −0.25 is computed as follows (Fig. 4.6):

```
# Compute beta-flexible clustering using cluster::agnes()
# beta = -0.25
spe.ch.beta2 <- agnes(spe.ch, method = "flexible",
  par.method = 0.625)
# Change the class of agnes object
class(spe.ch.beta2)      # [1] "agnes" "twins"
spe.ch.beta2 <- as.hclust(spe.ch.beta2)
class(spe.ch.beta1)      # [1] "hclust"
plot(spe.ch.beta2,
   labels = rownames(spe),
   main = "Chord - Beta-flexible (beta=-0.25)")
```

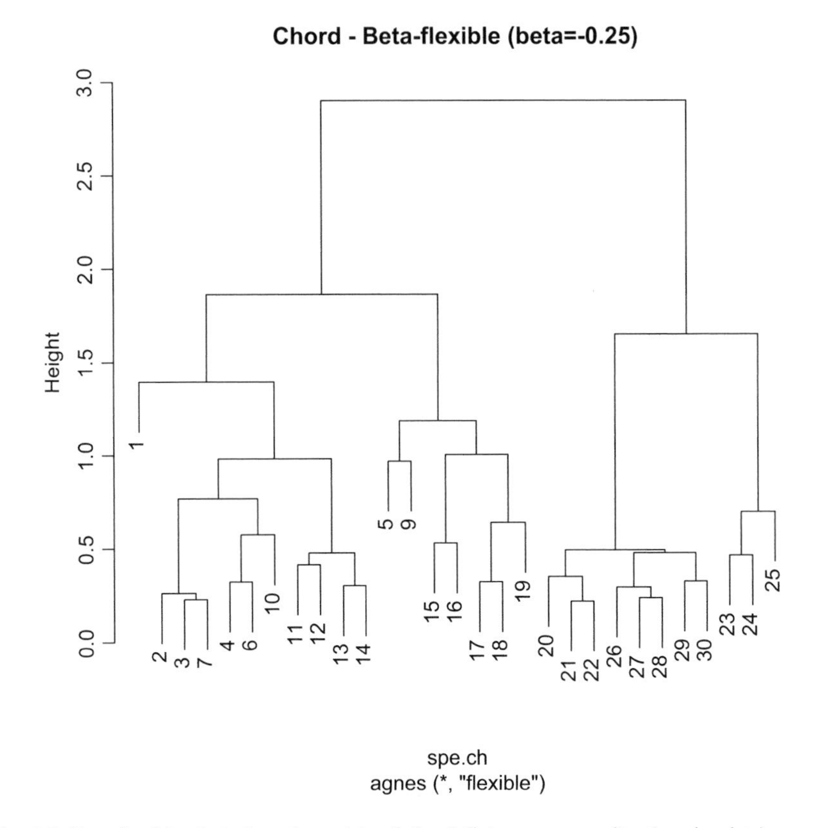

Fig. 4.6 Beta-flexible clustering of a matrix of chord distance among sites (species data)

Hint *To see the influence of beta on the shape of the dendrogram, apply the complete version of the R code (online material).*

> *Compare this dendrogram to the previous one (Ward method).*

4.7 Interpreting and Comparing Hierarchical Clustering Results

4.7.1 Introduction

Remember that unconstrained clustering is a heuristic procedure, not a statistical test. The choices of an association coefficient and a clustering method influence the result. This stresses the importance of choosing a method that is consistent with the aims of the analysis. The objects produced by **`hclust()`** contain the information necessary to fully describe the clustering results and draw the dendrogram. To display the list of items available in the output object, type **summary(object)** where "object" is the clustering result.

This information can also be used to help interpret and compare clustering results. We will now explore several possibilities offered by **R** for this purpose.

4.7.2 Cophenetic Correlation

The cophenetic distance between two objects in a dendrogram is the distance where the two objects become members of the same group. Locate any two objects, start from one, and "climb up the tree" to the first node leading down to the second object: the level of that node along the distance scale is the cophenetic distance between the two objects. A cophenetic matrix is a matrix representing the cophenetic distances among all pairs of objects. A Pearson's r correlation, called the *cophenetic correlation* in this context, can be computed between the original dissimilarity matrix and the cophenetic matrix. The method with the highest cophenetic correlation may be seen as the one that produces the clustering model that retains most of the information contained in the dissimilarity matrix. This does not necessarily mean, however, that this clustering model is the most adequate for the researcher's goal.

Of course, the cophenetic correlation cannot be tested for significance, since the cophenetic matrix is derived from the original dissimilarity matrix. The two sets of distances are not independent. Furthermore, the cophenetic correlation depends strongly on the clustering method used, in addition to the data.

As an example, let us compute the cophenetic matrix and correlation of four clustering results presented above, by means of the function **cophenetic()** of package **stats**.

```
# Single linkage clustering
spe.ch.single.coph <- cophenetic(spe.ch.single)
cor(spe.ch, spe.ch.single.coph)
# Complete linkage clustering
spe.ch.comp.coph <- cophenetic(spe.ch.complete)
cor(spe.ch, spe.ch.comp.coph)
# Average clustering
spe.ch.UPGMA.coph <- cophenetic(spe.ch.UPGMA)
cor(spe.ch, spe.ch.UPGMA.coph)
# Ward clustering
spe.ch.ward.coph <- cophenetic(spe.ch.ward)
cor(spe.ch, spe.ch.ward.coph)
```

Which dendrogram retains the closest relationship to the chord distance matrix? Cophenetic correlations can also be computed using Spearman or Kendall correlations:

```
cor(spe.ch, spe.ch.ward.coph, method = "spearman")
```

To illustrate the relationship between a dissimilarity matrix and a set of cophenetic matrices obtained from various methods, one can draw Shepard-like diagrams (Legendre and Legendre 2012 p. 414) by plotting the original dissimilarities against the cophenetic distances (Fig. 4.7):

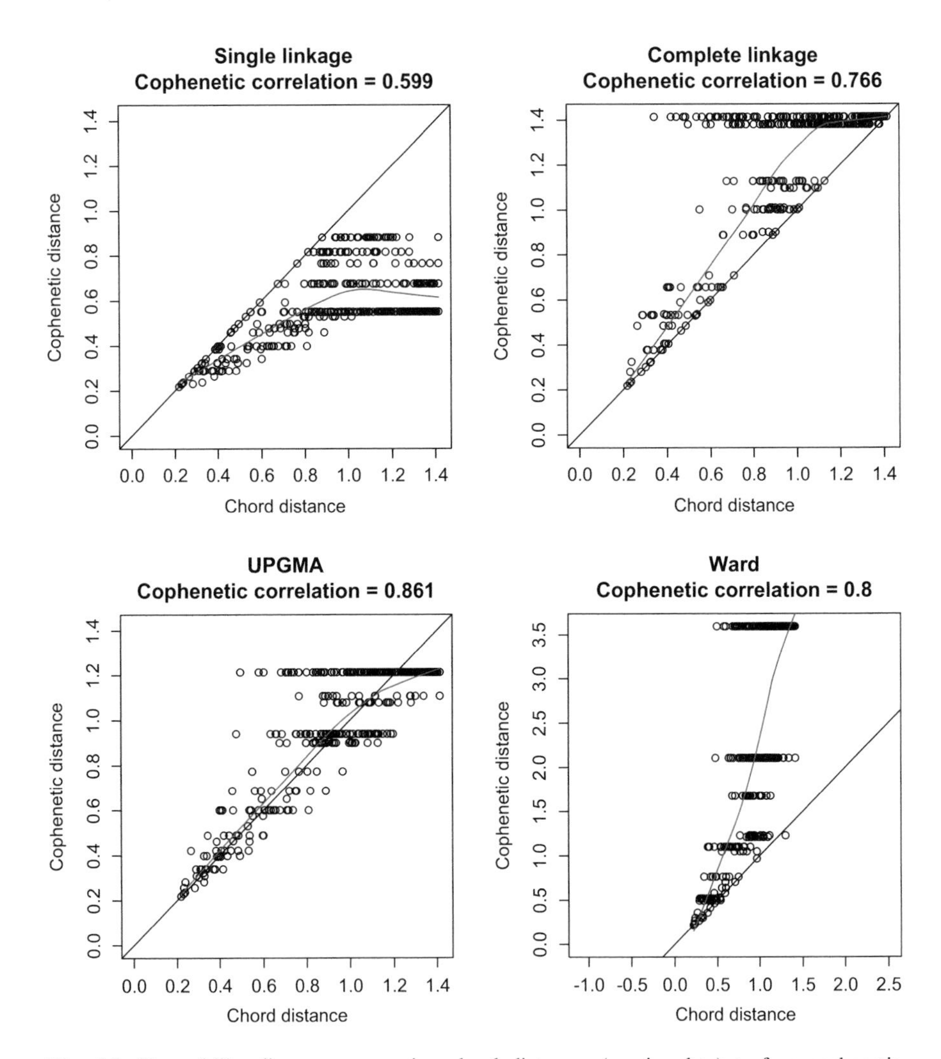

Fig. 4.7 Shepard-like diagrams comparing chord distances (species data) to four cophenetic distances. A LOWESS smoother shows the trend in each plot

```
# Shepard-like diagrams
par(mfrow = c(2, 2))
plot(
  spe.ch,
  spe.ch.single.coph,
  xlab = "Chord distance",
  ylab = "Cophenetic distance",
  asp = 1,
  xlim = c(0, sqrt(2)),
  ylim = c(0, sqrt(2)),
  main = c("Single linkage", paste("Cophenetic correlation =",
           round(cor(spe.ch, spe.ch.single.coph), 3)))
)
abline(0, 1)
lines(lowess(spe.ch, spe.ch.single.coph), col = "red")
plot(
  spe.ch,
  spe.ch.comp.coph,
  xlab = "Chord distance",
  ylab = "Cophenetic distance",
  asp = 1,
  xlim = c(0, sqrt(2)),
  ylim = c(0, sqrt(2)),
  main = c("Complete linkage", paste("Cophenetic correlation =",
           round(cor(spe.ch, spe.ch.comp.coph), 3)))
)
abline(0, 1)
lines(lowess(spe.ch, spe.ch.comp.coph), col = "red")
plot(
  spe.ch,
  spe.ch.UPGMA.coph,
  xlab = "Chord distance",
  ylab = "Cophenetic distance",
  asp = 1,
  xlim = c(0, sqrt(2)),
  ylim = c(0, sqrt(2)),
  main = c("UPGMA", paste("Cophenetic correlation =",
           round(cor(spe.ch, spe.ch.UPGMA.coph), 3)))
)
abline(0, 1)
lines(lowess(spe.ch, spe.ch.UPGMA.coph), col = "red")
plot(
  spe.ch,
  spe.ch.ward.coph,
  xlab = "Chord distance",
  ylab = "Cophenetic distance",
  asp = 1,
  xlim = c(0, sqrt(2)),
  ylim = c(0, max(spe.ch.ward$height)),
  main = c("Ward", paste("Cophenetic correlation =",
           round(cor(spe.ch, spe.ch.ward.coph), 3)))
)
abline(0, 1)
lines(lowess(spe.ch, spe.ch.ward.coph), col = "red")
```

Which method produces cophenetic distances (ordinate) that are well linearly related to the original distances (abscissa)?

Another possible statistic for the comparison of clustering results is the Gower (1983) distance[1], computed as the sum of squared differences between the original dissimilarities and cophenetic distances. The clustering method that produces the smallest Gower distance may be seen as the one that provides the best clustering model of the dissimilarity matrix. The cophenetic correlation and Gower distance criteria do not always designate the same clustering result as the best.

```
# Gower (1983) distance
(gow.dist.single <- sum((spe.ch - spe.ch.single.coph) ^ 2))
(gow.dist.comp <- sum((spe.ch - spe.ch.comp.coph) ^ 2))
(gow.dist.UPGMA <- sum((spe.ch - spe.ch.UPGMA.coph) ^ 2))
(gow.dist.ward <- sum((spe.ch - spe.ch.ward.coph) ^ 2))
```

Hint *Enclosing in brackets a command line producing an object induces the immediate screen display of the object.*

4.7.3 Looking for Interpretable Clusters

To interpret and compare clustering results, users generally look for interpretable clusters. This means that a decision must be made: at what level should the dendrogram be cut? Although it is not mandatory to select a single cutting level for a whole dendrogram (some parts of the dendrogram may be interpretable at finer levels than others), it is often practical to find one or a few levels where interpretations are made. These levels can be defined subjectively by visual examination of the dendrogram, or they can be chosen to fulfil some criteria, such as a predetermined number of groups for instance. In any case, adding information on the dendrograms or plotting additional information about the clustering results can be very useful.

4.7.3.1 Graph of the Fusion Level Values

The fusion level values of a dendrogram are the dissimilarity values where a fusion between two branches of a dendrogram occurs. Plotting the fusion level values may

[1]This Gower distance is not to be confused with the Gower dissimilarity presented in Chap. 3.

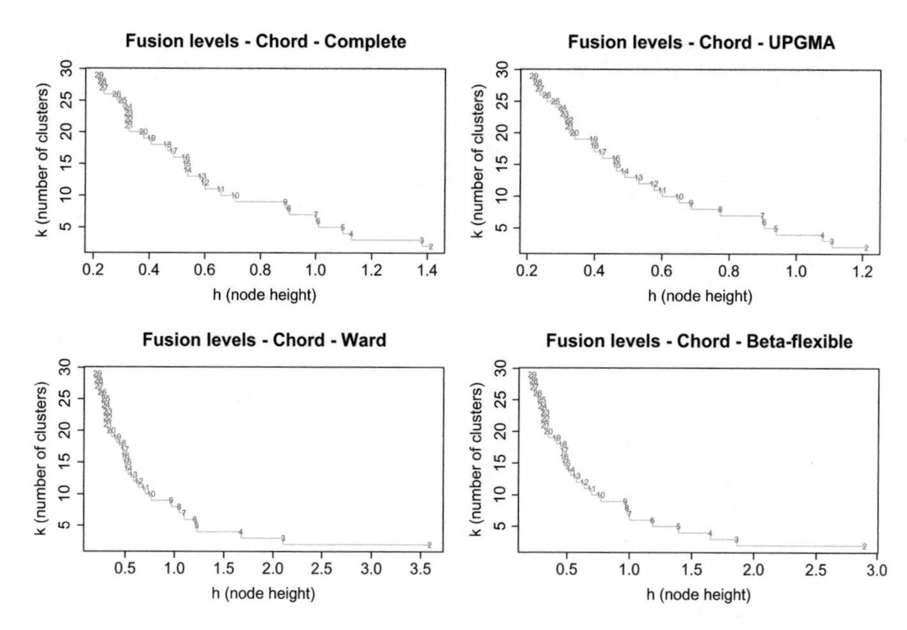

Fig. 4.8 Graphs of the fusion level values of four dendrograms. When read from right to left, long horizontal lines preceding steep increases suggest cutting levels, e.g. the 9-group solution in the complete linkage dendrogram or the 4-group solution in the Ward dendrogram

help define cutting levels. Let us plot the fusion level values for some of the dendrograms produced above (Fig. 4.8).

```
par(mfrow = c(2, 2))
# Plot the fusion level values of the complete linkage clustering
plot(
  spe.ch.complete$height,
  nrow(spe):2,
  type = "S",
  main = "Fusion levels - Chord - Complete",
  ylab = "k (number of clusters)",
  xlab = "h (node height)",
  col = "grey"
)
text(spe.ch.complete$height,
     nrow(spe):2,
     nrow(spe):2,
     col = "red",
     cex = 0.8)
# Plot the fusion level values of the UPGMA clustering
plot(
  spe.ch.UPGMA$height,
  nrow(spe):2,
  type = "S",
  main = "Fusion levels - Chord - UPGMA",
  ylab = "k (number of clusters)",
  xlab = "h (node height)",
  col = "grey"
)
```

```
text(spe.ch.UPGMA$height,
     nrow(spe):2,
     nrow(spe):2,
     col = "red",
     cex = 0.8)
# Plot the fusion level values of the Ward clustering
plot(
  spe.ch.ward$height,
  nrow(spe):2,
  type = "S",
  main = "Fusion levels - Chord - Ward",
  ylab = "k (number of clusters)",
  xlab = "h (node height)",
  col = "grey"
)
text(spe.ch.ward$height,
     nrow(spe):2,
     nrow(spe):2,
     col = "red",
     cex = 0.8)
# Plot the fusion level values of the beta-flexible clustering
# (beta = -0.25)
plot(
  spe.ch.beta2$height,
  nrow(spe):2,
  type = "S",
  main = "Fusion levels - Chord - Beta-flexible",
  ylab = "k (number of clusters)",
  xlab = "h (node height)",
  col = "grey"
)
text(spe.ch.beta2$height,
     nrow(spe):2,
     nrow(spe):2,
     col = "red",
     cex = 0.8)
```

Hint *Observe how the function* **`text()`** *is used to print the number of groups (clusters) directly on the graph.*

What is the suggested number of groups for each method? Go back to the dendrograms and cut them at the corresponding distances. Do the groups obtained always make sense? Do you obtain enough groups containing a substantial number of sites?
Remember that there is no single "truth" among these solutions. Each one may provide some insight into the data.

As it is obvious from the dendrograms and the graphs of the fusion levels, the four analyses tell different stories.

Now, if you want to set a common number of groups and compare the group contents among dendrograms, you can use the **cutree()** function and compute contingency tables:

```
# Choose a common number of groups
k <- 4  # Number of groups where at least a small jump is present
        # in all four graphs of fusion levels
# Cut the dendrograms and create membership vectors
spech.single.g <- cutree(spe.ch.single, k = k)
spech.complete.g <- cutree(spe.ch.complete, k = k)
spech.UPGMA.g <- cutree(spe.ch.UPGMA, k = k)
spech.ward.g <- cutree(spe.ch.ward, k = k)
spech.beta.g <- cutree(spe.ch.beta2, k = k)

# Compare classifications by constructing contingency tables
# Single vs complete linkage
table(spech.single.g, spech.complete.g)
# Single linkage vs UPGMA
table(spech.single.g, spech.UPGMA.g)
# Single linkage vs Ward
table(spech.single.g, spech.ward.g)
# Complete linkage vs UPGMA
table(spech.complete.g, spech.UPGMA.g)
# Complete linkage vs Ward
table(spech.complete.g, spech.ward.g)
# UPGMA vs Ward
table(spech.UPGMA.g, spech.ward.g)
# beta-flexible vs Ward
table(spech.beta.g, spech.ward.g)
```

If two classifications had provided the same group contents, the contingency tables would have shown only one non-zero frequency value in each row and each column. This was almost never the case here. For instance, the 26 sites of the second group of the single linkage clustering are distributed over the four groups of the Ward clustering.
However, one of the tables created above shows that two classifications resemble each other quite closely. Which ones?

4.7.3.2 Compare Two Dendrograms to Highlight Common Subtrees

To help select a partition found by several algorithms, it is useful to compare dendrograms and seek common clusters. The **tangelgram()** function from the **dendextend** package does this job nicely. Here we compare the dendrograms

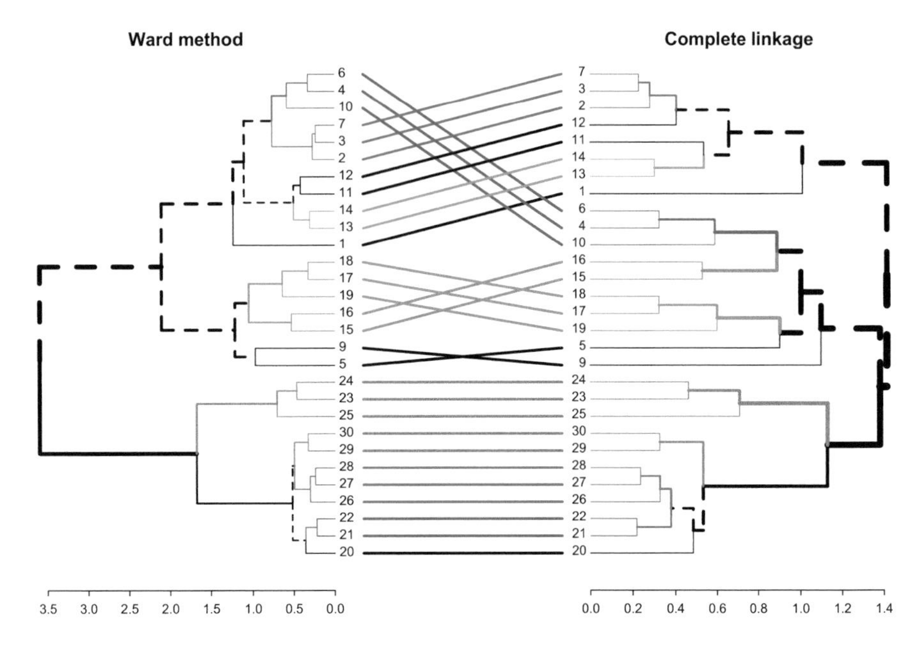

Fig. 4.9 Pairwise comparison of two dendrograms

previously produced by the Ward and complete linkage clustering methods (Fig. 4.9):

```
# Objects of class "hclust" must be first converted into objects of
# class "dendrogram"
class(spe.ch.ward)      # [1] "hclust"
dend1 <- as.dendrogram(spe.ch.ward)
class(dend1)            # [1] "dendrogram"
dend2 <- as.dendrogram(spe.ch.complete)
dend12 <- dendlist(dend1, dend2)
tanglegram(
  untangle(dend12),
  sort = TRUE,
  common_subtrees_color_branches = TRUE,
  main_left = "Ward method",
  main_right = "Complete linkage"
)
```

Hint *In order for the* **tanglegram()** *result to display the site names (as opposed to their number in the data set), the code in Sect. 4.3.1 shows how to incorporate the site labels into an object of class "dist" (here* **spe.ch***) used to compute the clusterings, by means of function* **attr()**.

Sites are ordered to match the two dendrogramsas well as possible. Colours highlight common clusters, whereas sites printed in black have different positions in the two trees. Can you recognize particularly "robust" clusters?

4.7.3.3 Multiscale Bootstrap Resampling

Unconstrained cluster analysis is a family of heuristic methods that are not aimed at testing *a priori* hypotheses. However, natural variation leads to sampling variability, and the results of a classification are likely to reflect it. It is therefore legitimate to assess the uncertainty (or its counterpart the robustness) of a classification. This has been done abundantly in phylogenetic analysis.

The choice approach for such validation procedures is bootstrap resampling (e.g. Efron 1979; Felsenstein 1985), which consists in randomly sampling subsets of the data and computing the clustering on these subsets. After having done this a large number of times, one counts the proportion of the replicate clustering results where a given cluster appears. This proportion is called the bootstrap probability (BP) of the cluster. Multiscale bootstrap resampling has been developed as an enhancement to answer some criticisms about the classical bootstrap procedure (Efron et al. 1996; Shimodaira 2002, 2004). In this method bootstrap samples of several different sizes are used to estimate the p-value of each cluster. This improvement produces "approximately unbiased" (AU) p-values. Readers are referred to the original publications for more details.

The **pvclust** package (Suzuki and Shimodaira 2006) provides functions to plot a dendrogram with bootstrap p-values associated to each cluster. AU p-values are printed in red. The less accurate BP values are printed in green. Clusters with high AU values (e.g. 0.95 or more) can be considered as strongly supported by the data. Let us apply this analysis to the dendrogram obtained with the Ward method (Fig. 4.10). **Beware**: in function **pvclust()** the data object must be transposed with respect to our usual layout (rows are variables).

```
# Compute p-values for all clusters (edges) of the dendrogram
spech.pv <-
  pvclust(t(spe.norm),
          method.hclust = "ward.D2",
          method.dist = "euc",
          parallel = TRUE)

# Plot dendrogram with p-values
plot(spech.pv)
# Highlight clusters with high AU p-values
pvrect(spech.pv, alpha = 0.95, pv = "au")
lines(spech.pv)
pvrect(spech.pv, alpha = 0.91, border = 4)
```

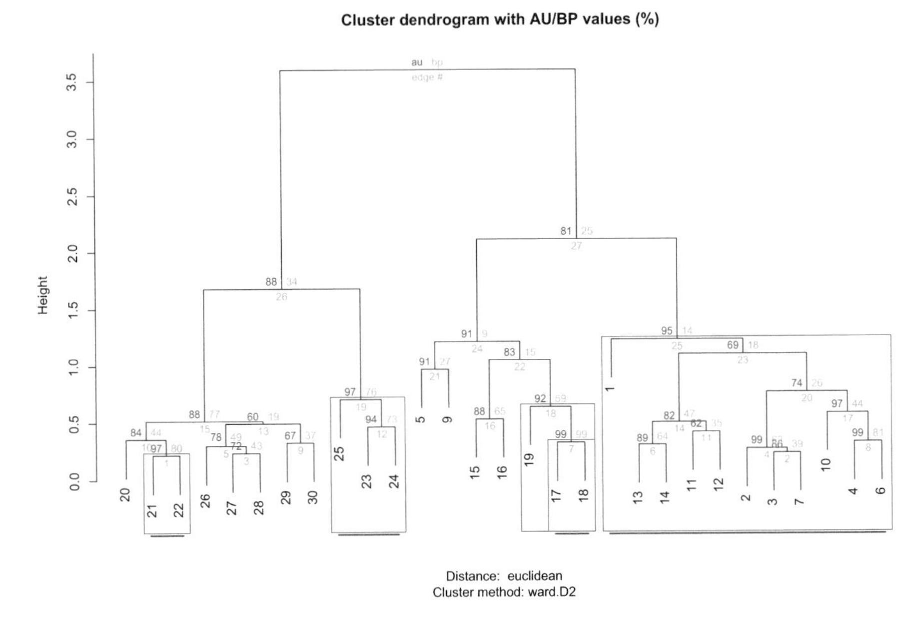

Fig. 4.10 Multiscale bootstrap resampling applied to the Chord-Ward dendrogram

Hint *The argument* `parallel = TRUE` *is used to greatly accelerate the computation with large data sets. It calls upon package* **parallel**.

Clusters enclosed in red boxes and underlined correspond to significant AU p-values (p ≥ 0.95), whereas clusters enclosed in blue rectangles show less significant values (p ≥ 0.91 in this example). This analysis is useful to highlight the most "robust" groups of sites. Note, however, that for the same data the results may vary from run to run because this is a procedure based on random resampling

Once we have chosen a dendrogram and assessed the robustness of its clusters, let us examine three methods that can help us identify an appropriate number of groups: silhouette widths, matrix comparison and diagnostic species.

4.7.3.4 Average Silhouette Widths

The *silhouette width* is a measure of the degree of membership of an object to its cluster, based on the average dissimilarity between this object and all objects of the cluster to which it belongs, compared to the same measure computed for the next closest cluster (see Sect. 4.7.3.7). Silhouette widths range from −1 to 1 and can be averaged over all objects of a partition.

We shall use the function **silhouette()** of package **cluster**. The documentation file of this function provides a formal definition of a silhouette width. In short, the larger the value is, the better the object is clustered. Negative values suggest that the corresponding objects may have been placed in the wrong cluster.

At each fusion level, the average silhouette width can be used as a measure of the quality of the partition (Rousseeuw quality index): compute silhouette widths measuring the intensity of the link of the objects to their groups, and choose the level where the within-group mean intensity is the highest, that is, the largest average silhouette width. A barplot is drawn (Fig. 4.11).

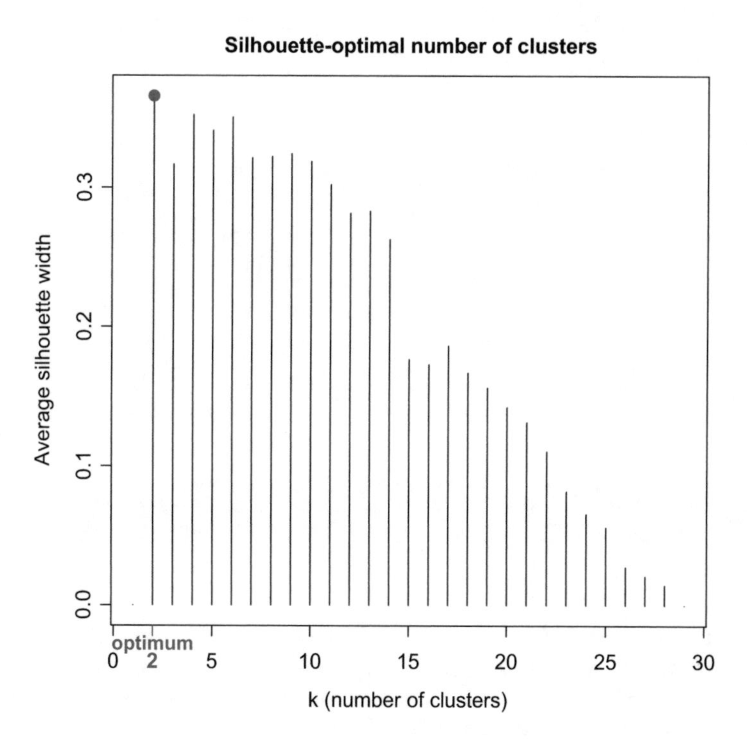

Fig. 4.11 Barplot showing the average silhouette widths for $k = 2$ to 29 groups. The best partition by this criterion is the one with the largest average silhouette width, i.e., in two groups. The second best, in 4 groups, and the third best, in 6 groups, are more interesting from an ecological point of view

```
# Choose and rename the dendrogram ("hclust" object)
hc <- spe.ch.ward

# Plot average silhouette widths (using Ward clustering) for all
# partitions except for the trivial partitions (k = 1 or k = n)
Si <- numeric(nrow(spe))
for (k in 2:(nrow(spe) - 1))
{
  sil <- silhouette(cutree(hc, k = k), spe.ch)
  Si[k] <- summary(sil)$avg.width
}
k.best <- which.max(Si)
plot(
  1:nrow(spe),
  Si,
  type = "h",
  main = "Silhouette-optimal number of clusters",
  xlab = "k (number of clusters)",
  ylab = "Average silhouette width"
)
axis(
  1,
  k.best,
  paste("optimum", k.best, sep = "\n"),
  col = "red",
  font = 2,
  col.axis = "red"
)
points(k.best,
       max(Si),
       pch = 16,
       col = "red",
       cex = 1.5)
```

Hint *Observe how the repeated computation of the average silhouette widths is done by a* **`for()`** *loop.*

As it often happens, this criterion has selected two groups as the optimal number. Considering the site numbers (reflecting their location along the river) in the dendrogram (Fig. 4.10), what is the explanation of this partitionin two groups? Is it interesting from an ecological point of view?

4.7.3.5 Comparison Between the Dissimilarity Matrix and Binary Matrices Representing Partitions

This technique compares the original dissimilarity matrix to binary matrices computed from the dendrogram cut at various levels (and representing group allocations). The idea is to choose the level where the matrix correlation between the two is the highest. Tests are impossible here since the matrices are not independent of one another; indeed, the matrices corresponding to the partitions are all derived from the original dissimilarity matrix.

To compute the binary dissimilarity matrices representing group membership, we will use our homemade function **grpdist()** to compute a binary dissimilarity matrix from a vector defining groups. The results are shown in Fig. 4.12.

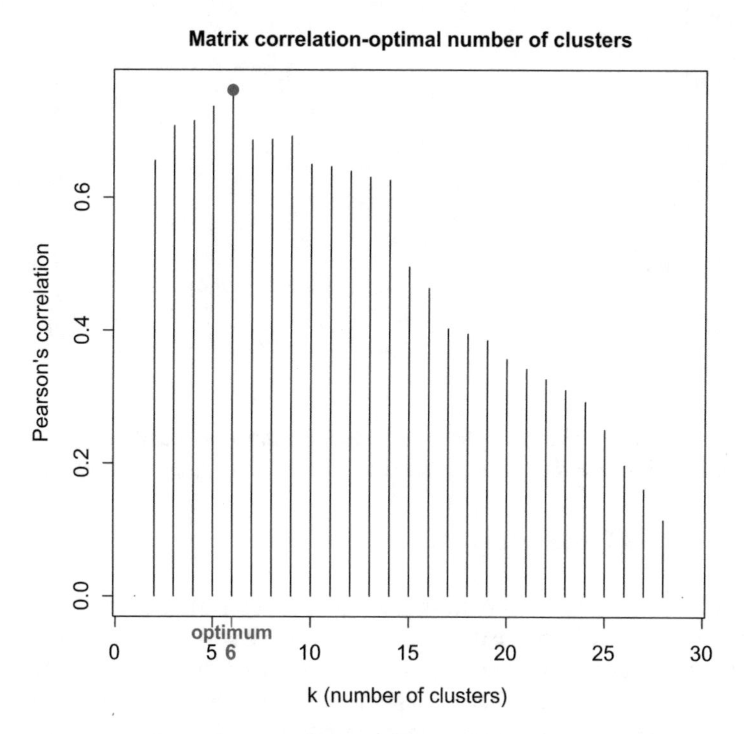

Fig. 4.12 Barplot showing the matrix correlations between the original dissimilarity matrix and binary matrices computed from the dendrogram cut at various levels

```
# Optimal number of clusters according to matrix correlation
# statistic (Pearson)
kt <- data.frame(k = 1:nrow(spe), r = 0)
for (i in 2:(nrow(spe) - 1)) {
  gr <- cutree(hc, i)
  distgr <- grpdist(gr)
  mt <- cor(spe.ch, distgr, method = "pearson")
  kt[i, 2] <- mt
}
k.best <- which.max(kt$r)
plot(
  kt$k,
  kt$r,
  type = "h",
  main = "Matrix correlation-optimal number of clusters",
  xlab = "k (number of clusters)",
  ylab = "Pearson's correlation"
)
axis(
  1,
  k.best,
  paste("optimum", k.best, sep = "\n"),
  col = "red",
  font = 2,
  col.axis = "red"
)
points(k.best,
       max(kt$r),
       pch = 16,
       col = "red",
       cex = 1.5)
```

The barplot shows that a partition in 3 to 6 clusters would achieve a high matrix correlation between the chord distance matrix and the binary allocation matrix.

4.7.3.6 Species Fidelity Analysis

Another internal criterion for assessing the quality of a partition is based on species fidelity analysis. The basic idea is to retain clusters that are best characterized by a set of diagnostic species, also called “indicator”, “typical”, “characteristic” or “differential” species, i.e. species that are significantly more frequent and abundant in a given group of sites. Specifically, the best partition would be the one that maximizes both (i) the sum of indicator values and (ii) the proportion of clusters with significant indicator species. Here, we shall anticipate on Sect. 4.11.2 and use index IndVal (Dufrêne and Legendre 1997), which integrates a specificity and a fidelity measure (Fig. 4.13).

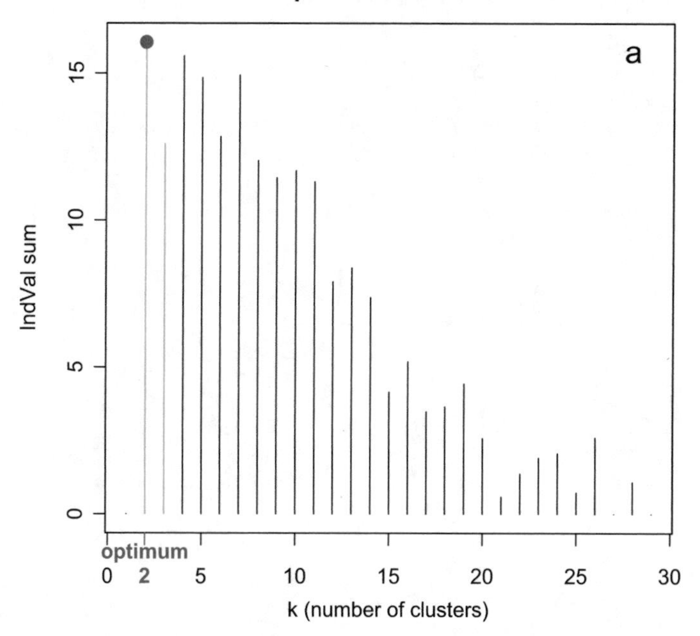

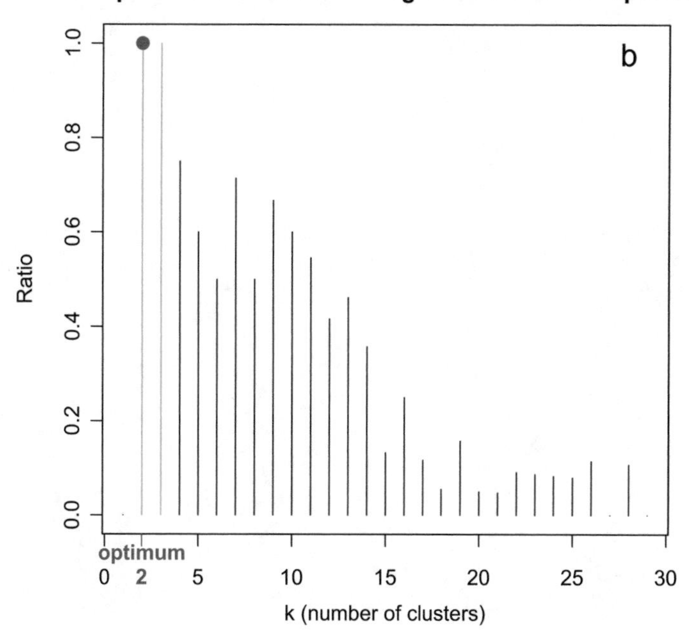

Fig. 4.13 Barplots showing the sum of species indicator values (IndVal) (**a**) and the proportion of significant indicator species (**b**) for all partitions obtained after cutting the dendrogram at various levels. The solutions where there are significant indicator species in all k clusters are highlighted in green

```
# Optimal number of clusters as per indicator species analysis
# (IndVal, Dufrene-Legendre; package: labdsv)
IndVal <- numeric(nrow(spe))
ng <- numeric(nrow(spe))
for (k in 2:(nrow(spe) - 1)) {
  iva <- indval(spe, cutree(hc, k = k), numitr = 1000)
  gr <- factor(iva$maxcls[iva$pval <= 0.05])
  ng[k] <- length(levels(gr)) / k
  iv <- iva$indcls[iva$pval <= 0.05]
  IndVal[k] <- sum(iv)
}
k.best <- which.max(IndVal[ng == 1]) + 1
col3 <- rep(1, nrow(spe))
col3[ng == 1] <- 3

par(mfrow = c(1, 2))
plot(
  1:nrow(spe),
  IndVal,
  type = "h",
  main = "IndVal-optimal number of clusters",
  xlab = "k (number of clusters)",
  ylab = "IndVal sum",
  col = col3
)
axis(
  1,
  k.best,
  paste("optimum", k.best, sep = "\n"),
  col = "red",
  font = 2,
  col.axis = "red"
)
points(
  which.max(IndVal),
  max(IndVal),
  pch = 16,
  col = "red",
  cex = 1.5
)
plot(
  1:nrow(spe),
  ng,
  type = "h",
  xlab = "k (number of groups)",
  ylab = "Ratio",
  main = "Proportion of clusters with significant indicator
          species",
  col = col3
)
```

```
axis(1,
     k.best,
     paste("optimum", k.best, sep = "\n"),
     col = "red",
     col.axis = "red")
points(k.best,
       max(ng),
       pch = 16,
       col = "red",
       cex = 1.5)
```

Hint *In addition to the search for the solutions meeting the two criteria, the code above highlights in green the solutions where there are significant indicator species in the* k *clusters (object* `ng`*). In the Doubs data set, it is the case only in the 2- and 3-cluster solutions.*

Only the partition in two clusters, simply contrasting the upstream and downstream sites, fully meets the two criteria. However, a partition in three or four clusters, allowing the discovery of more subtle structures, would be an acceptable choice despite the absence of positive differential species (i.e., a species that is more frequent in a given group than in others) for some groups.

4.7.3.7 Silhouette Plot of the Final Partition

In our example, the silhouette-based, matrix correlation-based and IndVal-based criteria do not return the same solution; these are, ranging from $k = 2$ to $k = 6$. A good compromise seems to be $k = 4$. Let us select this number for our final group diagnostics. We can select the Ward clustering as our final choice, since this method produced four well-balanced (not equal-sized, but without outliers) and well-delimited groups.

We can now proceed to examine if the group memberships are appropriate (i.e., no or few objects apparently misclassified). A silhouette plot is useful here (Fig. 4.14).

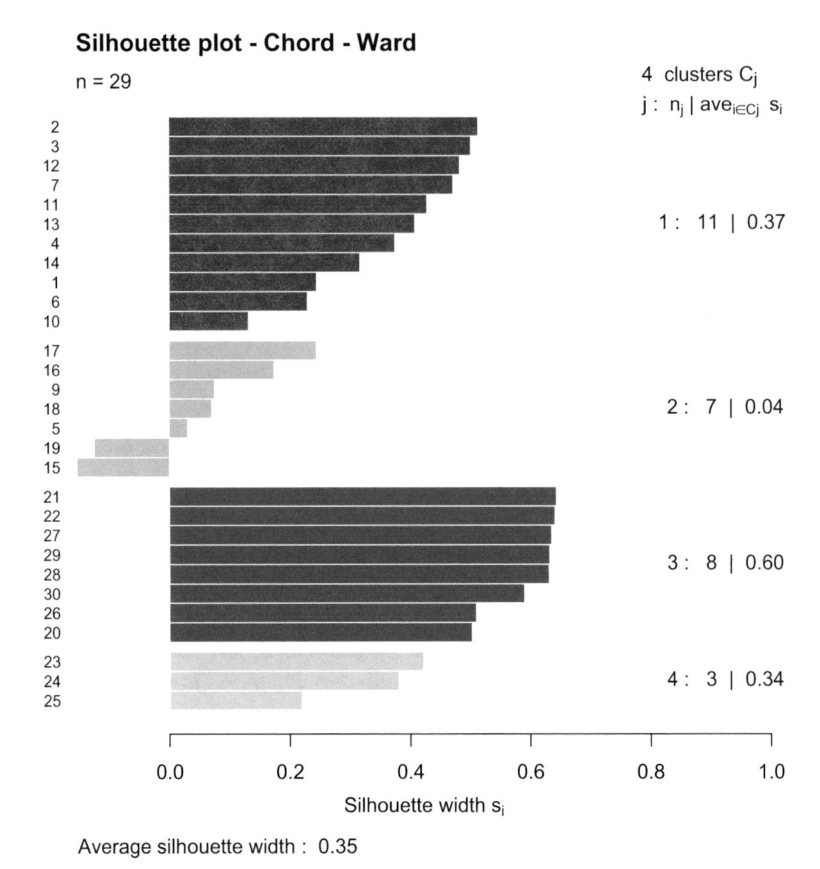

Fig. 4.14 Silhouette plot of the final, four-group partition from Ward clustering

```
# Choose the number of clusters
k <- 4
# Silhouette plot of the final partition
spech.ward.g <- cutree(spe.ch.ward, k = k)
sil <- silhouette(spech.ward.g, spe.ch)
rownames(sil) <- row.names(spe)
plot(
  sil,
  main = "Silhouette plot - Chord - Ward",
  cex.names = 0.8,
  col = 2:(k + 1),
  nmax = 100
)
```

Clusters 1 and 3 are the most coherent, while cluster 2 contains misclassified objects.

4.7.3.8 Final Dendrogram with Graphical Options

Now it is time to produce the final dendrogram, where we can represent the four groups and improve the overall appearance with several graphical options (Fig. 4.15). Try the code, compare the results with it, and use the documentation files to understand the arguments used. Note also the use of a homemade function **`hcoplot()`**.

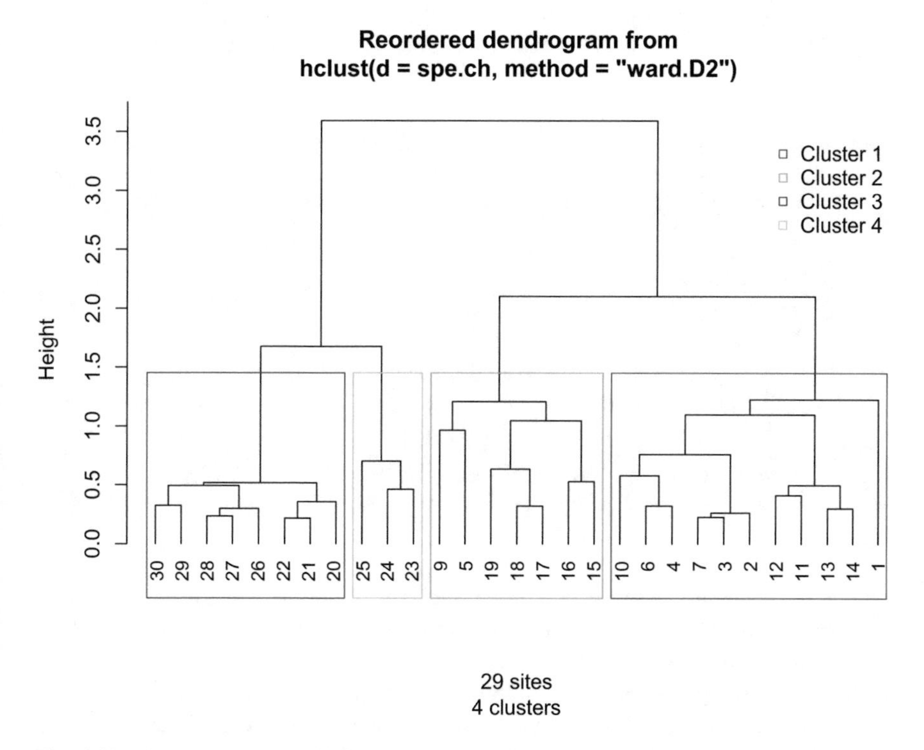

Fig. 4.15 Final dendrogram with boxes around the four selected clusters of sites from the fish species data

```
# Reorder clusters
spe.chwo <- reorder.hclust(spe.ch.ward, spe.ch)

# Plot reordered dendrogram with group labels
plot(
  spe.chwo,
  hang = -1,
  xlab = "4 groups",
  sub = "",
  ylab = "Height",
  main = "Chord - Ward (reordered)",
  labels = cutree(spe.chwo, k = k)
)
rect.hclust(spe.chwo, k = k)

# Plot the final dendrogram with group colors (RGBCMY...)
# Fast method using the additional hcoplot() function:
hcoplot(spe.ch.ward, spe.ch, lab = rownames(spe), k = 4)
```

Hints *Function* **`reorder.hclust()`** *reorders objects so that their original order in the dissimilarity matrix is respected as much as possible. This does not affect the topology of the dendrogram. Its arguments are (1) the clustering object and (2) the dissimilarity matrix.*

When using **`rect.hclust()`** *to draw boxes around clusters, as in the* **`hcoplot()`** *function used here, one can specify a fusion level (argument* `h`*) instead of a number of groups (argument* `k`*).*

Another function called **`identify.hclust()`** *allows the interactive cut of a tree at any position where one left-clicks with the mouse. It makes it possible to extract a list of objects from any given subgroup.*

The argument `hang = -1` *specifies that the branches of the dendrogram will all reach the value 0 and the labels will hang below that value.*

Let us conclude with several other representations of the clustering results obtained above. The usefulness of these representations depends on the context.

The **dendextend** package provides various functions to improve the representation of a dendrogram. First, the `hclust` object must be converted to a `dendrogram` object. Then, you can apply colours and line options to the branches of the dendrogram, e.g. based on the final partition (Fig. 4.16).

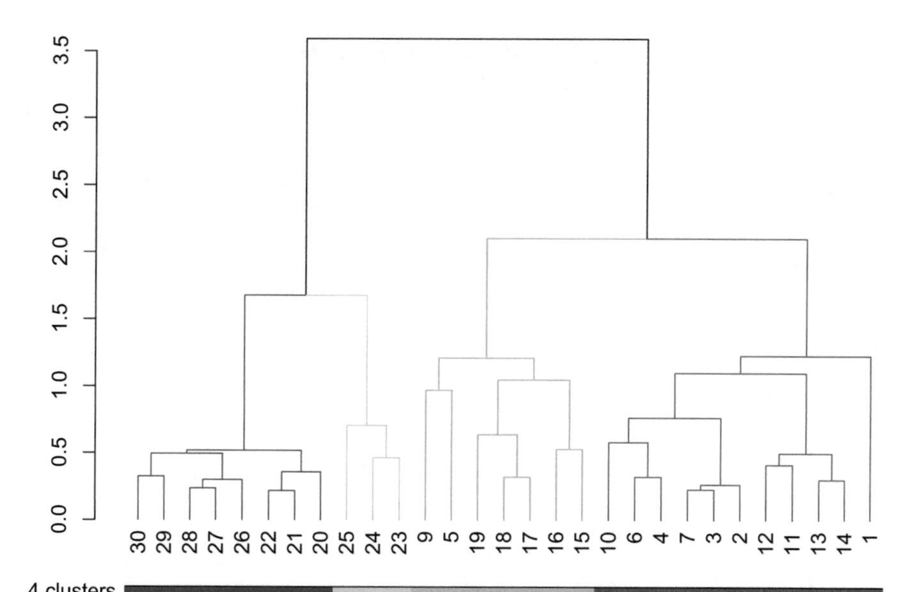

Fig. 4.16 Final dendrogram with coloured branches highlighting the four selected clusters of sites from the fish species data

```
# Convert the "hclust" object into a "dendrogram" object
dend <- as.dendrogram(spe.chwo)

# Plot the dendrogram with coloured branches
dend %>% set("branches_k_color", k = k) %>% plot

# Use standard colours for clusters
clusters <- cutree(dend, k)[order.dendrogram(dend)]
dend %>%
set("branches_k_color", k = k, value = unique(clusters) + 1) %>%
plot
# Add a coloured bar
colored_bars(clusters + 1,
             y_shift = -0.5,
             rowLabels = paste(k, "clusters"))
```

Hint *Note the non-conventional syntax used by several recent packages, including* **dendextend***, based on a chaining of instructions separated by the operator* `%>%`.

4.7.3.9 Spatial Plot of the Clustering Result

The following code allows us to plot the clusters on a map representing the river. This is a very useful way of representing results for spatially explicit data (Fig. 4.17).

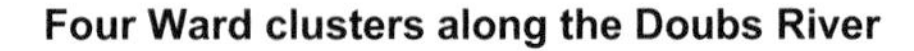

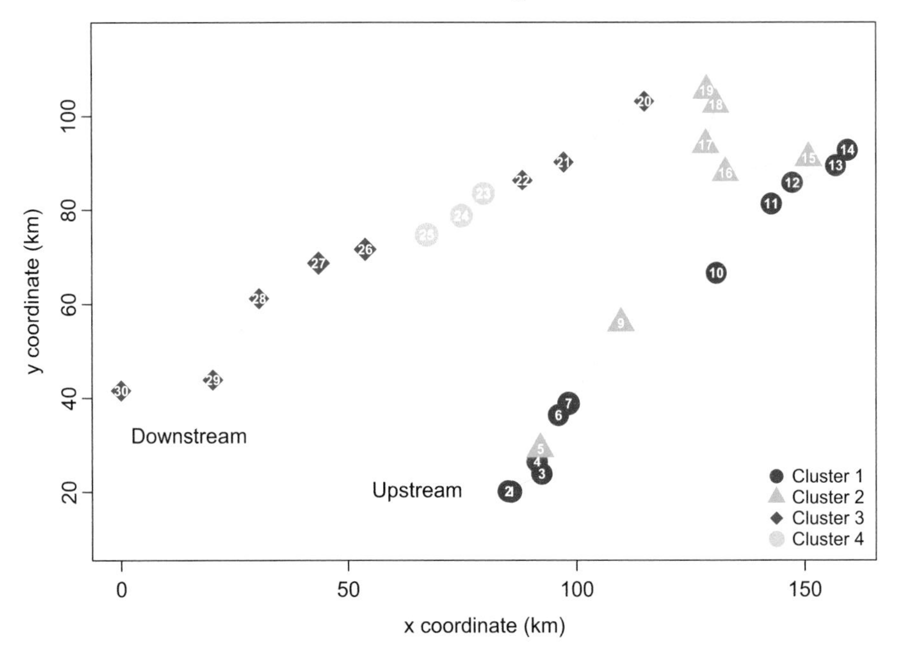

Fig. 4.17 The four ward clusters on a map of the Doubs River. Site 1 is hidden underneath site 2

Since we shall draw comparable maps later on, we assemble the code in a homemade generic function called **drawmap()**:

```
# Plot the Ward clusters on a map of the Doubs River
# (see Chapter 2)
drawmap(xy = spa,
        clusters = spech.ward.g,
        main = "Four Ward clusters along the Doubs River")
```

Go back to the maps of four fish species (Chap.2). Compare these to the map created above and shown in Fig. 4.17.

4.7.3.10 Heat Map and Ordered Community Table

See how to represent the dendrogram in a square matrix of coloured pixels, where the colour intensity represents the similarity among the sites (Fig. 4.18):

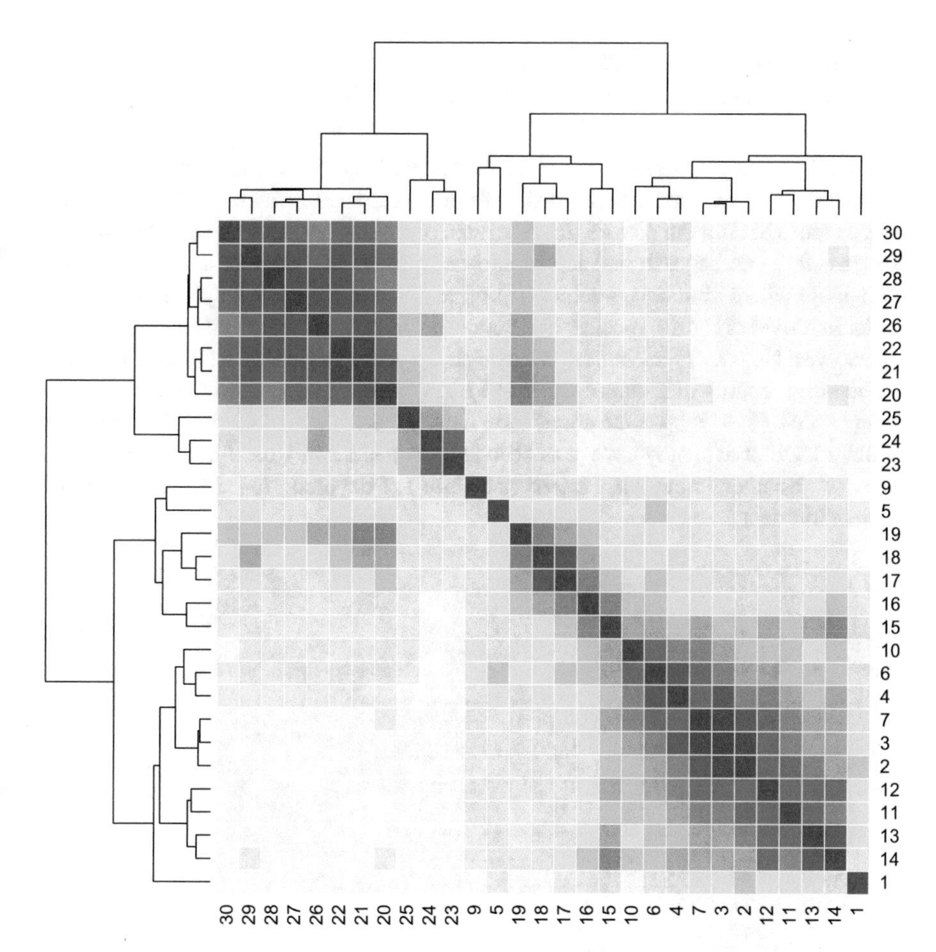

Fig. 4.18 Heat map of the chord *D* matrix reordered following the dendrogram

```
# Heat map of the dissimilarity matrix ordered with the dendrogram
heatmap(
  as.matrix(spe.ch),
  Rowv = dend,
  symm = TRUE,
  margin = c(3, 3)
)
```

Hint *Note that the* `hclust` *object has been converted to an object of class* `dendrogram`. *This allows many advanced graphical manipulations of tree-like structures. See the documentation file of the* **`as.dendrogram()`** *function, where examples are given.*

Observe how, thanks to the ordering, most "hot" (red) values representing high similarities are located close to the diagonal. The diagonal itself is trivial.

Finally, it may be useful to explore the clusters' species contents directly, i.e. to reorder the original data table according to the group memberships. In order to avoid large values to clog the picture, Jari Oksanen proposes **vegemite()**, a function of **vegan** that can use external information to reorder and display a site-by-species data table where abundance values can be recoded in different ways. If no specific order is provided for the species, these are ordered by their weighted averages on the site scores. Note that the abundance scale accepted by **vegemite()** must be made of one-digit counts only (e.g. from 0 to 9). Any two- or more digit value will cause **vegemite()** to stop. Our abundance data, expressed on a 0–5 scale, cause no problem. Other data may have to be recoded. **vegemite()** itself makes an internal use of another function, **coverscale()**, devoted to the rescaling of vegetation data.

```
# Ordered community table
# Species are ordered by their weighted averages on site scores.
# Dots represent absences.
or <- vegemite(spe, spe.chwo)
```

```
     32222222222  111111      1111
     09876210543959876506473221341
Icme 5432121......................
Abbr 54332431.....1...............
Blbj 54542432.1...1...............
Anan 54432222.....111.............
Gyce 5555443212...11..............
Scer 522112221...21...............
Cyca 53421321.....1111............
Rham 55432333.....221.............
Legi 35432322.1...1111............
Alal 55555555352..322.............
Chna 12111322.1...211.............
Titi 53453444...1321111.21........
Ruru 55554555121455221..1.........
Albi 53111123.....2341............
Baba 35342544.....23322.........1.
Eslu 453423321...41111..12.1....1.
Gogo 5544355421..242122111......1.
Pefl 54211432....41321..12........
Pato 2211.222.....3344............
Sqce 3443242312152132232211..11.1.
Lele 332213221...52235321.1.......
Babl .1111112...32534554555534124.
Teso .1...........11254........23.
Phph .1....11...13334344454544455.
Cogo ..............1123......2123.
Satr .1..........2.123413455553553
Thth .1............11.2......2134.
sites species
   29      27
```

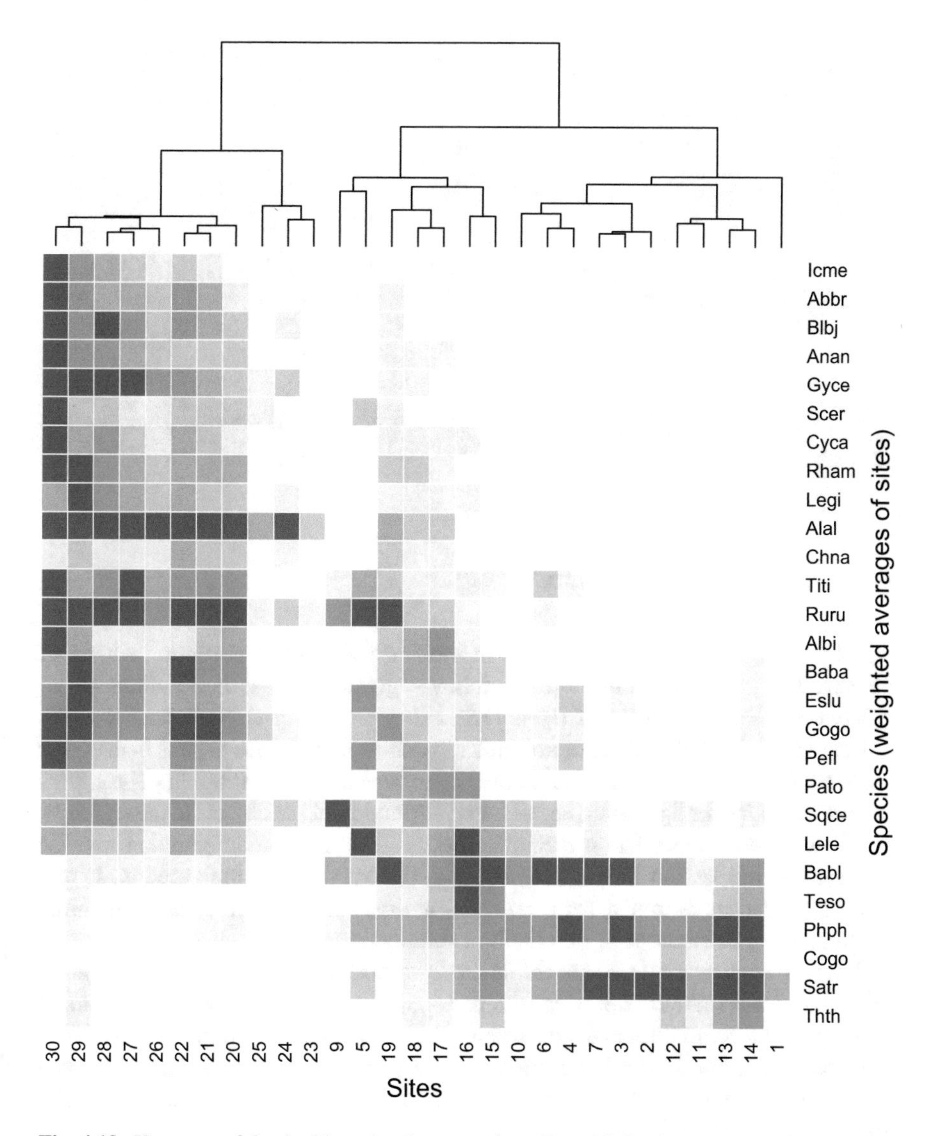

Fig. 4.19 Heat map of the doubly ordered community table, with dendrogram

This way of ordering species can also be used in a heat map with colour intensities proportional to the species abundances (Fig. 4.19):

```
# Heat map of the doubly ordered community table, with dendrogram
heatmap(
  t(spe[rev(or$species)]),
  Rowv = NA,
  Colv = dend,
  col = c("white", brewer.pal(5, "Greens")),
  scale = "none",
  margin = c(4, 4),
  ylab = "Species (weighted averages of sites)",
  xlab = "Sites"
)
```

Hint *A similar result can be obtained using the* **tabasco()** *function of the* **vegan** *package.*

4.8 Non-hierarchical Clustering

Non-hierarchical partitioning consists in looking for a single partition of a set of objects. The problem can be stated as follows: given n objects in a p-dimensional space, determine a partition of the objects into k groups, or clusters, such that the objects within each cluster are more similar to one another than to objects in the other clusters. The user determines the number of groups, k. The partitioning algorithms require an initial configuration, i.e. an initial attribution of the objects to the k groups, which will be optimized in a recursive process. The initial configuration may be provided by theory, but users often choose a random starting configuration. In that case, the analysis is run a large number of times with different random initial configurations in the hope to find the best solution.

Here we shall present two related methods, k-means partitioning and partitioning around medoids (PAM). These algorithms operate in Euclidean space. An important note is that if the variables in the data table are not dimensionally homogeneous, they must be standardized prior to partitioning. Otherwise, the total variance of the data has dimension equal to the sum of the squared dimensions of the individual variables, which is meaningless.

4.8.1 k-means Partitioning

The k-means method uses the local structure of the data to delineate clusters: groups are formed by identifying high-density regions in the data. To achieve this, the method iteratively minimizes an objective function called the total error sum of squares (E^2_k or TESS or SSE), which is the sum of the within-group sums-of-squares. This quantity is the sum, over the k groups, of the sums of the squared (Euclidean) distances among the objects in the groups, each divided by the number of objects in the group. This is the same criterion as used in Ward's agglomerative clustering.

4.8.1.1 *k*-means with Random Starts

If one has a pre-determined number of groups in mind, the recommended function to use is **kmeans()** of the **stats** package. The analysis can be automatically repeated a large number of times (argument `nstart`) using different random initial configurations. The function finds the best solution (smallest SSE value) after repeating the analysis 'nstart' times.

k-means is a *linear* method, i.e. it is not appropriate for raw species abundance data with lots of zeros (see Sect. 3.2.2). One possibility is to use non-Euclidean dissimilarity matrices, like the percentage difference (aka Bray-Curtis); such matrices should be square-root transformed and submitted to principal coordinate analysis (PCoA, Sect. 5.5) in order to obtain a full representation of the objects in Euclidean space. The resulting PCoA axes are then subjected to *k*-means partitioning. Another solution is to pre-transform the species data. To remain coherent with the previous sections, we can use the chord-transformed or "normalized" species data created in Sect. 4.3.1. When applied in combination with the Euclidean distance implicit in *k*-means, the analysis preserves the chord distance among sites. To compare with the results of Ward's clustering computed above, we ask for $k = 4$ groups and compare the outcome with the four groups derived from the Ward hierarchical clustering.

```
# k-means partitioning of the pre-transformed species data
# With 4 groups
spe.kmeans <- kmeans(spe.norm, centers = 4, nstart = 100)
spe.kmeans
```

Note: running the function again may produce slightly different results as each of the 'nstart' runs starts with a different random configuration. In particular, numbers given by the function to clusters are arbitrary!

```
# Comparison with the 4-group partition derived from
# Ward clustering
table(spe.kmeans$cluster, spech.ward.g)
```

```
   spech.ward.g
     1  2  3  4
  1  0  0  0  3
  2 11  1  0  0
  3  0  0  8  0
  4  0  6  0  0
```

Are the two results fairly similar? Which object(s) is (or are) classified differently?

To compute k-means partitioning from dissimilarity indices that cannot be obtained by a transformation of the raw data followed by calculation of the Euclidean distance, for example the percentage difference (aka Bray-Curtis) dissimilarity, one has to compute first a rectangular data table with n rows by principal coordinate analysis (PCoA, Sect. 5.5) of the dissimilarity matrix, then use that rectangular table as input to k-means partitioning. For the percentage difference dissimilarity, one has to compute the PCoA on the square root of the percentage difference dissimilarities to obtain a fully Euclidean solution, or use a PCoA function that provides a correction for negative eigenvalues. These points are discussed in Chap. 5.

A partitioning yields a single partition with a predefined number of groups. If you want to try several solutions with different k values, you must rerun the analysis. But which solution is the best in terms of number of clusters? To answer this question, one has to state what "best" means. Many criteria exist; some of them are available in the function **clustIndex()** of the package **cclust**. Milligan and Cooper (1985) recommend maximizing the Calinski-Harabasz index (F-statistic comparing the among-group to the within-group sum of squares of the partition), although its value tends to be lower for unequal-sized partitions. The maximum of 'ssi' ("Simple Structure Index", see the documentation file of **clustIndex()** for details) is another good indicator of the best partition in the least-squares sense.

Fortunately, one can avoid running **kmeans()** many times by hand. **vegan**'s function **cascadeKM()** is a *wrapper* for the **kmeans()** function, that is, a function that uses a basic function, adding new properties to it. It creates several partitions forming a cascade from small (argument **inf.gr**) to large values of k (argument **sup.gr**). Let us apply this function to our dataset, asking for 2–10 groups and the simple structure index criterion for clustering quality, followed by a plot of the results (Fig. 4.20).

```
# k-means partitioning, 2 to 10 groups
spe.KM.cascade <-
  cascadeKM(
    spe.norm,
    inf.gr = 2,
    sup.gr = 10,
    iter = 100,
    criterion = "ssi"
  )
summary(spe.KM.cascade)
plot(spe.KM.cascade, sortg = TRUE)
```

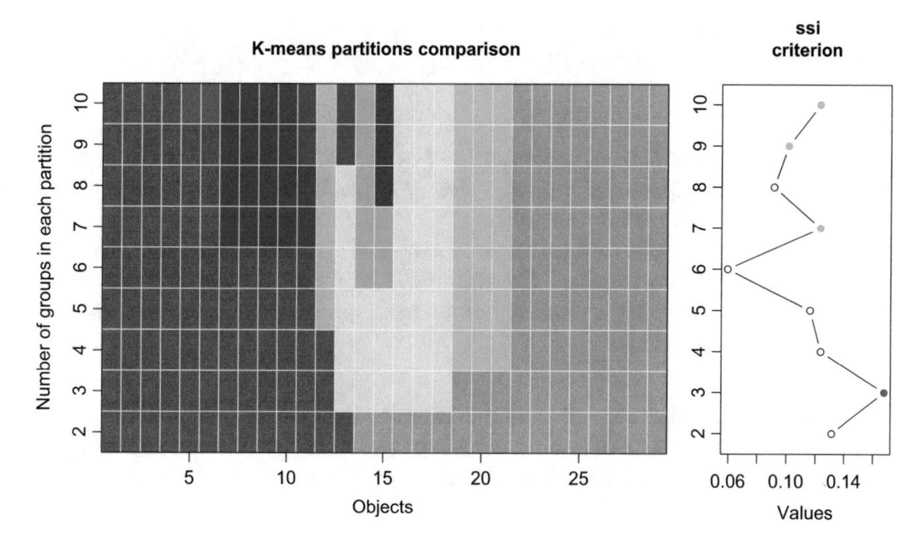

Fig. 4.20 *k*-means cascade plot showing the group attributed to each object for each partition

Hint *In the plot,* `sortg = TRUE` *reorders the objects in such a way as to put together, insofar as possible, the objects pertaining to each group. A more detailed explanation of this argument in provided in the function's documentation file.*

The plot shows the group attributed to each object for each partition (rows of the graph). The rows of the graph are the different values of k. The groups are represented by different colours; there are two colours for k = 2, three colours for k = 3, and so on. Another graph shows the values of the chosen stopping criterion for the different values of k. Since this is an iterative process, the results can vary from run to run.
How many groups does this cascade propose as the best solution? If one has reasons to prefer a larger number of groups, what would be the next best solution?

The function **`cascadeKM()`** provides numeric results as well. Among them, the element 'result' gives the TESS statistic and the value of the criterion (`calinski` or `ssi`) for each value of *k*. The element 'partition' contains a table showing the group attributed to each object. If the geographic coordinates of the objects are available, they can be used to plot a map of the objects, with symbols or colours representing the groups specified by one of the columns of this table.

```
summary(spe.KM.cascade)
spe.KM.cascade$results
```

The minimum of SSE is the criterion used by the algorithm to find the optimal grouping of the objects for a given value of k, while calinski and ssiare good criteria to find the optimal value of k.
Remember that the different partitions in k = {2, 3, ..., 10} groups are computed independently of one another. Examining the plot from bottom to top is NOT equivalent to examining a dendrogram because groups of successive partitions are not necessarily nested.

After defining site clusters, it is time to examine their contents. The simplest way is to define subgroups of sites on the basis of the typology retained and compute basic statistics. You can run the following example based upon the k-means 4-group partition.

```
# Reorder the sites according to the k-means result
spe.kmeans.g <- spe.kmeans$cluster
spe[order(spe.kmeans.g), ]

# Reorder sites and species using function vegemite()
ord.KM <- vegemite(spe, spe.kmeans.g)
spe[ord.KM$sites, ord.KM$species]
```

4.8.1.2 Use of k-means Partitioning to Optimize an Independently Obtained Classification

From another point of view, k-means partitioning can be used to optimise the result of a hierarchical clustering. Indeed, the nature of an agglomeration algorithm such as those explored in the previous sections prevents an object that has been included in a group to be translocated to another that appeared later in the agglomeration process, even if the latter would now be more appropriate. To overcome this potential problem, one can provide the k-means algorithm with prior information derived from the clustering to be optimized, either in the form of a $k \times p$ matrix of mean values of the p variables in the k groups obtained with the method, or as a list of k objects, one per group, considered to be "typical" for each group. In the latter case, these "typical" objects, called *medoids*, are used as seeds for the construction of groups in the technique presented in the next Section. Let us apply this idea to verify if the four groups obtained by a Ward clustering of the fish data are modified by a k-means partitioning.

k-means with Ward species means per group (centroids) as starting points:

```
# Mean species abundances on Ward site clusters
groups <- as.factor(spech.ward.g)
spe.means <- matrix(0, ncol(spe), length(levels(groups)))
row.names(spe.means) <- colnames(spe)
for (i in 1:ncol(spe)) {
  spe.means[i, ] <- tapply(spe.norm[, i], spech.ward.g, mean)
}
# Mean species abundances as starting points
startpoints <- t(spe.means)
# k-means on starting points
spe.kmeans2 <- kmeans(spe.norm, centers = startpoints)
```

A slightly different approach is to go back to the hierarchical clustering, identify the most "typical" object in each group (cf. silhouette plot), and provide these medoids as starting points to **kmeans()** *(argument* `centers`*):*

```
startobjects <- spe.norm[c(2, 17, 21, 23), ]
spe.kmeans3 <- kmeans(spe.norm, centers = startobjects)

# Comparison with the 4-group partition derived from
# Ward clustering:
table(spe.kmeans2$cluster, spech.ward.g)

# Comparison among the two optimized 4-group classifications:
table(spe.kmeans2$cluster, spe.kmeans3$cluster)

# Silhouette plot of the final partition
spech.ward.gk <- spe.kmeans2$cluster
k <- 4
sil <- silhouette(spech.ward.gk, spe.ch)
rownames(sil) <- row.names(spe)
plot(sil,
     main = "Silhouette plot - Ward & k-means",
     cex.names = 0.8,
     col = 2:(k + 1))
```

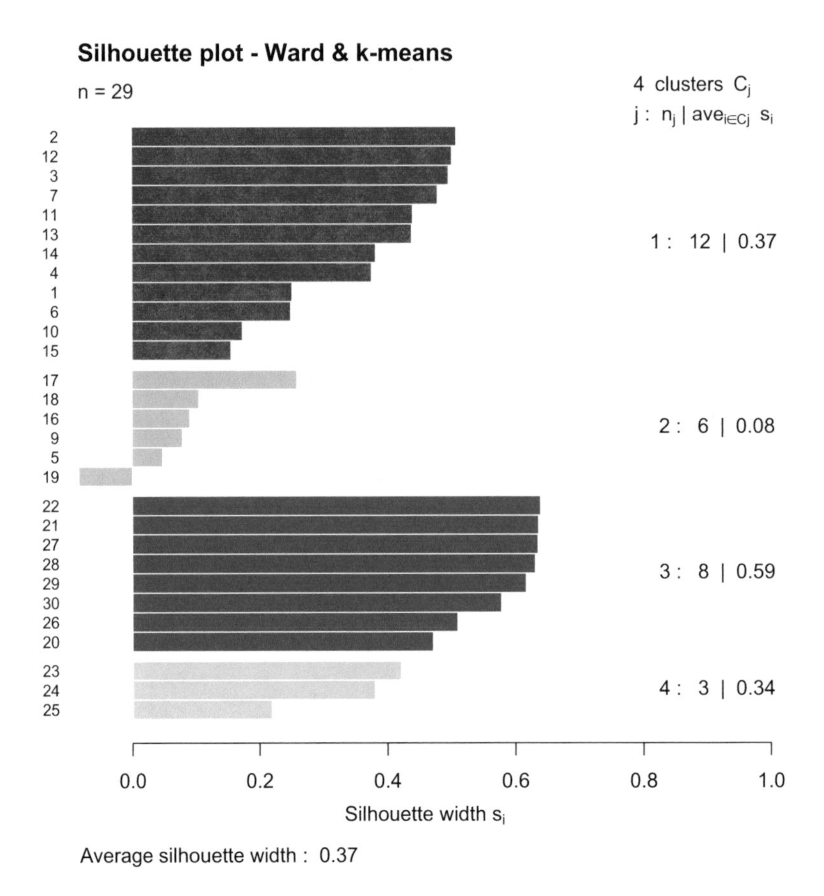

Fig. 4.21 Silhouette plot of the four-group partition from Ward clustering optimized by *k*-means

As can be seen in the code, comparison between the original and the optimized groups, or between the two optimization results, can be made by means of contingency tables (see Sect. 4.9.2). The two optimizations (based on species means and on starting objects) produce the same result for our choice of 4 "typical" objects. Note that this choice is critical. Comparison with the original Ward 4-group result shows that only one object, #15, has been moved from cluster 2 to cluster 1. This improves the silhouette plot (Fig. 4.21) and changes slightly the map of the groups along the river (Fig. 4.22). Note that the membership of site #19 remains unclear: its silhouette value remains negative.

```
# Plot the optimized Ward clusters on a map of the Doubs River
drawmap(xy = spa,
        clusters = spech.ward.gk,
        main = "Four optimized Ward clusters along the Doubs River"
)
```

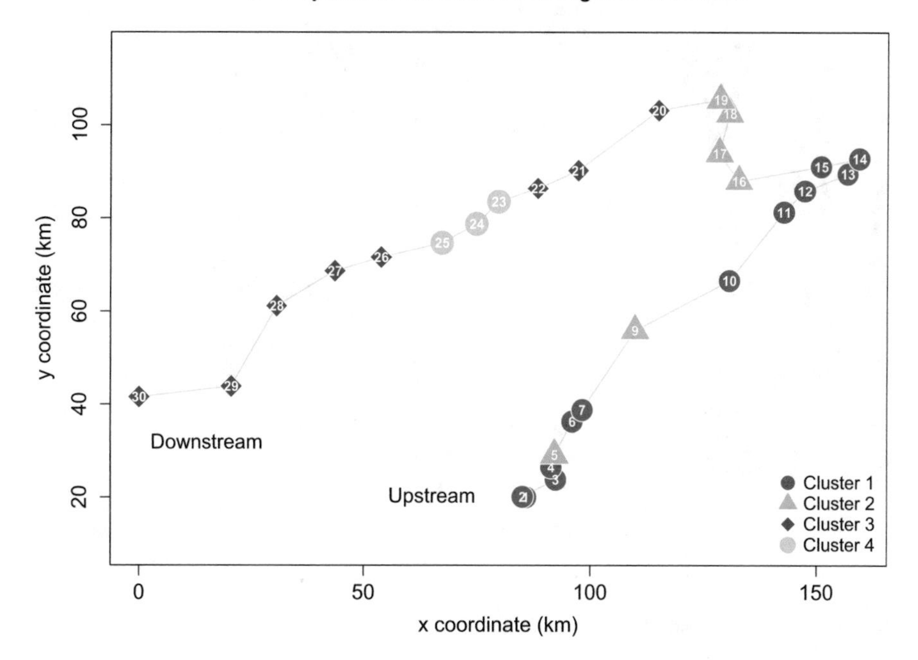

Fig. 4.22 The four optimized Ward clusters on a map of the Doubs River

4.8.2 *Partitioning Around Medoids (PAM)*

Partitioning around medoids (Chap. 2 *in* Kaufman and Rousseeuw 2005) *"searches for k representative objects or medoids among the observations of the dataset. These observations should represent the structure of the data. After finding a set of k medoids, k clusters are constructed by assigning each observation to the nearest medoid. The goal is to find k representative objects which minimize the sum of the dissimilarities of the observations to their closest representative object"* (excerpt from the **pam()** documentation file). By comparison, *k*-means minimizes the sum of the **squared** Euclidean distances within the groups. *k*-means is thus a traditional least-squares method, while PAM is not[2]. As implemented in **R**, **pam()** (package **cluster**) accepts raw data or dissimilarity matrices (an advantage over

[2]Many dissimilarity functions used in community ecology are non-Euclidean (for example the Jaccard, Sørensen and % difference indices). However, the square-root of these *D* functions is Euclidean. PAM minimizes the sum of the dissimilarities of the observations to their closest medoid. Since *D* is the square of the square-rooted (Euclidean) dissimilarities, PAM is a least-squares method when applied to these non-Euclidean *D* functions. See Legendre and De Cáceres (2013), Appendix S2.

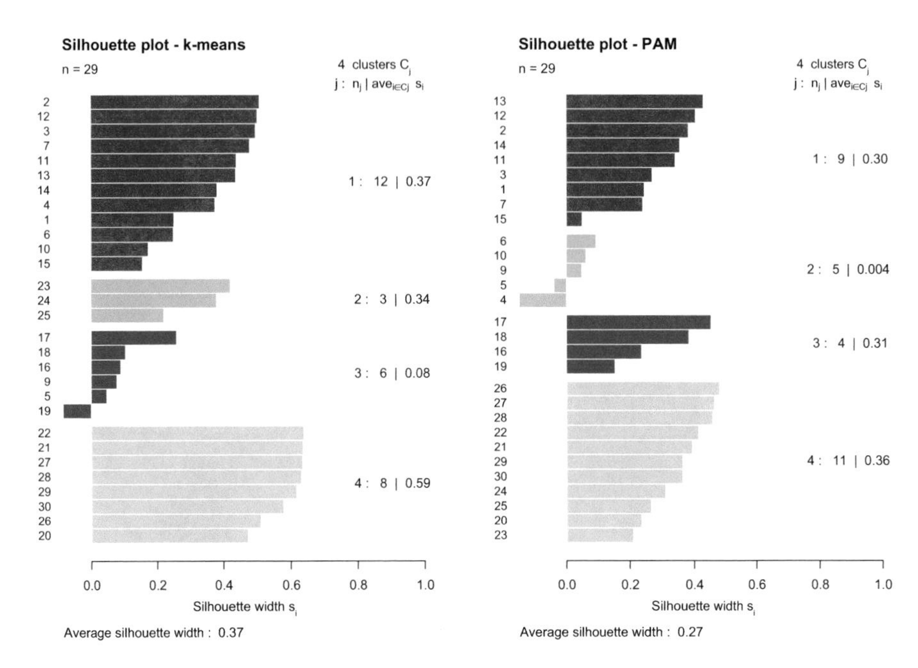

Fig. 4.23 Silhouette plots of the *k*-means and PAM results

`kmeans()` since it broadens the choice of association measures) and allows the choice of an optimal number of groups using the silhouette criterion. The code below ends with a double silhouette plot comparing the *k*-means and PAM results (Fig. 4.23).

Partitioning around medoids (PAM) computed from the chord distance matrix:

```
# Choice of the number of clusters
# Loop: obtain average silhouette widths (asw) for 2 to 28 clusters
asw <- numeric(nrow(spe))
for (k in 2:(nrow(spe) - 1))
  asw[k] <- pam(spe.ch, k, diss = TRUE)$silinfo$avg.width
k.best <- which.max(asw)
plot(
  1:nrow(spe),
  asw,
  type = "h",
  main = "Choice of the number of clusters",
```

```
  xlab = "k (number of clusters)",
  ylab = "Average silhouette width"
)
axis(
  1,
  k.best,
  paste("optimum", k.best, sep = "\n"),
  col = "red",
  font = 2,
  col.axis = "red"
)
points(k.best,
       max(asw),
       pch = 16,
       col = "red",
       cex = 1.5)
```

The not very interesting result k=2 is the best PAM solution with asw = 0.3841. Our previous choice of k=4 ends up with a poor performance in terms of silhouette width (asw = 0.2736). Nevertheless, let us compute a PAM for 4 groups:

```
# PAM for k = 4 clusters
spe.ch.pam <- pam(spe.ch, k = 4, diss = TRUE)
summary(spe.ch.pam)

spe.ch.pam.g <- spe.ch.pam$clustering
spe.ch.pam$silinfo$widths

# Compare with classification from Ward clustering and from k-means
table(spe.ch.pam.g, spech.ward.g)
table(spe.ch.pam.g, spe.kmeans.g)
```

The PAM result differs markedly from those of the Ward and k-means clusterings.

```
# Silhouette profile for k = 4 groups, k-means and PAM
par(mfrow = c(1, 2))
k <- 4
sil <- silhouette(spe.kmeans.g, spe.ch)
rownames(sil) <- row.names(spe)
plot(sil,
     main = "Silhouette plot - k-means",
     cex.names = 0.8,
     col = 2:(k + 1))
plot(
  silhouette(spe.ch.pam),
  main = "Silhouette plot - PAM",
  cex.names = 0.8,
  col = 2:(k + 1)
)
```

On the basis on this plot, you should be able to tell which solution (PAMor k-means) has a better silhouette profile.
You could also compare this result with the silhouette plot of the optimized Ward clustering produced earlier. Except cluster numbering, is there any difference between k-means and optimized Ward classifications? You can also address this question by examining a contingency table:

```
# Compare classifications from k-means and optimized Ward
# clustering
table(spe.kmeans.g, spech.ward.gk)
```

Hint *The PAM method is presented as "robust" because it minimizes a sum of dissimilarities instead of a sum of squared Euclidean distances. It is also robust in that it tends to converge to the same solution with a wide array of starting medoids for a given k value; this does not guarantee, however, that the solution is the most appropriate for a given research purpose.*

This example shows that even two methods that are devoted to the same goal and belong to the same general class (here non-hierarchical clustering) may provide diverging results. It is up to the user to choose the one that yields classifications that are bringing out more pertinent information or are more closely interpretable using environmental variables (next section).

4.9 Comparison with Environmental Data

All methods above have been presented with examples on species abundance data. They can actually be applied to any other type of data as well, particularly environmental data tables. Of course, care must be taken with respect to the choice of the proper coding and transformation for each variable (Chap. 2) and of the association measure (Chap. 3).

4.9.1 Comparing a Typology with External Data (ANOVA Approach)

We have seen that internal criteria, such as silhouette or other clustering quality indices, which rely on the species data only, were not always sufficient to select the "best" partition of the sites. The final choice of a typology should be based on the ecological interpretability of the groups. It could be seen as an external validation of the site typology.

Confronting clustering results (considered as response data) with external, independent explanatory data could be done by discriminant analysis (Sect. 6.5). From another point of view, the clusters obtained from the community composition data can be considered as a factor, or classification criterion, in the ANOVA sense. Here is a simplified example, showing how to perform quick assessments of the ANOVA assumptions (normality of residuals and homogeneity of variances) on several environmental variables separately, followed either by a parametric ANOVA or by a non-parametric Kruskal-Wallis test. Boxplots of the environmental variables (after some simple transformations to improve normality) for the four optimized Ward groups are also provided (Fig. 4.24). Note that, despite the fact that the clustering result based on the community composition data acts as an explanatory variable (factor) in the ANOVA, ecologically speaking we really look for an environmental interpretation of the groups of sites.

```
# Test of ANOVA assumptions
with(env, {
  # Normality of residuals
  shapiro.test(resid(aov(sqrt(ele) ~ as.factor(spech.ward.gk))))
  shapiro.test(resid(aov(log(slo) ~ as.factor(spech.ward.gk))))
  shapiro.test(resid(aov(oxy ~ as.factor(spech.ward.gk))))
  shapiro.test(resid(aov(sqrt(amm) ~ as.factor(spech.ward.gk))))
```

Residuals of `sqrt(ele)`, `log(slo)`, `oxy` *and* `sqrt(amm)` *are normally distributed, assuming that the power of the test is adequate. Try to find good normalizing transformations for the other variables.*

```
  # Homogeneity of variances
  bartlett.test(sqrt(ele), as.factor(spech.ward.gk))
  bartlett.test(log(slo), as.factor(spech.ward.gk))
  bartlett.test(oxy, as.factor(spech.ward.gk))
  bartlett.test(sqrt(amm), as.factor(spech.ward.gk))
```

Variable `sqrt(ele)` *has heterogeneous variances. It is not appropriate for parametric ANOVA.*

```
  # ANOVA of the testable variables
  summary(aov(log(slo) ~ as.factor(spech.ward.gk)))
  summary(aov(oxy ~ as.factor(spech.ward.gk)))
  summary(aov(sqrt(amm) ~ as.factor(spech.ward.gk)))

  # Kruskal-Wallis test of variable elevation
  kruskal.test(ele ~ as.factor(spech.ward.gk))
})
```

Are slope, dissolved oxygen and dissolved ammonium significantly different among species clusters?
Does elevation differ among clusters?

Hints *Note the use of* **`with()`** *at the beginning of the series of analyses to avoid the repetition of the name of the object* **`env`** *in each analysis. This is preferable to the use of* **`attach()`** *and* **`detach()`** *because the latter may lead to confusions if you have several datasets in your* **R** *console, and some happen to have variables with identical names.*

The null hypothesis for the Shapiro test is that the variable is normally distributed; in the Bartlett test, H_0 *states that the variances are equal among the groups. Therefore, for each of these tests, the p-value should be* <u>*larger*</u> *than the significance level, i.e.* $P > 0.05$*, for the ANOVA assumptions to be fulfilled.*

The parametric Bartlett test is sensitive to departures from normality. For non-normal data, we provide a function called **`bartlett.perm.R`** *that computes parametric, permutation and bootstrap (i.e., permutation with replacement) Bartlett tests.*

Two homemade generic functions will allow to perform post-hoc tests and show the results as letters on the boxplots of the environmental variables split by clusters (Fig. 4.24). Different letters denote significant differences among groups (in decreasing order). The **boxplert()** function should be used to perform ANOVA and LSD tests for multiple comparisons, whereas **boxplerk()** is used to perform a Kruskal-Wallis test and its corresponding post-hoc comparisons (both with Holm correction).

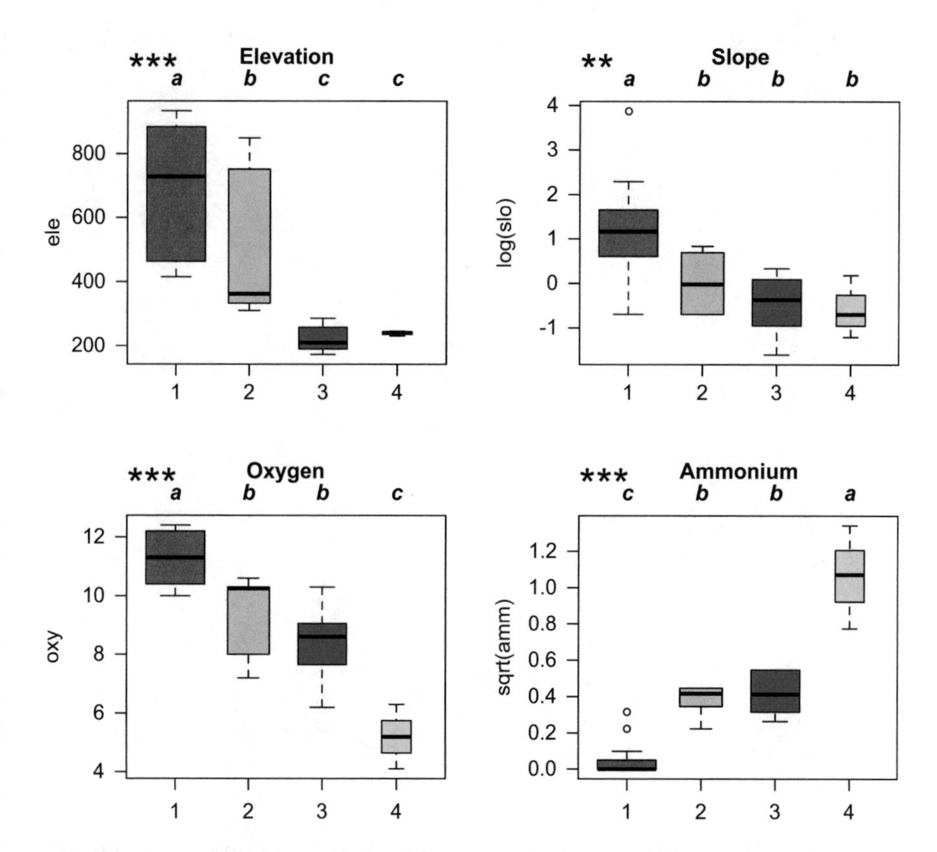

Fig. 4.24 Boxplots of four environmental variables grouped according to the four species-based groups from the optimized Ward clustering. Stars indicate the significance of the differences among groups for each environmental variable. The letters indicate which group means are significantly different

```
# Use boxplert() or boxplerk() to plot results with post-hoc tests
with(env, {
  boxplerk(
    ele,
    spech.ward.gk,
    xlab = "",
    ylab = "ele",
    main = "Elevation",
    bcol = (1:k) + 1,
    p.adj = "holm"
  )
  boxplert(
    log(slo),
    spech.ward.gk,
    xlab = "",
    ylab = "log(slo)",
    main = "Slope",
    bcol = (1:k) + 1,
    p.adj = "holm"
  )
  boxplert(
    oxy,
    spech.ward.gk,
    xlab = "",
    ylab = "oxy",
    main = "Oxygen",
    bcol = (1:k) + 1,
    p.adj = "holm"
  )
  boxplert(
    sqrt(amm),
    spech.ward.gk,
    xlab = "",
    ylab = "sqrt(amm)",
    main = "Ammonium",
    bcol = (1:k) + 1,
    p.adj = "holm"
  )
})
```

How would you qualify the ecology of the four fish community types based on this analysis?

Of course, the reverse procedure could be applied as well. One could cluster the environmental variables (to obtain a set of habitat types) and test if the species respond to these habitat types significantly through indicator species analysis (Sect. 4.11). In this approach, the species are tested one by one against the habitat types. Consider the question of multiple testing if several species are tested independently.

As an alternative, ordination-based multivariate approaches will be proposed in Chap. 6 to directly model and test the species-habitat relationships.

4.9.2 Comparing Two Typologies (Contingency Table Approach)

If you simply want to compare a typology generated from the species data to one independently obtained from the environmental variables, you can generate a table crossing the two typologies and test the relationship using a Fisher's exact test:

```
# Environment-based typology (see Chap. 2)
env2 <- env[, -1]
env.de <- vegdist(scale(env2), "euc")
env.kmeans <- kmeans(env.de, centers = 4, nstart = 100)
env.kmeans.g <- env.kmeans$cluster

# Table crossing the species and environment 4-group typologies
table(spe.kmeans.g, env.kmeans.g)
```

> *Do the two typologies tell the same story?*

```
# Test the relationship using a Fisher's exact test
fisher.test(table(spe.kmeans.g, env.kmeans.g))
```

Such tables could also be generated using categorical explanatory variables, which can be directly compared with the species typology.

4.10 Species Assemblages

Many approaches exist to address the problem of identifying species associations in a data set. Here are some examples.

4.10.1 Simple Statistics on Group Contents

The preceding sections immediately suggest a way to define crude assemblages: compute simple statistics (for instance mean abundances) from typologies obtained through a clustering method and look for species that are more present, abundant or specific in each cluster of sites. Here is an example.

```
# Compute mean species abundances in the four groups from the
# optimized Ward clustering
groups <- as.factor(spech.ward.gk)
spe.means <- matrix(0, ncol(spe), length(levels(groups)))
row.names(spe.means) <- colnames(spe)
for (i in 1:ncol(spe)) {
  spe.means[i, ] <- tapply(spe[, i], spech.ward.gk, mean)
}
group1 <- round(sort(spe.means[, 1], decreasing = TRUE), 2)
group2 <- round(sort(spe.means[, 2], decreasing = TRUE), 2)
group3 <- round(sort(spe.means[, 3], decreasing = TRUE), 2)
group4 <- round(sort(spe.means[, 4], decreasing = TRUE), 2)
# Species with abundances greater than group mean species abundance
group1.domin <- which(group1 > mean(group1))
group1
group1.domin
#... same for other groups
```

4.10.2 *Kendall's* W *Coefficient of Concordance*

Legendre (2005) proposed to use Kendall's *W* coefficient of concordance, together with permutation tests, to identify species assemblages in abundance data (this method cannot be applied to presence-absence data): *"An overall test of independence of all species is first carried out. If the null hypothesis is rejected, one looks for groups of correlated species and, within each group, tests the contribution of each species to the overall statistic, using a permutation test."* In this method, the search for species associations is done without any reference to a typology of the sites known *a priori* or computed from other data, for example environmental. The method aims at finding the most encompassing assemblages, i.e., the smallest number of groups containing the largest number of positively and significantly associated species.

The package **kendall.W** has been written to carry out these computations. Its functions are now part of **vegan**. The simulation results accompanying the Legendre (2005) paper show that *"when the number of judges [= species] is small, which is the case in most real-life applications of Kendall's test of concordance, the classical* χ^2 *test is overly conservative, whereas the permutation test has correct Type I error; power of the permutation test is thus also higher."* The **kendall.global()** function also includes a parametric *F*-test which does not suffer from the problems of the χ^2 test and has correct Type I error (Legendre 2010).

As a simple example, let us classify the fish species into several groups using *k*-means partitioning (Fig. 4.25), and run a global test (**kendall.global()**) to

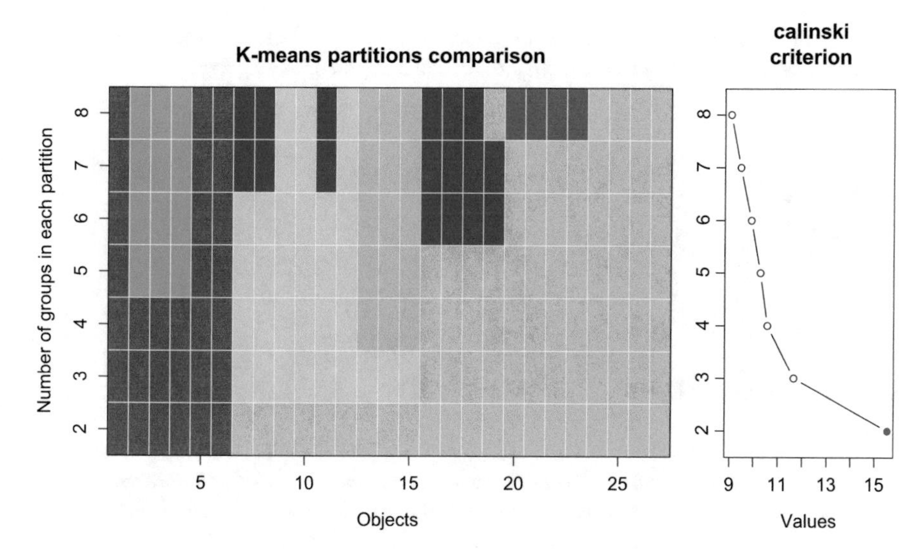

Fig. 4.25 *k*-means cascade plot showing the R-mode partitioning of the fish species. The Calinski-Harabasz criterion points to an optimum of 2 groups

know if all groups of species (called "judges" in the original paper) are globally significantly associated. If it is the case, we shall run post hoc tests (**`kendall.post()`**) on the species of each group to verify if all species within a group are concordant[3].

```
# Transformation of species data and transposition
spe.hel <- decostand(spe, "hellinger")
spe.std <- decostand(spe.hel, "standardize")
spe.t <- t(spe.std)
```

We can run a first test of Kendall concordance involving all species:

```
(spe.kendall.global1 <- kendall.global(spe.hel))
```

[3]Technical note: in the code that follows, the species data are first Hellinger-transformed (see Sect. 3.5). Then they are standardized. Standardization, which makes the variables dimensionless, is necessary because Kendall's W is based on correlation coefficients. For coherence reasons, a clustering of the species data, before Kendall's concordance analysis, must also be computed in a space where the variables are dimensionless.

The null hypothesis of absence of concordance is rejected. Thus, we can look for groups of species, then proceed with the Kendall analysis of these groups

```
# k-means partitioning of species
spe.t.kmeans.casc <- cascadeKM(
  spe.t,
  inf.gr = 2,
  sup.gr = 8,
  iter = 100,
  criterion = "calinski"
)
plot(spe.t.kmeans.casc, sortg = TRUE)
```

This result indicates that two groups may be a good choice. Avoid solutions with groups containing a single species except when it is clear that this species belongs to no other group. One group has 6 species and the other has 21. Three or four groups would also be fine at this point of the analysis: all groups would have 3 species or more.

```
# The partition into 2 groups is found in column 1 of the
# object $partition
(clusters2 <- spe.t.kmeans.casc$partition[, 1])
```

Partitions into three or four groups:

```
(clusters3 <- spe.t.kmeans.casc$partition[, 2])
(clusters4 <- spe.t.kmeans.casc$partition[, 3])
```

We will now examine the division of the species in two groups. Let us run a global Kendall W test on each group. This is done with a single call to the **kendall.global()** function.

```
# Concordance analysis
(spe.kendall.global2 <- kendall.global(spe.hel, clusters2))
```

Look at the corrected permutational p-values. If all values are equal to or smaller than 0.05, you can consider that all groups are globally significant, i.e. that on the whole they contain species that are concordant; this does not mean that all species in a globally significant group are concordant, only that at least some species are. If the corrected p-values for some groups were not significant (it is not the case with this example), it would indicate that these groups include non-concordant species and should be subdivided into smaller groups. In other words, a partition into more than 2 groups would be in order.

Now let us run *a posteriori* tests to identify the significantly concordant species within each group:

```
# A posteriori tests
(spe.kendall.post2 <- kendall.post(spe.hel, clusters2,
                                   nperm = 9999))
```

Look at the mean Spearman correlation coefficients of the individual species. A group contains concordant species if each of its species has a positive mean correlation with all the other species of its group. If a species has a negative mean correlation with all other members of its group, this indicates that this species should be left out of the group. Try a finer division of the groups and see if that species finds itself in a group for which it has a positive mean correlation. This species may also form a singleton, i.e. a group with a single species.

With 2 groups, we have in the largest group one species (`Sqce`) that has a negative mean correlation with all members of its group. This indicates that we should look for a finer partition of the species. Let us carry out a posteriori tests with 3 groups. Readers can also try with 4 groups and examine the results.

```
(spe.kendall.post3 <- kendall.post(spe.hel, clusters3,
                                   nperm = 9999))
```

Now all species in the three groups have positive mean Spearman correlations with the other members of their group. `Sqce` finds itself in a new group of 9 species with which it has a mean positive correlation, although its contribution to the concordance of that group is not significant. All the other species in all three groups contribute significantly to the concordance of their group. So we can stop the analysis and consider that three groups of species, with respectively 12, 9 and 6 species, adequately describe the species associations in the Doubs River.

Ecological theory predicts nested structures in ecological relationships. Within communities, subgroups of species can be more or less loosely or densely associated. One can explore such avenues by investigating smaller species groups within the large species associations revealed by the Kendall *W* test results.

The groups of species defined here may be further interpreted ecologically by different means. For instance, mapping their abundances along the river and computing summary statistics on the sites occupied by the species assemblages can help in assessing their ecological roles. Another avenue towards interpretation is to compute a redundancy analysis (RDA, Sect. 6.3) of the significantly associated species with respect to a set of explanatory environmental variables.

4.10.3 Species Assemblages in Presence-Absence Data

A method exists for presence-absence data (Clua et al. 2010). It consists in computing the *a* component of Jaccard's S_7 coefficient (as a measure of co-occurrence among species) in R-mode and assessing its probability by means of a permutation test in the spirit of the Raup and Crick (1979) coefficient. The p-values act as dissimilarities: they have very small values for highly co-occurring species. We provide an **R** function called **`test.a()`** to compute this coefficient. Readers are invited to apply it to the species of the Doubs fish data transformed to presence-absence form. A critical point is to specify enough permutations for the probabilities to survive a correction for multiple testing (see Sect. 7.2.6). There are 27 species, and thus $27 \times 26/2 = 351$ tests will be run. A Bonferroni correction requires a p-value of $0.05/351 = 0.0001425$ to remain significant at the 0.05 level. This requires at least 9999 permutations, since the smallest p-value would be $1/(9999 + 1) = 0.0001$. 99,999 permutations provide a finer estimation of the p-value but beware: computation can take several minutes.

```
# Transform the data to presence-absence
spe.pa <- decostand(spe, "pa")
# Test the co-occurrence of species
res <- test.a(spe.pa, nperm = 99999)
summary(res)
```

In the output object, `res$p.a.dist` *contains a matrix of p-values of class dist. The next step is to compute a Holm correction (see Sect. 7.2.6) on the matrix of p-values unfolded as a vector.*

```
# Compute a Holm correction on the matrix of p-values
res.p.vec <- as.vector(res$p.a.dist)
adjust.res <- p.adjust(res.p.vec, method = "holm")
range(adjust.res)
```

Among the corrected Holm p-values, find 0.05 or the closest value smaller than 0.05:

```
(adj.sigth <- max(adjust.res[adjust.res <= 0.05]))
```

Now find the uncorrected p-value corresponding to `adj.sigth`*:*

```
(sigth <- max(res.p.vec[adjust.res <= 0.05]))
```

In this run it is 0.00017. The significant values thus have an uncorrected probability of 0.00017 or less. Replace all larger values in the probability matrix by 1:

```
res.pa.dist <- res$p.a.dist
res.pa.dist[res.pa.dist > sigth] <- 1
```

How many (unadjusted) values are equal to or smaller than `sigth`*?*

```
length(which(res.p.vec <= sigth))
```

The dissimilarity matrix made of the p-values can be displayed as a heat map (see Chap. 3):

```
# Heat map of significant p-values
coldiss( res.pa.dist,
         nc = 10,
         byrank = TRUE,
         diag = TRUE)
```

4.10.4 Species Co-occurrence Network

Co-occurrence network analysis becomes more and more popular in community ecology, especially to investigate ecological interactions among species or among communities. The principle is to analyse associations among species based on their co-occurrence in ecological meta-communities or multitrophic species assemblages. Based on the topology of the co-occurrence network, sociological groups of species are defined, called "modules". Network structure is characterized by two main properties, namely "modularity" (the extent to which species co-occurrences are organized into modules, i.e. densely connected, non-overlapping subsets of species) and "nestedness" (the tendency of the network to show a nested pattern wherein the species composition of small assemblages is a nested subset of larger assemblages; see Sect. 8.4.3). The role of a species is defined by its position compared with other species in its own module (its "standardized within-module degree", i.e., the number of links that a node has with other nodes in the same module, standardized by the mean and standard deviation of the number of links per node in the module) and how well it connects to species in other modules (its among-module "connectivity"). For more details, see e.g. Olesen et al. (2007) and Borthagaray et al. (2014).

Basically, the network is built from an adjacency matrix, which may be binary (species significantly co-occur or not) or numeric (links among species are weighted). Various metrics of positive, neutral or negative association among species can be used to compute the adjacency matrix, including Spearman correlations, Jaccard similarity coefficients or the p-values obtained in the previous section. These metrics can be computed from a species presence-absence or abundance matrix. A threshold is often applied to restrict the non-zero links to the most important positive associations to build undirected networks. The network is drawn in such a way that frequently co-occurring species are located close to one another in the graph, using various algorithms.

Several **R** packages are devoted to network analysis. Restricting ourselves to a basic introduction to this approach, we shall use **`igraph`** for network processing and **`picante`** for computing co-occurrence distances (Hardy 2008).

Let us begin by computing several adjacency matrices from the fish species dataset used above. You can choose one of these symmetric matrices to build the undirected co-occurrence network, here from Jaccard dissimilarity (Fig. 4.26).

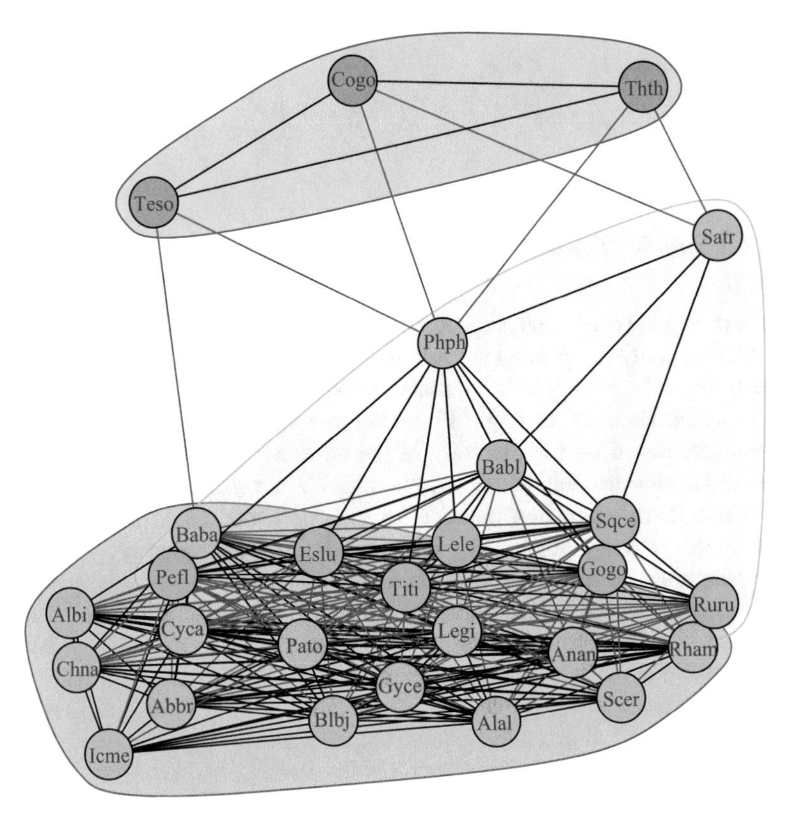

Fig. 4.26 Graph of the co-occurrence network of the fish species based on Jaccard similarity, showing three modules (species bubble colours). Positive intra-module associations are indicated by black lines, positive inter-module ones by red lines

```
# Attach supplementary packages
library(igraph)
library(rgexf)

# Adjacency matrix from the binary matrix of significant
# co-occurrences (results of the previous section)
adjm1 <- 1 - as.matrix(res.pa.dist)
diag(adjm1) <- 0

# Adjacency matrix from the "a" distance matrix
adjm2 <- 1 - as.matrix(res$p.a.dist)
adjm2[adjm2 < 0.5] <- 0
diag(adjm2) <- 0

# Adjacency matrix from the Spearman rank correlation matrix
adjm3 <- cor(spe, method = "spearman")
# Only positive associations (rho >= 0.25)
adjm2[adjm3 < 0.25] <- 0
adjm2[adjm3 >= 0.25] <- 1  # binary co-occurrences
diag(adjm3) <- 0

# Species co-occurrence dissimilarities
# (picante package, Hardy 2008)
adjm4 <- species.dist(spe.pa, metric = "jaccard")
adjm4 <- as.matrix(adjm4)
adjm4[adjm4 < 0.4] <- 0

# Select an adjacency matrix
adjm <- adjm4
summary(as.vector(adjm))

# Plot histogram of adjacency values
hist(adjm)

# Build graph
go <- graph_from_adjacency_matrix(adjm, weighted = TRUE,
                                  mode = "undirected")
plot(go)

# Network structure detection: find densely connected subgraphs
# (modules) in a graph
wc <- cluster_optimal(go)
modularity(wc)
membership(wc)
plot(wc, go)

# Detach package rgexf
detach("package:rgexf", unload = TRUE)
# If not sufficient:
unloadnamespace("rgexf")
```

4.11 Indicator Species

4.11.1 Introduction

The search for indicator or characteristic species in groups of sites is a particularly important question in applied ecology, where conservation and management specialists look for methods to quickly assess or predict the type of environment characterizing a given location, to assess the type of living community (e.g., vegetation type) in a given area, or to monitor temporal changes in some ecological variables (e.g., mean annual temperature). In fundamental ecology, one often looks for ecological preferences of species among a set of site groups, either to gain knowledge about the species themselves or to improve the knowledge of the habitats sampled in the groups of sites. These approaches are complementary, and different methods exist to address them. De Cáceres and Legendre (2009) presented the two complementary approaches and their related indices, classified into indicator value and correlation indices, respectively.

Indicator value indices are based on the concepts of *specificity* (highest when the species is present in the target group but not elsewhere) and *fidelity* (highest when the species is present in all sites of the target group). A high indicator value is obtained by a combination of high specificity and fidelity.

Correlation indices rely on the computation of statistics related to Pearson's *r* correlation coefficient, measuring the relationship between a vector indicating the presence or abundance of a species at different sites and a vector describing to which predefined group each site belongs.

De Cáceres and Legendre (2009) describe 12 indices belonging to the two groups. They state that indicator value indices are more useful for assessing the species predictive values as bioindicators, e.g. for field determination of community types or for ecological monitoring, whereas correlation indices should be used to determine the ecological preference of a given species among a set of alternative site groups (De Cáceres and Legendre 2009, p. 3573).

4.11.2 IndVal: Species Indicator Values

The Dufrêne and Legendre (1997) IndVal index is computed as the product of the specificity of a species to the targeted group by its fidelity to the targeted group. Specificity is defined by the mean abundance of the species within the targeted group compared to its mean abundance across the groups; fidelity is the proportion of sites of the targeted group where the species is present. Dave Roberts, the author of the package `labdsv` used hereunder, summarized the concept as follows (pers. comm.): "*The indval approach looks for species that are both necessary and sufficient, i.e. if you find that species you should be in that type, and if you are in that type you should find that species*".

The groups of sites can be defined in various ways. A simple, although circular way is to use the result of a cluster analysis based on the species data. The indicator species are then simply the most prominent members of these groups. Of course, one cannot rely on statistical tests to identify the indicator species in this case, since the classification and the indicator species do not derive from independent data. Another approach, which is conceptually and statistically more fruitful, consists in clustering the sites on the basis of independent data (environmental variables, for instance). The indicator species can then be considered *indicator* in the true sense of the word, i.e. species closely related to the ecological conditions of their group. The statistical significance of the indicator values (i.e., the probability of obtaining by chance as high an indicator value as observed) is assessed by means of a permutation test.

The Dufrêne and Legendre index is available in function **`indval()`** of package **`labdsv`**. Let us apply it to our fish data. As an example of a search for indicator species related to one explanatory variable, we could for instance divide the data set into groups of contiguous sites on the basis of the variable `dfs` (distance from the source), acting as a surrogate of the overall river gradient conditions, and look for indicator species in the groups.

```
# Divide the sites into 4 groups depending on the distance from
# the source of the river
dfs.D1 <- dist(data.frame(dfs = env[, 1],
                row.names = rownames(env)))
dfsD1.kmeans <- kmeans(dfs.D1, centers = 4, nstart = 100)
# Cluster delimitation and numbering
dfsD1.kmeans$cluster
# Cluster labels are in arbitrary order
# See Hint below
grps <- rep(1:4, c(8, 10, 6, 5))
# Indicator species for this typology of the sites
(iva <- indval(spe, grps, numitr = 10000))
```

Hint ***BEWARE***: *in the output table, the cluster numbering corresponds to the cluster numbers fed to the program, which do not necessarily follow a meaningful order. The k-means analysis produces arbitrary group labels in the form of numbers. The column headers of the* **`indval()`** *output are the ones provided to the program, but the columns have been reordered according to the labels. Consequently, the groups do not follow the order along the river. This is why we have manually constructed an object (*`grps`*) with sequential group numbers.*

The resulting object contains the following tables:

- `relfrq` = relative frequency of the species in each group = number of sites where species is present/number of sites in the group
- `relabu` = relative abundance of the species across groups = total abundance in group/grand total abundance
- `indval` = indicator value (IndVal) of each species
- `maxcls` = cluster where the species has highest IndVal
- `indcls` = highest IndVal of the species
- `pval` = permutational p-value of IndVal

Correct the p-values for multiple testing:

```
pval.adj <- p.adjust(iva$pval)
```

The following lines of code extract the significant indicator species with their group with highest IndVal, the corresponding IndVal, the species' indicator p-value and their total frequency in the data set.

```
# Table of the significant indicator species
gr <- iva$maxcls[pval.adj <= 0.05]
iv <- iva$indcls[pval.adj <= 0.05]
pv <- iva$pval[pval.adj <= 0.05]
fr <- apply(spe > 0, 2, sum)[pval.adj <= 0.05]
fidg <- data.frame(
  group = gr,
  indval = iv,
  pvalue = pv,
  freq = fr
)
fidg <- fidg[order(fidg$group, -fidg$indval), ]
fidg

# Export the result to a CSV file (to be opened in a spreadsheet)
write.csv(fidg, "IndVal-dfs.csv")
```

On this basis, what do you think of the results? How do they relate to the four ecological zones presented in Chap.1?
Note that the indicator species identified here may differ from the members of the species assemblages identified in Sect. 4.10. The indicator species are linked to predefined groups, whereas the species assemblages are identified without any prior classification or reference to environmental conditions.

In Sect. 4.12 you will find another application in relationship with the MRT method. Some detailed results will be presented.

Package **indicspecies**, which was produced as a companion package to the De Cáceres and Legendre (2009) paper, computes various indicator species indices, including IndVal (or, actually, the square root of IndVal). Two especially interesting features of this package are the possibility of pooling two or more groups of a typology in turn to look for species that may be indicators of pooled groups (in function **multipatt()**) and the computation of bootstrapped confidence intervals around indicator values (argument nboot of function **strassoc()**).

Let us first compute the same analysis as above, using function **multipatt()** this time and taking advantage of the possibility of pooling the groups by two.

```
# Indval with indicspecies::multipatt() with search for indicator
# species of pooled groups
(iva2 <- multipatt(
  spe,
  grps,
  max.order = 2,
  control = how(nperm = 999)
))
```

In function **multipatt()**, the default indicator measure is func = IndVal.g. The ".g" indicates that the measure is corrected for unequal group sizes. This corresponds to the original Dufrêne and Legendre (1997) IndVal measure and is recommended.

The output object contains several matrices:

- comb = matrix of the belonging of the sites to all possible combinations up to the order requested in the analysis;
- str = the "strength of association", i.e., the value of the chosen index;
- A = if the function is the IndVal index, value of its A component (specificity); otherwise NULL;
- B = if the function is the IndVal index, value of its B component (fidelity); otherwise NULL;
- sign = results of the best patterns, i.e., the group or combination of groups where the statistic is the highest, and permutation test results for that result. No correction for multiple testing is provided.

Here again, the p-values should be corrected for multiple testing:

```
(pval.adj2 <- p.adjust(iva2$sign$p.value))
```

One can also request a simpler output with the **summary()** function:

```
summary(iva2, indvalcomp = TRUE)
```

The summary displays the groups and combinations of groups (if any has been requested) with significant indicator species. In each of these groups it displays the species' IndVal value (stat) and the p-value of the permutational test. With argument indvalcomp = TRUE, the A (specificity) and B (fidelity) components of the IndVal index are displayed as well.

The results of this second analysis are slightly different from the previous ones because for some species the highest IndVal is found for a combination of groups instead of a single group. For instance, for the brown trout (Satr) the highest value is found for the combination of groups 1 + 2 whereas, if the analysis is conducted on separate groups only, the highest IndVal is in group 1.

Our next analysis consists in computing confidence intervals around IndVal values using function **strassoc()**. The number of iterations is provided by argument nboot.

```
# Indval with bootstrap confidence intervals of indicator values
(iva2.boot <- strassoc(spe, grps, func = "IndVal.g", nboot = 1000))
```

The output object consists in a list of three elements: the indicator value ($stat), and the lower ($lowerCI) and upper ($upperCI) limits of the confidence intervals. For the first three species, the results are the following (they vary from run to run since they result from bootstrapping):

```
$stat
         1           2           3           4
Cogo     0.0000000   0.89442719  0.00000000  0.00000000
Satr     0.6437963   0.67667920  0.00000000  0.05923489
Phph     0.5784106   0.75731725  0.09901475  0.05423261
(...)
$lowerCI
         1           2           3           4
Cogo     0.00000000  0.70710678  0.00000000  0.00000000
Satr     0.39086798  0.45825757  0.00000000  0.00000000
Phph     0.30618622  0.58177447  0.00000000  0.00000000
(...)
$upperCI
         1           2           3           4
Cogo     0.0000000   1.0000000   0.0000000   0.0000000
Satr     0.8277591   0.8537058   0.0000000   0.1924501
Phph     0.7585133   0.9004503   0.2348881   0.1721326
(...)
```

When interpreting confidence intervals, keep in mind that any indicator value whose lower limit is equal to 0 can be considered non-significant, even if the value itself is greater than 0. For instance, in groups 3 and 4, the IndVal values of the Eurasian minnow (Phph) are 0.09 and 0.054 respectively, but the lower limit of the confidence intervals is 0. Also, two values whose confidence intervals are overlapping cannot be considered different. For example, in group 1, the CI limits for the brown trout (Satr) and the Eurasian minnow (Phph) are [0.391; 0.828] and

[0.306; 0.758]. Therefore, there is a large probability that the true IndVal values for these two species reside somewhere within the common part of these ranges, i.e., [0.391; 0.758], but there is no way of determining that one value is larger than the other.

4.11.3 Correlation-Type Indices

As mentioned above, correlation indices were devised to help identify the ecological preferences of species among a set of groups. De Cáceres and Legendre (2009) pointed out that this approach is "*probably more useful* [for this purpose] *than the indicator value approach, because the former naturally allows the detection of negative preferences*". Beware, however, of the problems relative to the absence of species from a set of sites. The absence of a species may be due to different reasons in each site, and thus it should not be interpreted in ecological terms without due caution (see Sect. 3.2.2). Species with high ecological preferences for a given group of sites representing well-defined ecological conditions are often called "diagnostic" species in plant ecology and are useful to identify vegetation types in field surveys (Chytrý et al. 2002 *in* De Cáceres and Legendre 2009).

The simplest correlation-type index for presence-absence data is called Pearson's ϕ (phi) coefficient of association (Chytrý et al. 2002 *in* De Cáceres and Legendre 2009). It consists in the correlation between two binary vectors. For all sites, one vector gives the presence or absence of the species, and the other indicates that the site belongs (1) or not (0) to a given group. With quantitative (abundance) data, the phi coefficient is called the point biserial correlation coefficient and a vector of abundances replaces the presence-absence vector. In function **strassoc()**, the default indicator measure is the phi coefficient, requested by `func = "r"`, but an `"r.g"` option is available that corrects the measure for unequal group sizes. We recommend the latter option, following De Cáceres' **indicspecies** tutorial.

Let us compute the phi coefficient (corrected for unequal group sizes) on the fish abundance data.

```
# Phi correlation index
iva.phi <- multipatt(
  spe,
  grps,
  func = "r.g",
  max.order = 2,
  control = how(nperm = 999)
)
summary(iva.phi)
```

The results of this analysis look quite similar to that of the IndVal analysis. Note that the statistical tests highlight the highest *positive* phi values (see the summary). Therefore, it is natural that both approaches (species indicator values and correlation) highlight the same strongest associations between species and selected groups. However, it may be interesting to display all the phi values to identify possible sets of environmental conditions (represented by the groups, if these have been defined on the basis of environmental variables) that are *avoided* by some species (with the caveat mentioned above):

```
round(iva.phi$str, 3)
```

For instance, the bleak (`Alal`) shows a very strong negative association ($\phi = -0.92$) with the group of sites "1 + 2", which mirrors its strongest positive association ($\phi = 0.92$) with the group of sites "3 + 4". Indeed, the bleak is absent from all but 3 sites of groups 1 + 2, and is present in all sites of groups 3 + 4. Obviously, the bleak prefers the lower regions of the river and avoids the higher ones. The ecological characteristics of the various sections of the Doubs River are described and analysed in various sections of this book.

It is also possible to compute bootstrap confidence intervals around the phi values, using the same **`strassoc()`** function as was used to compute the IndVal coefficient:

```
iva.phi.boot <- strassoc(spe, grps, func = "r.g", nboot = 1000)
```

Since the phi values can be negative or positive, you can use this opportunity to obtain a bootstrap test of significance of the phi values: all values where both confidence limits have the same sign (i.e., CI not encompassing 0) are deemed significant. This complements the permutational tests for the negative values. For instance, the avoidance of the bullhead (`Cogo`) of the conditions prevailing in group 1 is significant in that sense, the CI interval being [–0.313, –0.209].

4.12 Multivariate Regression Trees (MRT): Constrained Clustering

4.12.1 Introduction

Multivariate regression trees (MRT; De'ath 2002) are an extension of univariate regression trees, a method allowing the recursive partitioning of a quantitative response variable under the control of a set of quantitative or categorical explanatory variables (Breiman et al. 1984). Such a procedure is sometimes called constrained or supervised clustering. The result is a tree whose "leaves" (terminal groups of sites) are composed of subsets of sites chosen to minimize the within-group sums of squares (as in a *k*-means clustering), but where each successive partition is defined by a threshold value or a state of one of the explanatory variables. Among the numerous potential solutions in terms of group composition and number of leaves, one usually retains the one that has the best *predictive* power. This stresses the fact that, contrary to most constrained ordination methods described in Chap. 6, where the selection of explanatory variables is made on the basis of *explanatory* power, MRT focuses on prediction, making it a very interesting tool for practical applications, as in environmental management. The focus on prediction is embedded in the method, as will become clear below.

MRT is a powerful and robust method, which can handle a wide variety of situations, even those where some values are missing, and where the relationships between the response and explanatory variables are nonlinear, or where high-order interactions among explanatory variables are present.

4.12.2 Computation (Principle)

The computation of a MRT consists in two procedures running together: (1) constrained partitioning of the data and (2) cross-validation of the results. Let us first briefly explain the two procedures. After that we will see how they are applied together to produce a model that has the form of a decision tree.

4.12.2.1 Constrained Partitioning of the Data

- For each explanatory variable, produce all possible partitions of the sites into two groups. For a quantitative variable, this is done by sorting the sites according to the ordered values of the variable, and repeatedly splitting the series after the first, second... $(n - 1)^{\text{th}}$ object. For a categorical variable, allocate the objects to two groups, screening all possible combinations of levels. In all cases, compute the resulting sum of within-group sums of squared distances to the group means (within-group SS) for the response data. Retain the solution minimizing this quantity, along with the identity and value of the explanatory variable or the level of the categorical variable producing the partition retained.
- Repeat the same procedure within each of the two subgroups retained above; in each group, retain the best partition along with the corresponding explanatory variable and its threshold value.
- Continue within all partitions until all objects form their own group or until a preselected smallest number of objects per group is reached. At that point, select the tree with size (number of groups) appropriate to the aims of the study. For studies with a predictive objective, cross-validation, which is a procedure to identify the best predictive tree, is developed below.
- Apart from the number and composition of the leaves, an important characteristic of a tree is its *relative error* (RE), i.e., the sum of the within-group SS over all leaves divided by the overall SS of the data. In other words, this is the fraction of variance not explained by the tree. Without cross-validation, among the successive partitioning levels, one would retain the solution minimizing RE; this would

be equivalent to retaining the solution maximizing the R^2. However, this would be an explanatory rather than predictive approach. De'ath (2002) states that "*RE gives an over-optimistic estimate of how accurately a tree will predict for new data, and predictive accuracy is better estimated from the cross-validated relative error (CVRE)*".

4.12.2.2 Cross-Validation of the Partitions and Pruning of the Tree

At which level should we prune a tree, i.e., cut each of its branches so as to retain the most sensible partition? To answer this question in a prediction-oriented manner, one uses a subset of the objects (training set) to construct the tree, and the remaining objects (test set) to validate the result by allocating them to the constructed

The measure of predictive error is the cross-validated relative error (CVRE). The function is:

$$CVRE = \frac{\sum_{k=1}^{\nu}\sum_{i=1}^{n}\sum_{j=1}^{p}\left(y_{ij(k)} - \widehat{y}_{j(k)}\right)^2}{\sum_{i=1}^{n}\sum_{j=1}^{p}\left(y_{ij} - \bar{y}_j\right)^2} \tag{4.1}$$

where $y_{ij(k)}$ is one observation of the test set k, $\widehat{y}_{j(k)}$ is the predicted value of one observation in one leaf (centroid of the sites of that leaf), and the denominator represents the overall dispersion (sum of squares) of the response data.

The CVRE can thus be defined as the ratio between the dispersion unexplained by the tree (summed over the k test sets) divided by the overall dispersion of the response data. Of course, the numerator changes after every partitioning event. CVRE is 0 for perfect predictors and close to 1 for a poor set of predictors.

4.12.2.3 MRT Procedure

Now that we have both components of the methods, let us put them together to explain the sequence of events of a cross-validated MRT run:

- Randomly split the data into k groups; by default $k = 10$.
- Leave one of the k groups out and build a tree by constrained partitioning, the decisions being made on the basis of the minimal within-group SS.
- Rerun the step above $k - 1$ times, leaving out each of the test groups in turn.
- In each of the k solutions above, and for each possible partition size (number of groups) within these solutions, reallocate the test set. Compute the CVRE for all partition sizes of the k solutions (one CVRE value per size). Eq. 4.1 encompasses the computation for one level of partitioning and all k solutions.
- Pruning of the tree: retain the partition size for which the CVRE is smallest. An alternative solution is to retain a smallest size for which the CVRE value is the

minimal CVRE value plus one standard error of the CVRE values. This is called "the 1 SE rule".

- To obtain an estimate of the error of this process, run it a large number of times (100 or 500 times) with other random assignments of the objects to *k* groups.
- The final tree retained is the one showing most of the smallest CVRE values over all permutations, or the one respecting most often the 1 SE rule.
- In **mvpart**, the smallest possible number of objects in one group is equal to `ceiling(log2(n))` when argument `minauto = TRUE`. Setting `minauto = FALSE` allows the partitioning to proceed until the data set is fully split into individual observations.

In MRT analysis, the computations of sums-of-squares (SS) are done in Euclidean space. To account for special characteristics of the data, they can be pre-transformed prior to being submitted to the procedure. Pre-transformations for species data prior to their analysis by Euclidean-based methods are presented in Sect. 2.2.4. For response data that are environmental descriptors with different physical units, the variables should be standardized before they are used in MRT.

4.12.3 Application Using Packages **mvpart** *and* **MVPARTwrap**

As of this writing, the only package implementing a complete and handy version of MRT is **mvpart**. Unfortunately, this package is no longer supported by the R Core Team, so that no updates are available for **R** versions posterior to R 3.0.3. Nevertheless, **mvpart** can still be installed on more recent versions of **R** by applying the following code.

```
# On Windows machines, Rtools (3.4 and above) must be installed
# first. Go to: https://cran.r-project.org/bin/windows/Rtools/
# After that (for Windows and MacOS), type:
install.packages("devtools")
library(devtools)
install_github("cran/mvpart", force = TRUE)
install_github("cran/MVPARTwrap", force = TRUE)
```

Package **mvpart** has been written to compute MRT, using univariate regression trees computed by a function called **rpart()**. Its use requires that the response data belong to class 'matrix' and the explanatory variables to class 'data frame'. The relationship is written as a formula of the same type as those used in regression functions (see **?lm**). The example below shows the simplest implementation, where one uses all the variables contained in the explanatory data frame.

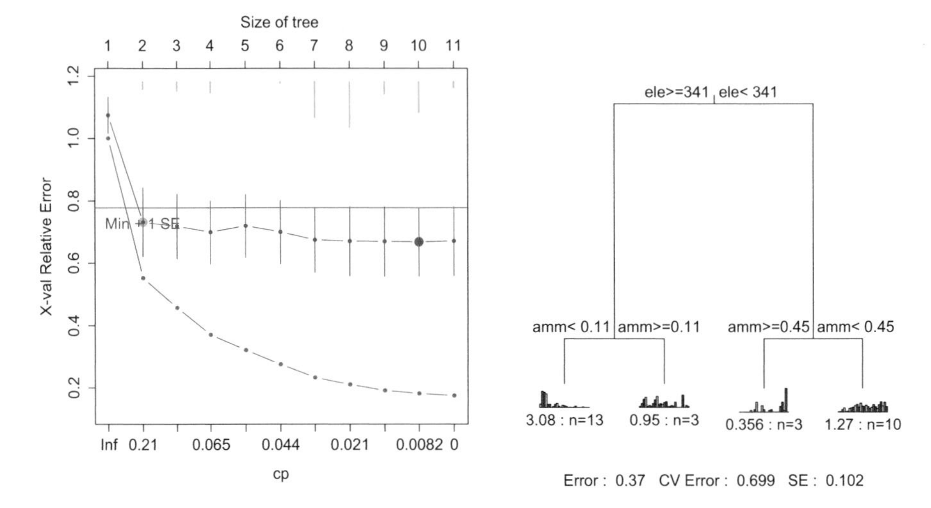

Fig. 4.27 Left: graph of the (steadily decreasing) relative error RE and the cross-validated relative error CVRE. The solution with the smallest CVRE is indicated (red point), as well as CVRE error bars. The green vertical bars indicate the number of times that the solution was selected as the best one during the cross-validation iterations. Right: Multivariate regression tree of the Doubs fish species explained by their environmental variables. Interpretation: see text

Let us build a multivariate regression tree of the Doubs fish data constrained by the environmental variables. We will use the fish species data with site vectors normalized to length 1 (chord transformation in Sect. 2.2.4) so that the distance actually preserved is the chord distance. Among the arguments, we will use `xval` = 29 cross-validation groups (i.e., as many groups as rows in the data, instead of the usual 10, owing to the small number of observations), 100 iterations, and we will allow ourselves to interactively pick a solution on a graph provided by **`mvpart()`** (Fig. 4.27, left). The graph displays the relative error (RE), which is steadily decreasing when the number of groups increases, and the cross-validated relative error (CVRE), which usually drops first sharply and then goes to a minimum before increasing again. Prediction is optimal for a given number of clusters, but its quality decreases when the data are exaggeratedly fragmented into many small groups.

The best solution is not always obvious, however. Sometimes it is the simple two-group solution that shows the smallest CVRE. The graph provides error bars representing one standard error for the CVRE, and an orange horizontal line located one standard error above the minimal CVRE solution (large red spot). According to De'ath (2002), one may select that tree as the best predictive tree or, following the rule proposed by Breiman et al. (1984) for univariate regression trees, one may select the smallest tree within one standard error of the best tree; this is the tree with k = 2 in our example, identified by "Min + 1 SE". That tree is more parsimonious and only slightly worse than the best predictive tree.

```
par(mfrow = c(1, 2))
spe.ch.mvpart <-
  mvpart(
    data.matrix(spe.norm) ~ .,
    env,
    margin = 0.08,
    cp = 0,
    xv = "pick",
    xval = nrow(spe),
    xvmult = 100
  )

summary(spe.ch.mvpart)
printcp(spe.ch.mvpart)
```

Hint *Argument* `xv = "pick"` *allows one to interactively pick a tree among those proposed. If one prefers that the tree with the minimum CVRE be automatically chosen, then use* `xv = "min"`.

If argument `xv = "pick"` has been used, which we recommend, one left-clicks on the point representing the desired number of groups. A tree is then drawn. Here we decided to pick the 4-group solution. While not the absolute best, it still ranks among the good ones and avoids producing too many small groups (Fig. 4.27, right).

The tree produced by this analysis is rich in information. Apart from the general statistics appearing at the bottom of the plot (residual error, i.e. the one-complement of the R^2 of the model; cross-validated error; standard error), the following features are important:

- Each node is characterized by a threshold value of an explanatory variable. For instance, the first node splits the data into two groups of 16 and 13 sites on the basis of elevation. The critical value (here 341 m) is often not found among the data; it is the mean of the two values delimiting the split. If two or more explanatory variables lead to equal results, an arbitrary choice is made among them. In this example, for instance, variable `dfs` with value 204.8 km would yield the same split.
- Each leaf (terminal group) is characterized by its number of sites and its RE as well as by a small barplot representing the abundances of the species (in the same order as is the response data matrix). Although difficult to read if there are many species, these plots show that the different groups are indeed characterized by different species. A more formal statistical approach is to search for characteristic or indicator species (Sect. 4.11). See below for an example.
- The tree can be used to allocate a new observation to one of the groups on the basis of the values of the relevant environmental variables. "Relevant" means here that the variables needed to allocate an object may differ depending on the

branch of the tree. Observe that a given variable may be used several times along the course of the successive binary partitions.

As proposed in the code below, apart from the residuals, one can retrieve the objects of each node and examine the node's characteristics at will:

```
# Residuals of MRT
par(mfrow = c(1, 2))
hist(residuals(spe.ch.mvpart), col = "bisque")
plot(predict(spe.ch.mvpart, type = "matrix"),
     residuals(spe.ch.mvpart),
     main = "Residuals vs Predicted")
abline(h = 0, lty = 3, col = "grey")

# Group composition
spe.ch.mvpart$where

# Group identity
(groups.mrt <- levels(as.factor(spe.ch.mvpart$where)))

# Fish composition of first leaf
spe.norm[which(spe.ch.mvpart$where == groups.mrt[1]), ]

# Environmental variables of first leaf
env[which(spe.ch.mvpart$where == groups.mrt[1]), ]
```

One can also use the MRT results to produce pie charts of the fish composition of the leaves (Fig. 4.28):

```
# Table and pie charts of fish composition of leaves
leaf.sum <- matrix(0, length(groups.mrt), ncol(spe))
colnames(leaf.sum) <- colnames(spe)
for (i in 1:length(groups.mrt))
{
  leaf.sum[i, ] <-
    apply(spe.norm[which(spe.ch.mvpart$where == groups.mrt[i]), ],
          2, sum)
}
leaf.sum

par(mfrow = c(2, 2))
for (i in 1:length(groups.mrt)){
  pie(which(leaf.sum[i, ] > 0),
      radius = 1,
      main = paste("leaf #", groups.mrt[i]))
}
```

Unfortunately, extracting other numerical results from an **`mvpart()`** object is no easy task. This is why our colleague Marie-Hélène Ouellette wrote a wrapper doing just that and providing a wealth of additional, useful information. The package

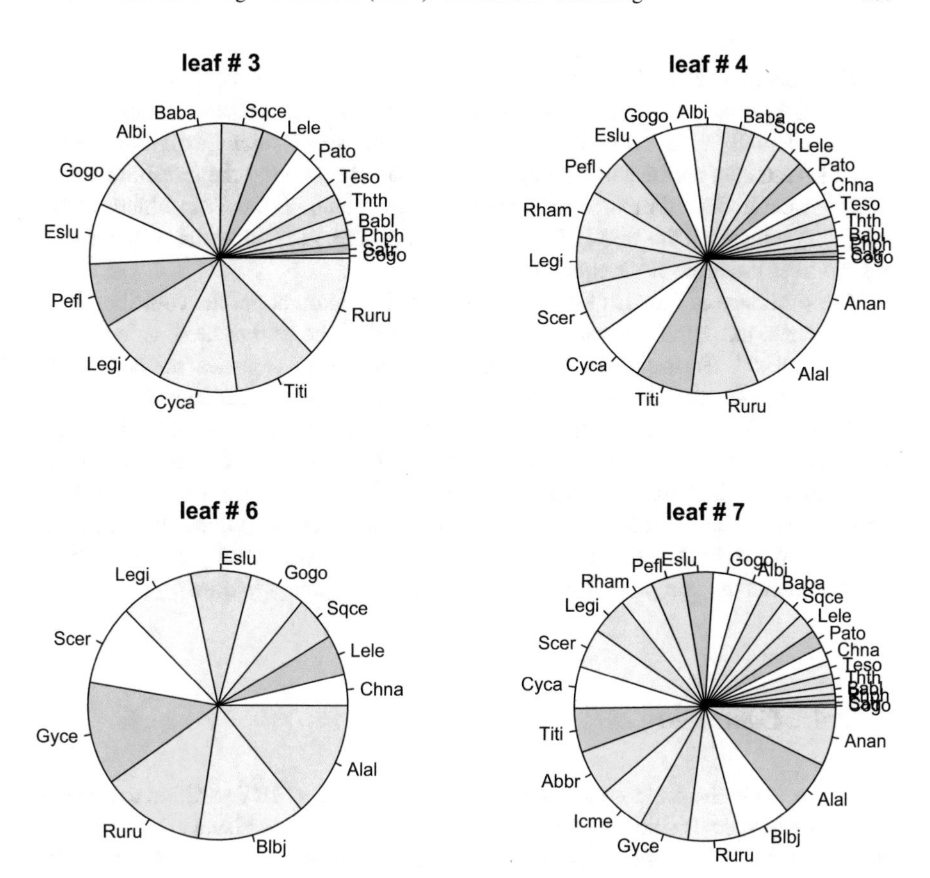

Fig. 4.28 Pie chart of the fish composition of the four leaves retained in the MRT analysis

is called **MVPARTwrap** and the function is **MRT()**. Its output comprises two graphs and a lot of numerical results. Let us apply it to our previous result.

```
# Extracting MRT results from an mvpart object
spe.ch.mvpart.wrap <-
  MRT(spe.ch.mvpart, percent = 10, species = colnames(spe))
summary(spe.ch.mvpart.wrap)
```

Hint *Argument* `percent` *indicates the smallest percentage of variance explained by a species at a node that one considers interesting. No test is made here; it is an arbitrarily chosen value.*

The function displays some results on-screen during execution. These are presented in a form close to canonical ordination results (see Chap. 6), because the function actually runs a redundancy analysis on the species data, explained by the variables retained by the MRT analysis, and recoded following the threshold values of the tree nodes. Among the results, let us mention the R^2, which is the one-complement of the tree RE. In our example, Fig. 4.27 gives an error of 0.37 for the tree; therefore R^2 is equal to 0.63.

The summary of the result object provides information about the contribution of each node to the explained variance ("Complexity"). The sum of these values gives the overall R^2. Furthermore, it identifies "discriminant" species at each node, selecting the species that contribute most to the explained variance (down to a minimum arbitrarily set by the user with the argument `percent`). The mean (transformed) abundances are given for both branches of the nodes, so one can see the branch for which the species are discriminant. For instance, our example analysis shows species `Satr`, `Phph` and `Babl` for the left branch (higher altitude), and `Alal` for the right. The summary lists the sites present in each leaf. The result object itself contains the complete results from which the summary is drawn.

4.12.4 Combining MRT and IndVal

As suggested in the previous section, one could submit the MRT partition to a search for indicator species (IndVal, Sect. 4.11.2). This is better than visual examination of the results for the identification of "discriminant" species: one can test the significance of the indicator values and correct the p-values for multiple testing using the Holm correction.

```
# Indicator species search on the MRT result
spe.ch.MRT.indval <- indval(spe.norm, spe.ch.mvpart$where)
pval.adj3 <- p.adjust(spe.ch.MRT.indval$pval)      # Corrected prob.
```

```
 Cogo  Satr  Phph  Babl  Thth  Teso  Chna  Pato  Lele  Sqce  Baba  Albi
1.000 0.027 0.036 0.330 1.000 1.000 0.330 0.860 1.000 1.000 0.027 0.504
 Gogo  Eslu  Pefl  Rham  Legi  Scer  Cyca  Titi  Abbr  Icme  Gyce  Ruru
0.594 1.000 0.860 0.027 0.027 0.860 0.027 0.068 0.027 0.144 0.330 1.000
 Blbj  Alal  Anan
0.027 0.027 0.027
```

The following list gives the leaf number for the significant indicator species, followed by the list of the indicator values:

```
# For each significant species, find the leaf with the highest
# IndVal
spe.ch.MRT.indval$maxcls[which(pval.adj3 <= 0.05)]
```

```
Satr Phph Baba Rham Legi Cyca Abbr Blbj Alal Anan
   1    1    4    4    4    4    4    4    4    4
```

```
# IndVal value in the best leaf for each significant species
spe.ch.MRT.indval$indcls[which(pval.adj3 <= 0.05)]
```

```
     Satr      Phph      Baba      Rham      Legi      Cyca      Abbr
0.7899792 0.5422453 0.6547659 0.8743930 0.6568224 0.7964617 0.9000000
     Blbj      Alal      Anan
0.7079525 0.6700303 0.8460903
```

One sees that not all groups harbour indicator species, and that most of these are in the fourth (rightmost) group. Individually, the result for the brown trout (`Satr`), for instance, shows a significant indicator value of 0.79 in the first leaf.

Then, it would be interesting to compare the constrained typology of sites brought by MRT with the one obtained by the unconstrained optimized Ward clustering. For this purpose, we transform the leaf numbers to sequential numbers, i.e. the levels of the membership factor.

```
# Partition of objects based on MRT
spech.mvpart.g <- factor(spe.ch.mvpart$where)
levels(spech.mvpart.g) <- 1:length(levels(spech.mvpart.g))
# Compare with partition from unconstrained clustering
table(spech.mvpart.g, spech.ward.g)
```

Finally, the MRT clusters can be mapped on the Doubs River (Fig. 4.29):

```
# Plot of the MRT clusters on a map of the Doubs River
drawmap(xy = spa,
        clusters = spech.mvpart.g,
        main = "Four MRT clusters along the Doubs River")
```

4.13 MRT as a Monothetic Clustering Method

Multivariate regression trees provide a solution for monothetic clustering, which is a form of (unconstrained) clustering where a single response variable is selected for each split of the tree. The application here consists of building the tree using the species data as both the response and explanatory variables. This allows the definition of groups of sites based on a homogeneous composition and explained by the presence (or the relative abundance) of indicator species added sequentially. This method is related to the "association analysis" proposed by Williams and Lambert (1959). It can be applied to presence-absence or pre-transformed abundance response data; the explanatory data may be binary (presence-absence) or quantitative (untransformed or transformed), the same data matrix being used, possibly after

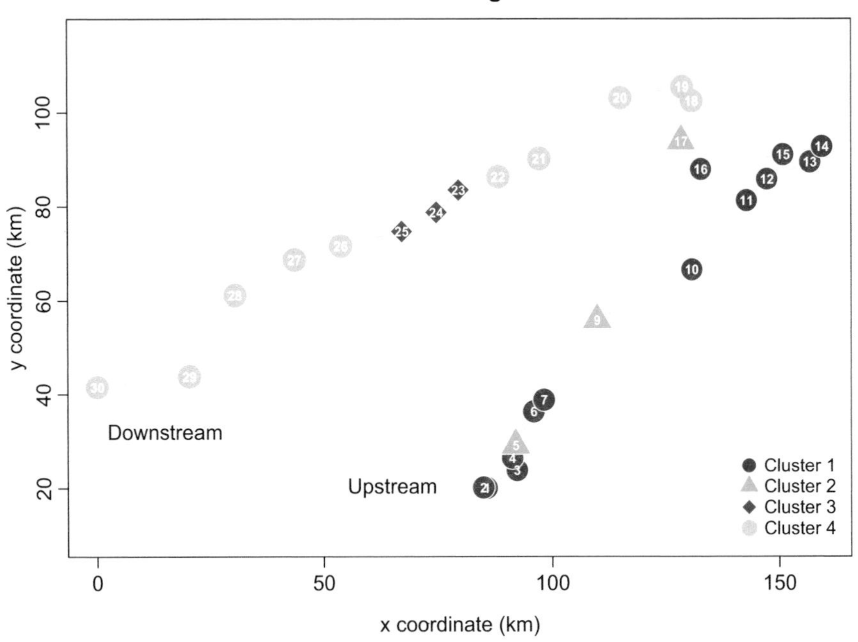

Fig. 4.29 Four MRT clusters along the Doubs River

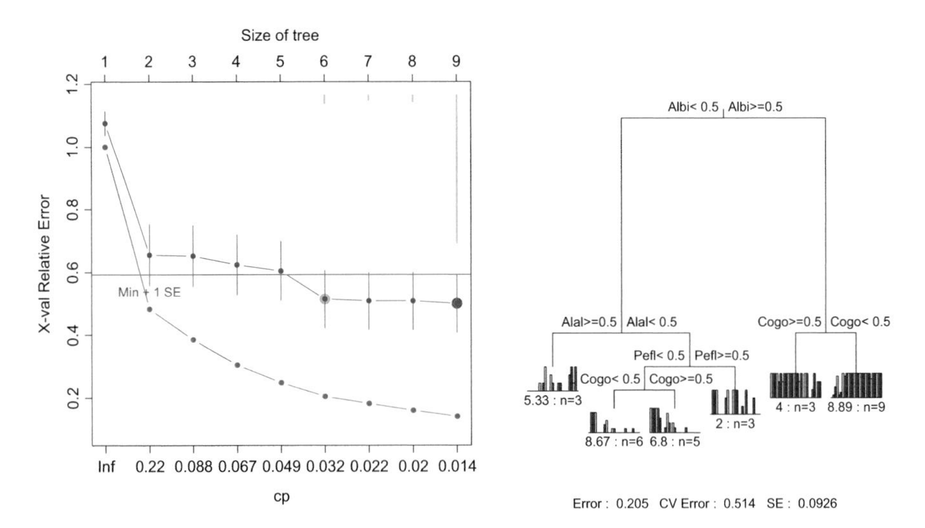

Fig. 4.30 Monothetic clustering of the fish species data from MRT with the presence-absence dissimilarity matrix used both as response and explanatory matrix

different transformations, for both the explanatory and response data. Here we apply this method to the Doubs fish data (Figs. 4.30 and 4.31).

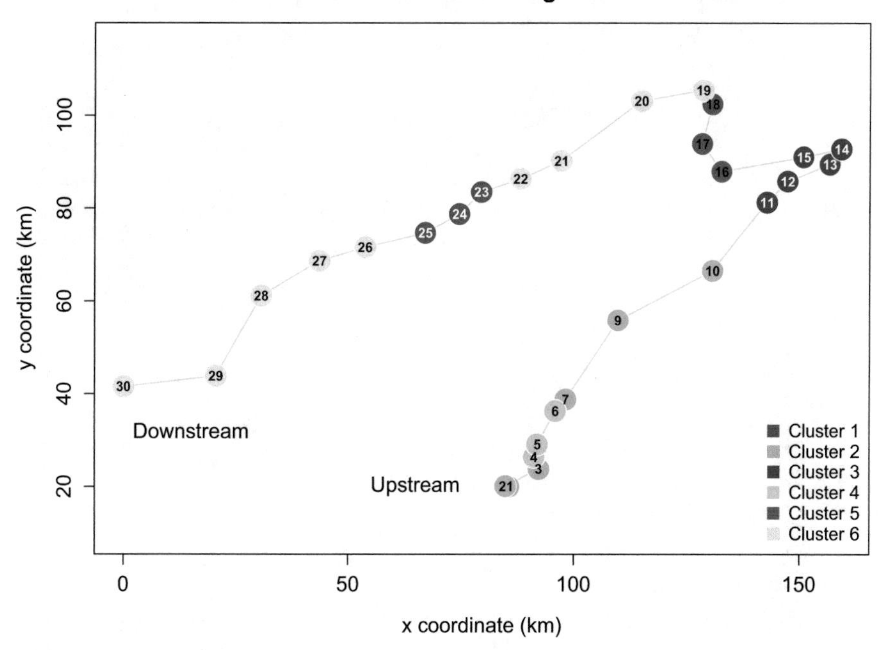

Fig. 4.31 Six monothetic clusters along the Doubs River. The response and "explanatory" matrices are the presence-absence data

```
# spe.pa (presence-absence) is the response and the explanatory
# matrix
par(mfrow = c(1, 2))
spe.pa <- decostand(spe, "pa")
res.part1 <-
  mvpart(
    data.matrix(spe.pa) ~ .,
    data = spe.pa,
    margin = 0.08,
    xv = "p",
    xvmult = 100
  )
# spe.norm is the response, spe.pa is the explanatory matrix
res.part2 <-
  mvpart(
    data.matrix(spe.norm) ~ .,
    data = spe.pa,
    margin = 0.08,
    xv = "p",
    xvmult = 100
  )
```

```
# spe.norm is the response matrix and spe (untransformed) is the
# explanatory matrix
res.part3 <-
  mvpart(
    data.matrix(spe.norm) ~ .,
    data = spe,
    margin = 0.08,
    xv = "p",
    cp = 0,
    xvmult = 100
  )

# Membership of objects to groups - presence-absence on both sides
res.part1$where

res.part1.g <- factor(res.part1$where)
levels(res.part1.g) <- 1:length(levels(res.part1.g))
# Compare with groups from unconstrained clustering
table(res.part1.g, spech.ward.g)
table(res.part1.g, spech.ward.gk)

# Plot of the MRT clusters on a map of the Doubs River
drawmap3(xy = spa,
         clusters = res.part1.g,
         main = "Six monothetic clusters along the Doubs River")
```

4.14 Sequential Clustering

In cases where the data present themselves in a spatial (transect) or temporal sequence, the contiguity information can be taken into account when looking for groups, and for discontinuities, along the series. Several methods have been proposed for that purpose in the temporal (e.g. Gordon and Birks 1972, 1974; Gordon 1973, Legendre et al. 1985) and spatial (Legendre et al. 1990) contexts. Sequential clustering can also be computed by MRT, the constraint being a variable representing the sampling sequence (Legendre and Legendre 2012 Sect. 12.6.4). For the 29 sites of the Doubs data, a vector containing the numbers 1 to 29 or, equivalently, the variable `dfs`, would be appropriate. Computation of this example is detailed in Borcard et al. (2011, Sect. 4.11.5). The code is presented in the accompanying material of the present book.

Here we will apply a method of clustering with contiguity constraint developed for stratigraphic research and called CONISS (Grimm 1987). It can be construed as a variant of Ward's minimum variance clustering with the constraint that sites can only join if they are contiguous along a spatial or temporal sequence. The CONISS algorithm is available through function **`chclust()`** of package **`rioja`**.

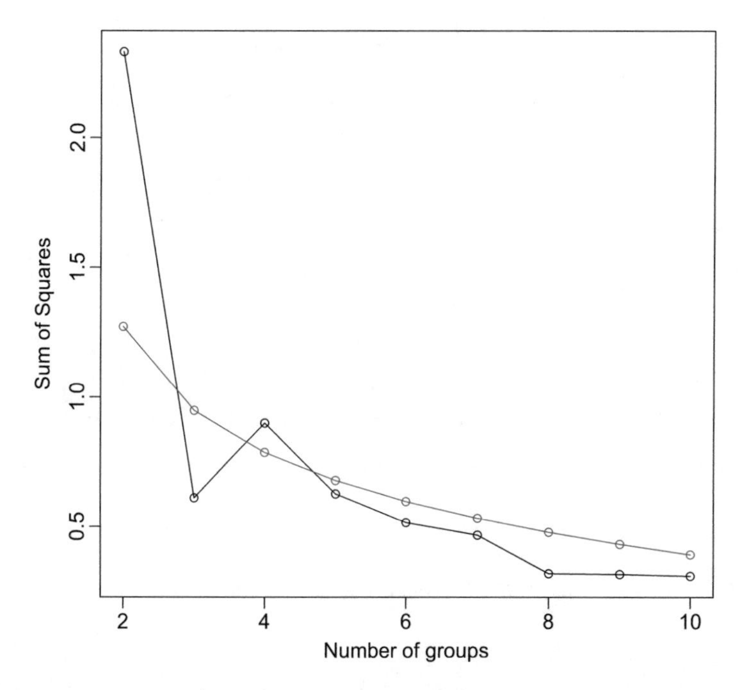

Fig. 4.32 Group-by-SS graph comparing the dispersion of the classification at different fusion levels to a null model. Red line: broken stick model

We apply it to a matrix of percentage difference (aka Bray-Curtis) dissimilarities computed from the fish species data. To choose the number of clusters, the dispersion (sum of squares, SS) of the hierarchical classification is compared to that obtained from a broken stick model (Fig. 4.32):

```
# Default method CONISS
# On the percentage difference dissimilarity matrix
spe.chcl <- chclust(vegdist(spe))

# Compare the fusion levels to a broken stick model
bstick(spe.chcl, 10)
```

The graph suggests cutting the dendrogram to retain two or four clusters (points above the red line obtained from the broken stick model).

For consistency with previous clustering results, we choose to cut the dendrogram in 4 groups. The result is displayed in Fig. 4.33 as a dendrogram and in the form of a map of the four clusters along the river.

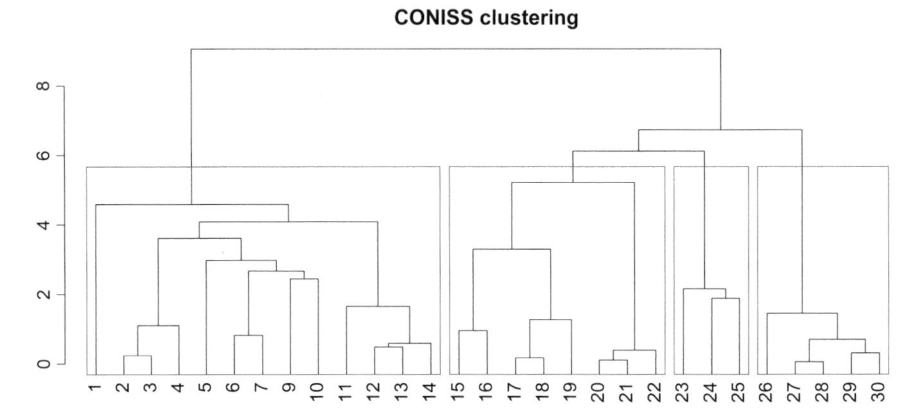

Fig. 4.33 Dendrogram of the clustering with contiguity constraint of the fish species data

```
# Cut the dendrogram in 4 clusters
k <- 4
(gr4 <- cutree(spe.chcl, k = k))

# Plot the dendrogram
plot(spe.chcl, hang = -1, main = "CONISS clustering")
rect.hclust(spe.chcl, k = k)
```

If you want, you can space out the branches of the dendrogram according to the distance from the source:

```
# Dendrogram with observations plotted according to dfs
plot(spe.chcl,
     xvar = env$dfs,
     hang = -1,
     main = "CONISS clustering",
     cex = 0.8)
```

See on the map of the Doubs River how sites are now clustered in a consistent sequence. The constraint of spatial contiguity has forced the separation of sites 15–22 and 26–30 into separate groups because the sequence is interrupted by the group 23–25 of polluted sites (Fig. 4.34):

```
# Plot the clusters on a map of the Doubs River
drawmap(xy = spa,
        clusters = gr4,
        main = "Sequential clusters along the river")
```

Compare this sequence with the traditional zonation of fish communities presented in Chap. 1.

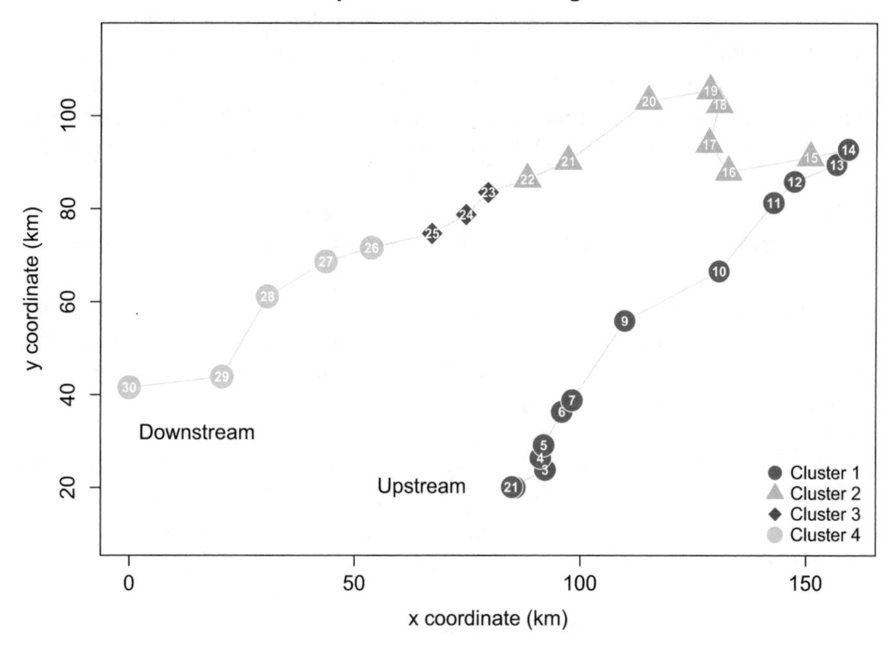

Fig. 4.34 Map of four groups along the Doubs River

4.15 A Very Different Approach: Fuzzy Clustering

At the beginning of this chapter, we defined clusters produced by clustering methods as non-overlapping entities. This definition is a natural consequence of the focus of most clustering methods on discontinuities. However, there is another approach to clustering that recognizes that sometimes cluster limits may not be so clear-cut as one would like them to be. In that approach, an object may be, to different degrees, a member of two or several groups. As an example, the colour green may be obtained by mixing blue and yellow paint, the proportion of each of the two primary colours determining the shade of green. Consequently, a different family of hierarchical and non-hierarchical methods has been developed, namely fuzzy clustering. We will not develop this family in detail, but briefly show one approach that is akin to non-hierarchical *k*-means partitioning. Its name is *c*-means clustering (Kaufman and Rousseeuw 2005).

4.15.1 Fuzzy c-means Using Package `cluster`'s Function `fanny()`

Instead of a classification where a given object belongs to one and only one cluster, *c*-means clustering associates to all objects a series of membership values measuring the strength of their memberships in the various clusters. An object that is clearly linked to a given cluster has a strong membership value for that cluster and weak (or null) values for the other clusters. The membership values add up to 1 for each object. To give an example from sensory evaluation, imagine the difficulty of applying the two-state descriptor {sweet, bitter} to beers. A beer may be judged by a panel of tasters to be, say, 30% sweet and 70% bitter. This is an example of a fuzzy classification of beers for that descriptor.

Fuzzy *c*-means clustering is implemented in several packages, e.g. **cluster** (function **fanny()**) and **e1071** (function **cmeans()**). The short example below uses the former.

Function **fanny()** accepts either site-by-species or dissimilarity matrices. In the former case, the default metric is `euclidean`. Here we will directly use as input the chord distance matrix '`spe.ch`' previously computed from the fish species data. An identical result would be obtained by using the chord-transformed species data '`spe.norm`' with the `metric = "euclidean"` argument.

The plot function can return two diagrams: an ordination (see Chap. 5) of the clusters and a silhouette plot. Here we present the latter (Fig. 4.35) and we replace the original ordination diagram by a principal coordinate analysis (PCoA, see Sect. 5.5) combined with star plots of the objects (Fig. 4.36). Each object is associated with a small star plot (resembling a little pie chart) whose segment radiuses are proportional to its membership coefficient.

```
k <- 4        # Choose the number of clusters
spe.fuz <- fanny(spe.ch, k = k, memb.exp = 1.5)
summary(spe.fuz)

# Site fuzzy membership
spe.fuz$membership
# Nearest crisp clustering
spe.fuz$clustering
spefuz.g <- spe.fuz$clustering

# Silhouette plot
plot(
  silhouette(spe.fuz),
  main = "Silhouette plot - Fuzzy clustering",
  cex.names = 0.8,
  col = spe.fuz$silinfo$widths + 1
)

# Ordination of fuzzy clusters (PCoA)
# Step 1: ordination (PCoA) of the fish chord distance matrix
dc.pcoa <- cmdscale(spe.ch)
dc.scores <- scores(dc.pcoa, choices = c(1, 2))
# Step 2: ordination plot of fuzzy clustering result
plot(dc.scores,
     asp = 1,
     type = "n",
     main = "Ordination of fuzzy clusters (PCoA)")
```

```
abline(h = 0, lty = "dotted")
abline(v = 0, lty = "dotted")
# Step 3: representation of fuzzy clusters
for (i in 1:k) {
  gg <- dc.scores[spefuz.g == i, ]
  hpts <- chull(gg)
  hpts <- c(hpts, hpts[1])
  lines(gg[hpts, ], col = i + 1)
}
stars(
  spe.fuz$membership,
  location = dc.scores,
  key.loc = c(0.6, 0.4),
  key.labels = 1:k,
  draw.segments = TRUE,
  add = TRUE,
  # scale = FALSE,
  len = 0.075,
  col.segments = 2:(k + 1)
)
```

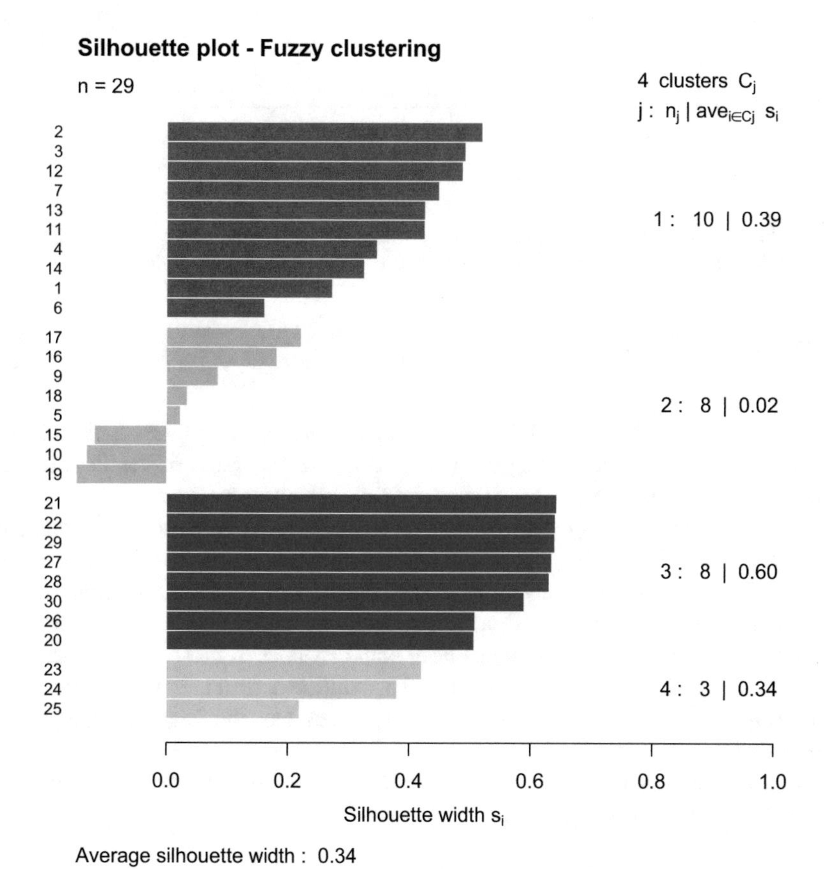

Fig. 4.35 Silhouette plot of the *c*-means fuzzy clustering of the fish data preserving the chord distance

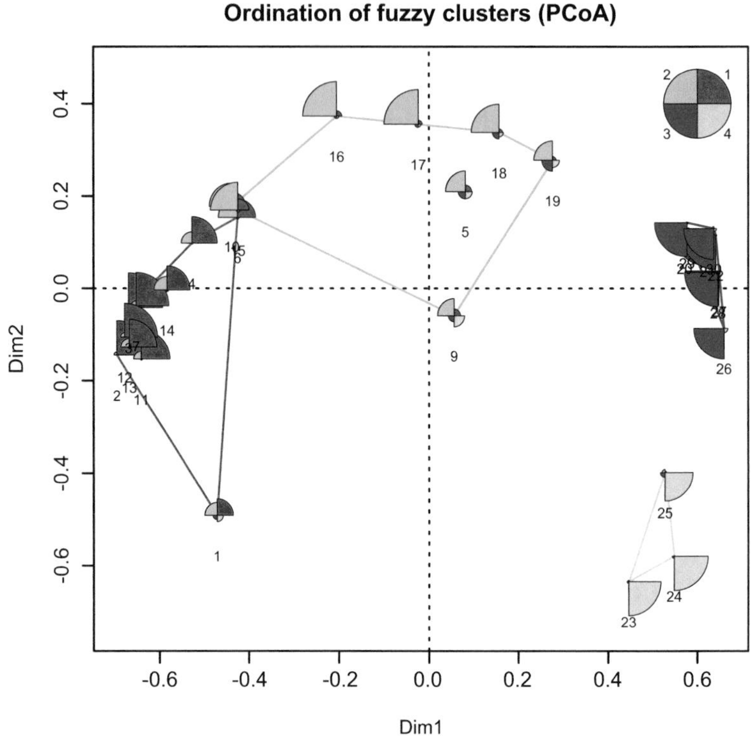

Fig. 4.36 *c*-means fuzzy clustering of the fish data preserving the chord distance. Principal coordinate ordination associated with star plots showing the memberships of the sites

Hints *Argument* `memb.exp` *is a kind of "fuzziness exponent" with values ranging from 1 (close to non-fuzzy clustering) to any large value.*

The silhouette plot (Fig. 4.35) shows, in particular, that cluster 2 is not well defined. On the ordination diagram (Fig. 4.36), the star plots of the ill-classified objects (10, 15, 19) also illustrate that their membership is unclear.

The numerical results give the membership coefficients of the objects. The sum of each row is equal to 1. Objects belonging unambiguously to one cluster, like sites 2, 21 or 23, have a high membership value for that cluster and correspondingly low values for the other clusters. Conversely, one can easily locate objects that are difficult to classify: their coefficients have similar values in most if not all clusters. Sites 5, 9 and 19 are good examples. An additional result is the nearest crisp clustering, i.e., the cluster to which each object has the highest membership coefficient.

Sectors representing fuzzy membership can also be added to the map of the sites along the Doubs River (Fig. 4.37):

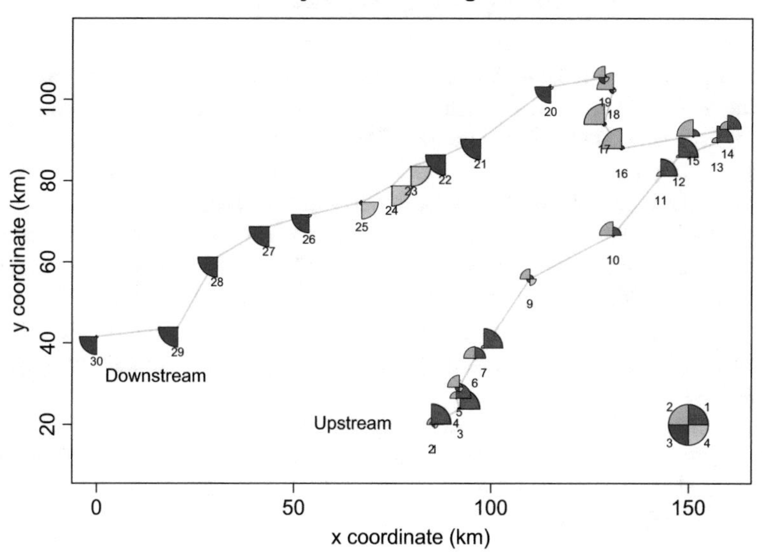

Fig. 4.37 Four fuzzy clusters along the Doubs River

```
# Plot the fuzzy clusters on a map of the Doubs River
plot(
  spa,
  asp = 1,
  type = "n",
  main = "Fuzzy clusters along the river",
  xlab = "x coordinate (km)",
  ylab = "y coordinate (km)"
)
lines(spa, col = "light blue")
text(65, 20, "Upstream", cex = 1.2)
text(15, 32, "Downstream", cex = 1.2)
# Add sectors to represent fuzzy membership
for (i in 1:k) {
  stars(
    spe.fuz$membership,
    location = spa,
    key.loc = c(150, 20),
    key.labels = 1:k,
    draw.segments = TRUE,
    add = TRUE,
    len = 5,
    col.segments = 2:(k + 1)
  )
}
```

4.15.2 Noise Clustering Using the `vegclust()` Function

A recent package called **vegclust,** developed by Miquel De Cáceres, provides a large range of options to perform non-hierarchical or hierarchical fuzzy clustering of community data under different models (De Cáceres et al. 2010). An interesting one is called "noise clustering" (Davé and Krishnapuram 1997). It allows the consideration of outliers, i.e. unclassified objects, denoted "N", beside fuzzy clusters labelled "M1", "M2"… The principle of the method consists in defining a cluster called "Noise" in addition to the regular clusters. This "Noise" cluster is represented by an imaginary point located at a constant distance δ from all observations. The effect of this cluster is to capture the "*objects that lie farther than* δ *from all the c "good" centroids*" (De Cáceres et al. 2010). A small δ results in a large membership in the "Noise" cluster.

To perform noise clustering, we apply the **`vegclust()`** function to the normalized species matrix with the argument `method = "NC"`. As before, we project the result of the noise clustering on a PCoA plot using sectors to represent fuzzy membership (Fig. 4.38).

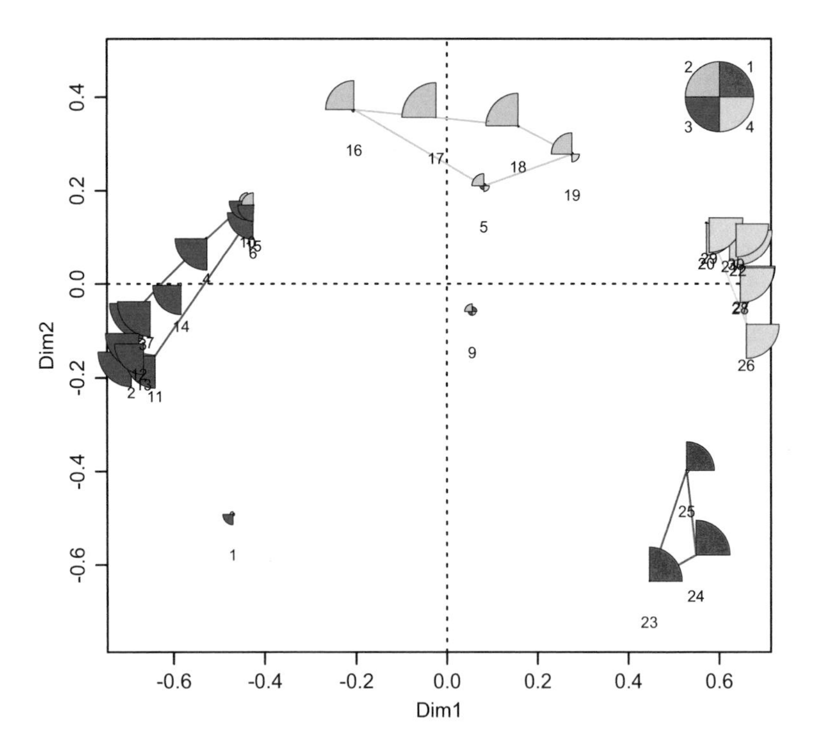

Fig. 4.38 Ordination plot of the noise clustering of the fish species data.

```
# Create noise clustering with four clusters. Perform 30 starts
# from random seeds and keep the best solution
k <- 4
spe.nc <- vegclust(
  spe.norm,
  mobileCenters = k,
  m = 1.5,
  dnoise = 0.75,
  method = "NC",
  nstart = 30
)
spe.nc

# Medoids of species
(medoids <- spe.nc$mobileCenters)

# Fuzzy membership matrix
spe.nc$memb

# Cardinality of fuzzy clusters (i.e., the number of objects
# belonging to each cluster)
spe.nc$size

# Obtain hard membership vector, with 'N' for objects that are
# unclassified
spefuz.g <- defuzzify(spe.nc$memb)$cluster
clNum <- as.numeric(as.factor(spefuz.g))

# Ordination of fuzzy clusters (PCoA)
plot(dc.scores,
     asp = 1,
     type = "n")
abline(h = 0, lty = "dotted")
abline(v = 0, lty = "dotted")
for (i in 1:k)
{
  gg <- dc.scores[clNum == i, ]
  hpts <- chull(gg)
  hpts <- c(hpts, hpts[1])
  lines(gg[hpts, ], col = i + 1)
}
stars(
  spe.nc$memb[, 1:4],
  location = dc.scores,
  key.loc = c(0.6, 0.4),
  key.labels = 1:k,
  draw.segments = TRUE,
  add = TRUE,
  len = 0.075,
  col.segments = 2:(k + 1)
  )
```

The hard partition contained in the **`clNum`** object is added to the ordination plot (Fig. 4.39).

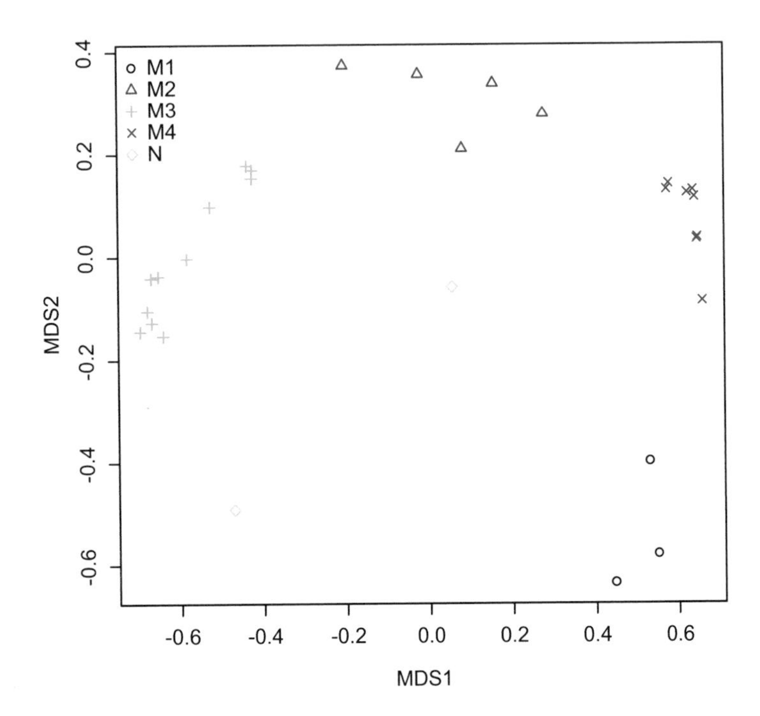

Fig. 4.39 Ordination plot of the defuzzified noise clustering of the fish species data, showing the two unclassified objects

```
# Defuzzified site plot
plot(
  dc.pcoa,
  xlab = "MDS1",
  ylab = "MDS2",
  pch = clNum,
  col = clNum
)
legend(
  "topleft",
  col = 1:(k + 1),
  pch = 1:(k + 1),
  legend = levels(as.factor(spefuz.g)),
  bty = "n"
)
```

Finally, let us plot the fuzzy clusters on the map of the river (Fig. 4.40).

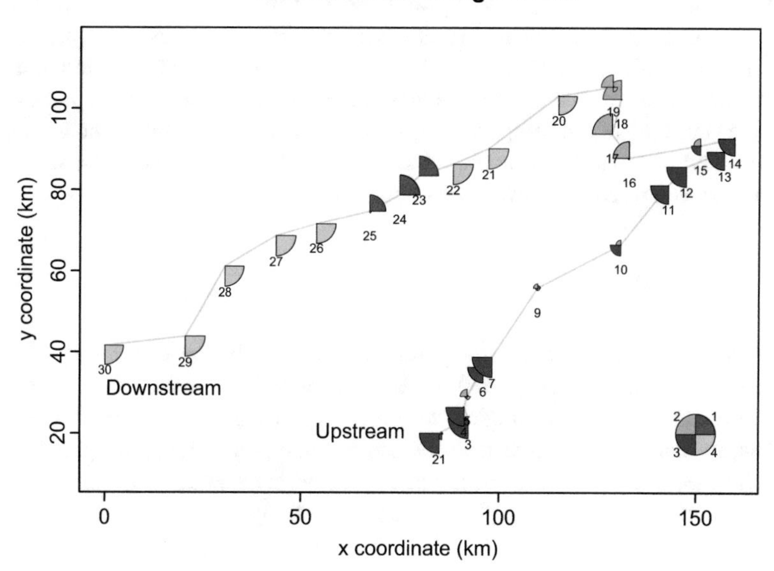

Fig. 4.40 Four noise clusters along the Doubs River

```
# Plot the fuzzy clusters on a map of the Doubs River
plot(
  spa,
  asp = 1,
  type = "n",
  main = "Noise clusters along the river",
  xlab = "x coordinate (km)",
  ylab = "y coordinate (km)"
)
lines(spa, col = "light blue")
text(65, 20, "Upstream", cex = 1.2)
text(15, 32, "Downstream", cex = 1.2)
# Add sectors to represent fuzzy membership
for (i in 1:k) {
  stars(
    spe.nc$memb[, 1:4],
    location = spa,
    key.loc = c(150, 20),
    key.labels = 1:k,
    draw.segments = TRUE,
    add = TRUE,
    # scale = FALSE,
    len = 5,
    col.segments = 2:(k + 1)
  )
}
```

While "hard" clustering may appear somewhat unrealistic in ecology, its application is of great help when one needs a typology or a decision-making tool requiring unambiguous allocation of the sites. Fuzzy clustering is a more nuanced, and therefore more realistic approach in most ecological situations if its purpose is to describe similarity relationships among sites. Other very powerful methods exist that are designed to reveal the structure of continuous data. These methods, simple and constrained ordination, are explored in the following chapters.

4.16 Conclusion

This chapter has not covered all the possibilities offered by the large family of cluster analyses, but you have explored its main avenues and seen how flexible this approach can be. Every ecological research project has its own features and constraints; in many cases cluster analysis can provide rich insights into the data. The clustering techniques themselves are numerous, as are also the ways of interpreting the results. It is up to you to exploit this lore to optimise the output of your research.

Chapter 5
Unconstrained Ordination

5.1 Objectives

While cluster analysis looks for discontinuities in a dataset, ordination extracts the main trends in the form of continuous axes. It is therefore particularly well adapted to analyse data from natural ecological communities, which are generally structured in gradients.

Practically, you will:

- learn how to choose among various ordination techniques (PCA, CA, MCA, PCoA and NMDS), compute them using the correct options, and properly interpret the ordination diagrams;
- apply these techniques to the Doubs River or the Oribatid mite data;
- overlay the result of a cluster analysis on an ordination diagram to improve the interpretation of both analyses;
- interpret the structures revealed by the ordination of the species data using the environmental variables from a second dataset;
- write your own PCA function.

5.2 Ordination Overview

5.2.1 Multidimensional Space

A multivariate data set can be viewed as a collection of sites positioned in a space where each variable defines one dimension. There are thus as many dimensions as variables. To reveal the structure of the data, it would be interesting to represent the main trends in the form of scatter plots of the sites. Since ecological data generally contain more than two variables, it is tedious and not very informative to draw the objects in a series of scatter plots defined by all possible pairs of descriptors. For

D. Borcard et al., *Numerical Ecology with R*, Use R!,
https://doi.org/10.1007/978-3-319-71404-2_5

instance, if the matrix contains 10 descriptors, the number of plots to draw would be equal to $(10 \times 9)/2 = 45$. Such a series of scatter diagrams would allow neither to bring out the most important structures of the data, nor to visualise the relationships among descriptors (which, in general, are not linearly independent of one another). The aim of ordination methods is to represent the data along a reduced number of orthogonal axes, constructed in such a way that they represent, in decreasing order, the main trends of variation of the data. These trends can then be interpreted visually or in association with other methods like clustering or regression. Here we shall describe five techniques. All these methods are descriptive: no statistical test is provided to assess the significance of the structures detected. That is the role of constrained ordination, a family of methods that will be presented in Chap. 6.

5.2.2 *Ordination in Reduced Space*

Most usual ordination methods (except NMDS) are based on the extraction of the eigenvectors of an association matrix. They can be classified according to the dissimilarity preserved among sites and to the type of variables that they can handle. Legendre and Legendre (2012, Table 9.1, p. 426) provide a table showing their domains of application.

The basic principle of ordination in reduced space is the following. Imagine an $n \times p$ data set containing n objects and p variables. The n objects can be represented as a cluster of points in the p-dimensional space. Now, this cluster is generally not spheroid: it is elongated in some directions and flattened in others. These directions are not necessarily aligned with a single dimension (= a single variable) of the multidimensional space. The direction where the cluster is most elongated corresponds to the direction of largest variance of the cluster. This is the first axis that an ordination will extract. Indeed, the direction of largest variance corresponds to the strongest gradient present in the data: this is where the most important information resides. The next axis to be extracted is the second most important in variance, provided that it is *orthogonal* (linearly independent, scalar product of 0) to the first one. The process continues until all axes have been computed.

When there are a few major structures in the data (gradients or groups) and the method has been efficient at extracting them, then the few first axes contain most of the useful information, i.e., they have extracted most of the variance of the data. In that case, the distances among sites in the projection in reduced space (most often two-dimensional) are relatively similar to the distances among objects in the multidimensional space. Note, however, that an ordination can be useful even when the first axes account for small proportions of the variance. This may happen when there are some interesting structures in an otherwise noisy data set. The question arising is then: how many axes should one retain and interpret? In other words, how many axes represent interpretable structures? The answer depends on the method and the data; several procedures will be explained in due course to help users answer this question.

The methods that will be presented in this chapter are:

- **Principal component analysis (PCA)**: the main eigenvector-based method. Analysis performed on raw, quantitative data. Preserves the Euclidean distance among sites in scaling 1, and the Mahalanobis distance in scaling 2. Scaling is explained in Sect. 5.3.2.2.
- **Correspondence analysis (CA)**: works on data that must be frequencies or frequency-like, dimensionally homogeneous, and non-negative. Preserves the χ^2 distance among rows (in scaling 1) or columns (in scaling 2). In ecology, almost exclusively used to analyse community composition data.
- **Multiple correspondence analysis MCA**: ordination of a table of categorical variables, i.e. a data frame where all variables are factors.
- **Principal coordinate analysis (PCoA)**: devoted to the ordination of dissimilarity matrices, most often in the Q mode, instead of site-by-variables tables. Hence, great flexibility in the choice of association measures (Chap. 3).
- **Nonmetric multidimensional scaling (NMDS)**: unlike the three others, this is not an eigenvector-based method. NMDS tries to represent the set of objects along a predetermined number of axes while preserving the ordering relationships among them. Operates from a dissimilarity matrix.

PCoA and NMDS can produce ordinations from any square dissimilarity matrix, which have been created in (transformed to) class "dist" in **R**.

5.3 Principal Component Analysis (PCA)

5.3.1 Overview

Imagine a data set whose variables are normally distributed. This data set will be said to show a multinormal distribution. The first principal axis (or principal-component axis) of a PCA of this data set is the straight line that goes through the greatest dimension of the concentration ellipsoid describing this multinormal distribution. The following axes, which are orthogonal to one another and successively shorter, go through the following greatest dimensions of the ellipsoid (Legendre and Legendre 2012). One can extract a maximum of p principal axes from a data set containing p variables.

Stated otherwise, PCA carries out a rotation of the original system of axes defined by the variables, such that the successive new axes (called principal components) are orthogonal to one another, and correspond to the successive dimensions of maximum variance of the scatter of points. The principal components give the positions of the objects in the new system of coordinates. PCA works on a *dispersion matrix* **S**, i.e. an association matrix among variables containing the variances and covariances of the variables (when these are dimensionally homogeneous), or the correlations computed from dimensionally heterogeneous variables. It is exclusively devoted to the analysis of quantitative variables. The dissimilarity preserved is the Euclidean distance and the relationships detected are linear. Therefore, it is not generally appropriate to the analysis of raw species abundance data. These can, however, be subjected to PCA after an appropriate pre-transformation (Sects. 3.5 and 5.3.3).

In a PCA ordination diagram, following the tradition of scatter diagrams in Cartesian coordinate systems, *objects* are represented as *points* and *variables* are displayed as *arrows*.

Later in this chapter (Sect. 5.7), you will see how to program a PCA in **R** using matrix equations. But for everyday users, PCA is available in several **R** packages. A convenient function for ecologists is **`rda()`** in package **`vegan`**. The name of the function refers to redundancy analysis, a method that will be presented in Chap. 6. Other possible functions are **`PCA.newr()`** provided with this book, or else **`dudi.pca()`** (package **`ade4`**) and **`prcomp()`** (package **`stats`**).

5.3.2 *PCA of the Environmental Variables of the Doubs River Data Using* `rda()`

Let us work again with the Doubs data. We have 11 quantitative environmental variables at our disposal. How are they correlated? What can we learn from the ordination of the sites?

Since the variables are expressed in different measurement scales (they are dimensionally heterogeneous), we shall compute PCA from a correlation matrix. Correlations are the covariances of standardized variables.

5.3.2.1 Preparation of the Data

```
# Load the required packages
library(ade4)
library(vegan)
library(gclus)
library(ape)
library(missMDA)
library(FactoMineR)

# Source additional functions that will be used later in this
# chapter. Our scripts assume that files to be read are in
# the working directory
source("cleanplot.pca.R")
source("PCA.newr.R")
source("CA.newr.R")

# Load the Doubs data
load("Doubs.RData")
# Remove empty site 8
spe <- spe[-8, ]
env <- env[-8, ]
spa <- spa[-8, ]

# Load the oribatid mite data
load("mite.RData")
```

5.3.2.2 PCA of a Correlation Matrix

```
# A reminder of the content of the env dataset
summary(env)

# PCA based on a correlation matrix
# Argument scale=TRUE calls for a standardization of the variables
env.pca <- rda(env, scale = TRUE)
env.pca
summary(env.pca) # Default scaling 2
summary(env.pca, scaling = 1)
```

Hint *If you don't want to see the site and species scores, add argument* `axes = 0` *to the* **`summary()`** *call.*

Note that the scaling (see below) is called at the step of the summary (or, below, for the drawing of biplots) and not for the analysis itself.

The "summary" output looks as follows for scaling 2 (scalings are explained below); some results have been deleted:

```
Call:
rda(X = env, scale = TRUE)

Partitioning of correlations:
              Inertia Proportion
Total              11          1
Unconstrained      11          1

Eigenvalues, and their contribution to the correlations

Importance of components:
                          PC1    PC2     PC3     PC4     PC5 ...
Eigenvalue             6.0979 2.1672 1.03761 0.70353 0.35174 ...
Proportion Explained   0.5544 0.1970 0.09433 0.06396 0.03198 ...
Cumulative Proportion  0.5544 0.7514 0.84571 0.90967 0.94164 ...
```

```
Scaling 2 for species and site scores
* Species are scaled proportional to eigenvalues
* Sites are unscaled: weighted dispersion equal on all
* dimensions
* General scaling constant of scores:  4.189264
Species scores

        PC1     PC2      PC3      PC4      PC5      PC6
dfs  1.0842  0.5150 -0.25749 -0.16168  0.21132 -0.09485
ele -1.0437 -0.5945  0.17984  0.12282  0.12464  0.14022
 (...)

Site scores (weighted sums of species scores)

        PC1      PC2      PC3      PC4      PC5      PC6
1  -1.41243 -1.47560 -1.74593 -2.95533  0.23051  0.49227
2  -1.04173 -0.81761  0.34075  0.54364  0.92835 -1.76876
```

The ordination output uses some vocabulary that requires explanations.

- **Inertia**: in **vegan's** language, this is the general term for "variation" in the data. This term comes from the world of correspondence analysis (Sect. 5.4). In PCA, the "inertia" is either the sum of the variances of the variables (PCA on a covariance matrix) or, as in this case (PCA on a correlation matrix), the sum of the diagonal values of the correlation matrix, i.e., the sum of all correlations of the variables with themselves, which corresponds to the number of variables (11 in this example).
- **Constrained and unconstrained**: see Sect. 6.1 (canonical ordination). In PCA, the analysis is unconstrained, i.e. *not* constrained by a set of explanatory variables, and so are the results.
- **Eigenvalues**, symbolized λ_j: these are measures of the importance (variance) of the PCA axes. They can be expressed as **Proportions Explained**, or proportions of variation accounted for by the axes, by dividing each eigenvalue by the "total inertia".
- **Scaling**: not to be confused with the argument "scale" calling for standardization of variables. "Scaling" refers to the way ordination results are projected in the reduced space for graphical display. There is no single way of optimally displaying objects and variables together in a PCA biplot, i.e., a plot showing two types of results, here the sites and the variables. Two main types of scaling are generally used. Each of them has properties that must be kept in mind for proper interpretation of the biplots. Here we give the essential features of each scaling. Please refer to Legendre and Legendre (2012, p. 443–445) for a complete account.

  – **Scaling 1** = distance biplot: the eigenvectors are scaled to unit length. (1) **Distances among objects in the biplot are approximations of their Euclidean distances in multidimensional space.** (2) *The angles among descriptor vectors do not reflect their correlations.*

  - **Scaling 2** = correlation biplot: each eigenvector is scaled to the square root of its eigenvalue. (1) *Distances among objects in the biplot are not approximations of their Euclidean distances in multidimensional space.* (2) **The angles between descriptors in the biplot reflect their correlations**.
  - In both cases, projecting an object at right angle on a descriptor approximates the position of the object along that descriptor.
  - Bottom line: if the main interest of the analysis is to interpret the relationships among **objects**, choose **scaling 1**. If the main interest focuses on the relationships among **descriptors**, choose **scaling 2**.
  - A compromise scaling, **Scaling 3**, also called "symmetric scaling", is also offered, which consists in scaling both the site and species scores by the square roots of the eigenvalues. This scaling is supposed to allow a simultaneous representation of sites and scores without emphasizing one or the other point of view. This compromise scaling has no clear interpretation rules, and therefore we will not discuss it further in this book.

- **Species scores**: coordinates of the arrowheads of the variables. For historical reasons, response variables are always called "species" in **vegan**, no matter what they represent, because **vegan** contains software for ***veg***etation ***an***alysis.
- **Site scores**: coordinates of the sites in the ordination diagram. Objects are always called "Sites" in vegan output files.

5.3.2.3 Extracting, Interpreting and Plotting Results from a **vegan** Ordination Output Object

vegan output objects are complex entities, and extraction of their elements does not follow the basic rules of **R**. Type **?cca.object** in the **R** console. This calls a documentation file explaining all features of an **rda()** or **cca()** output object. The examples at the end of that documentation file show how to access some of the ordination results directly. Here we shall access some important results as examples. Further results will be examined later when needed.

Eigenvalues
First, let us examine the eigenvalues. Are the first few clearly larger than the following ones? Here a question arises: how many ordination axes are meaningful to display and interpret?

PCA is not a statistical test, but a heuristic procedure; it aims at representing the major features of the data along a reduced number of axes (hence the expression "ordination in reduced space"). Usually, the user examines the eigenvalues, and decides how many axes are worth representing and displaying on the basis of the amount of variance explained. The decision can be completely arbitrary (for instance, interpret the number of axes necessary to represent 75% of the variance of the data), or assisted by one of several procedures proposed to set a limit between the axes that represent interesting variation of the data and axes that merely display the remaining, essentially random variance. One of these procedures consists in computing a **broken stick model**, which randomly divides a stick of unit length into the same number of pieces as there are PCA eigenvalues. The theoretical equation for the broken stick model is known. The pieces are then put in order of decreasing lengths and compared to the eigenvalues. One interprets only the axes whose eigenvalues are larger than the length of the corresponding piece of the stick, or, alternately, one may compare the sum of eigenvalues, from 1 to k, to the sum of the values from 1 to k predicted by the broken stick model. One can compute a scree plot for an ordination (i.e., a plot of the eigenvalues shown in decreasing order of importance). The **screeplot.cca()** function of **vegan** also shows the values predicted by the broken stick model, as follows (Fig. 5.1)[1].

```
# Examine and plot partial results from PCA output
?cca.object    # Explains how an ordination object
               # produced by vegan is structured and how to
               # extract its results.
# Eigenvalues
(ev <- env.pca$CA$eig)
# Scree plot and broken stick model
screeplot(env.pca, bstick = TRUE, npcs = length(env.pca$CA$eig))
```

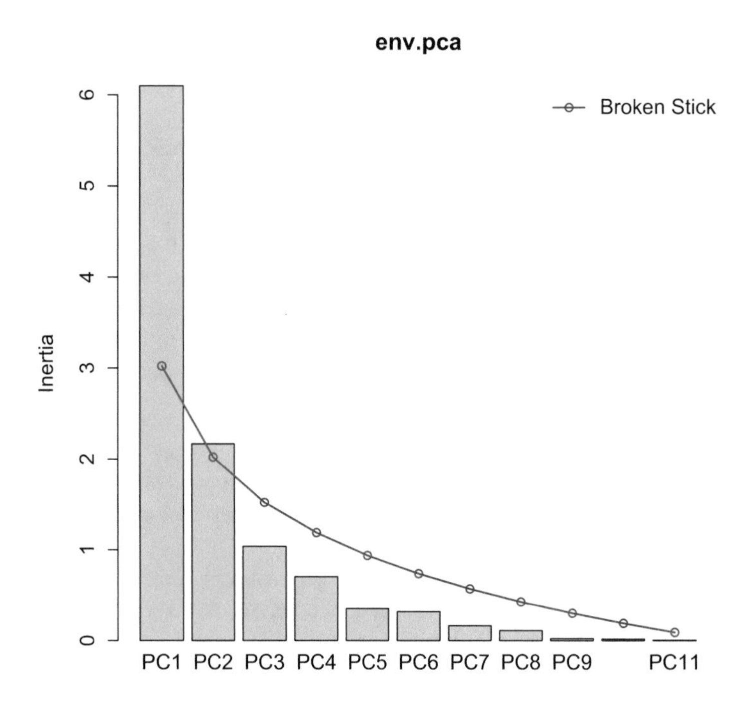

Fig. 5.1 Scree plot and broken stick model to help assess the number of interpretable axes in PCA. Application to the Doubs environmental data

[1]Comparison of a PCA result with the broken stick model can also be done by using function **PCAsignificance()** of package **BiodiversityR**.

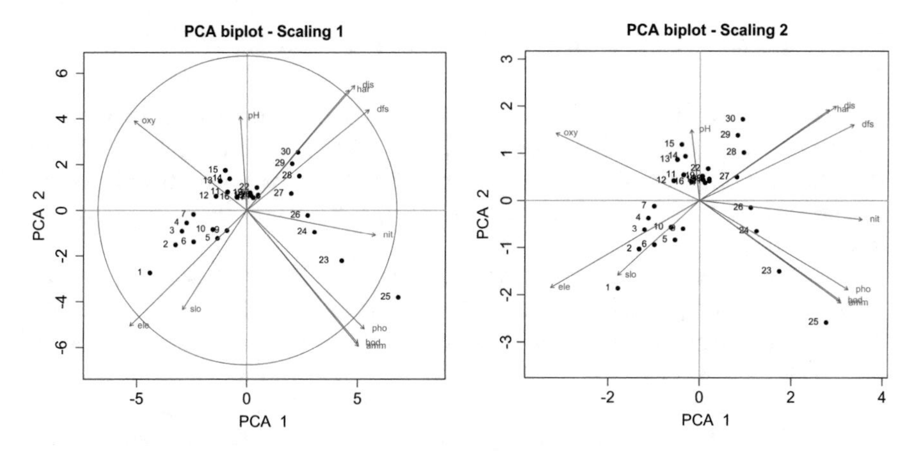

Fig. 5.2 PCA biplots of the Doubs environmental data, drawn with function `cleanplot.pca()`

Biplots of sites and variables

To plot PCA results in a proper manner, one has to show objects as points and variables as arrows. Two plots will be produced here, the first in scaling 1 (optimal display of distance relationships among objects), the second in scaling 2 (optimal display of covariances among variables) (Fig. 5.2). We will present two functions: **vegan**'s **biplot.rda()** and a function directly drawing scaling 1 and 2 biplots from **vegan** results: **cleanplot.pca()**.

```
# Plots using biplot.rda
par(mfrow = c(1, 2))
biplot(env.pca, scaling = 1, main = "PCA - scaling 1")
biplot(env.pca, main = "PCA - scaling 2") # Default scaling 2

# Plots using cleanplot.pca
# A rectangular graphic window is needed to draw the plots together
par(mfrow = c(1, 2))
cleanplot.pca(env.pca, scaling = 1, mar.percent = 0.08)
cleanplot.pca(env.pca, scaling = 2, mar.percent = 0.04)
```

Hints *One can also plot subsets of sites or variables, by creating* **R** *objects containing the biplot scores of the sites or variables of interest and using functions* **biplot()**, **text()** *and* **arrows()**. *Alternately, you can use our function* **cleanplot.pca()** *and directly select the variables that you want to plot by using argument* `select.spe`.

To help memorize the meaning of the scalings, **vegan** *now accepts argument* `scaling = "sites"` *for scaling 1 and* `scaling="species"` *for scaling 2. This is true for all* **vegan** *functions involving scalings.*

What does the circle in the left-hand plot mean? See below...

Now it is time to interpret the two biplots. First, the proportion of variance accounted for by the first two axes is 0.751, or 75.1%. This high value makes us confident that our interpretation of the first pair of axes will extract most of the relevant information from the data. Here is an example of how such a biplot can be interpreted.

First, the **scaling 1 biplot** displays a feature that must be explained. The circle is called a **circle of equilibrium contribution**. Its radius is equal to $\sqrt{d/p}$, where d is the number of axes represented in the biplot (usually $d = 2$) and p is the number of dimensions of the PCA space (i.e., usually the number of variables of the data matrix)[2]. The radius of this circle represents the length of the vector representing a variable that would contribute equally to all dimensions of the PCA space. Therefore, for any given pair of axes, the variables that have vectors longer than this radius make a higher contribution than average and can be interpreted with confidence. In scaling 2 it is not possible to draw a circle of equilibrium contribution, because scaling 2 is a projection into a Mahalanobis space, not a Euclidean space. Actually, a circle could only be drawn if all variables were standardized and strictly orthogonal to one another, which is never the case in practice.

The **scaling 1 biplot** shows a gradient from left to right, starting with a group formed by sites 1–10 which display the highest values of elevation (`ele`) and slope (`slo`), and the lowest values in river discharge (`dis`), distance from the source (`dfs`) and hardness (`har`). The second group of sites (11–16) has the highest values in oxygen content (`oxy`) and the lowest in nitrate concentration (`nit`). A third group of very similar sites (17–22) show intermediate values in almost all the measured variables; they are not spread out by the variables contributing to axes 1 and 2. Phosphate (`pho`) and ammonium (`amm`) concentrations, as well as biological oxygen demand (`bod`) show their maximum values around sites 23–25; the values decrease afterwards. Overall, the progression from oligotrophic, oxygen-rich to eutrophic, oxygen-deprived water is clear.

The **scaling 2 biplot** shows that the variables are organized in groups. The lower left part of the biplot shows that elevation and slope are very highly, positively correlated, and that these two variables are very highly, negatively correlated with another group comprising distance from the source, river discharge and hardness. Oxygen content is positively correlated with slope and elevation, but very negatively with phosphate and ammonium concentration and, of course, with biological oxygen demand. The right part of the diagram shows the variables associated with the lower

[2]Note, however, that **`vegan`** uses an internal constant to rescale its results, so that the vectors and the circle represented here are not equal but proportional to their original values. See the code of the **`cleanplot.pca()`** function.

section of the river, i.e. the group discharge and hardness, which are highly correlated with the distance from the source, and the group of variables linked to eutrophication, i.e. phosphate, ammonium concentration and biological oxygen demand. Positively correlated with these two groups is nitrate concentration. Nitrate and pH have nearly orthogonal arrows, indicating a correlation close to 0. pH displays a shorter arrow, showing its lesser importance for the ordination of the sites in the ordination plane. A plot of axes 1 and 3 would emphasize its contribution to axis 3.

This example shows how useful a biplot representation can be in summarizing the main features of a data set. Clusters and gradients of sites are obvious, as are the correlations among the variables. The correlation biplot (scaling 2) is far more informative than the visual examination of a correlation matrix among variables; the latter can be obtained by typing **`cor(env)`**).

Technical remark: **`vegan`** allows its output object to be plotted by the low-level plotting function **`plot(env.pca)`.** However, the use of this function provides PCA plots where sites as well as variables are represented by points. This is misleading, since the points representing the variables are actually the apices (tips) of vectors that must be drawn as arrows for the plot to be interpreted correctly. For further information about this function, type **`?plot.cca`.**

5.3.2.4 Projecting Supplementary Variables into a PCA Biplot

Supplementary variables can be added to a PCA plot through the function **`predict()`**. This function uses the ordination result to compute the ordination scores of new variables as a function of their values in the sites. The data frame containing the supplementary items must have the exact same row names as the original data.

Let us compute a PCA of the Doubs environment data, but without the two last variables, `oxy` and `bod`, then "predict" the position of the new variables in the ordination plot as an exercise:

```
# PCA of the environmental variables minus oxy and bod
env.pca2 <- rda(env[, -c(10, 11)], scale = TRUE)
# Create data frame with oxy and bod (our "new" variables)
new.var <- env[, c(10, 11)]
# Compute position of new variables (arrow tips)
new.vscores <-
  predict(env.pca2,
          type = "sp",
          newdata = new.var,
          scaling = 2)
# Plot of the result - scaling 2
biplot(env.pca2, scaling = 2)
arrows(
  0,
  0,
  new.vscores[, 1],
  new.vscores[, 2],
  length = 0.05,
  angle = 30,
  col = "blue"
)
text(
  new.vscores[, 1],
  new.vscores[, 2],
  labels = rownames(new.vscores),
  cex = 0.8,
  col = "blue",
  pos = 2
)
```

This example produces a result that is quite close to the one of the PCA computed with all environmental variables. This is because the two missing variables, `oxy` and `bod`, are well correlated to several others. Therefore, the PCA produced without them resembles the complete one, and the projection of the missing variables falls close to their position in the complete PCA. If we had, for the sake of demonstration, "predicted" the position of variables that were actually present in the original data, then the resulting "predicted" arrows would have been superimposed on the original arrows.

Important note: the projection of new variables into a PCA result works only if the supplementary variables have been treated in the same way as the ones that have been used in the PCA. In the cases above, the PCA has been computed on standardized variables (`scale = TRUE`), but the standardization has been done *within* the analysis, i.e., the data provided to the function **`rda()`** were untransformed. Since function **`predict()`** extracts its information from the PCA output (`env.pca2` in this example), the supplementary variables have automatically been treated in the same way as the original ones, i.e., standardized within the "prediction" process. By contrast, if the variables had been standardized *prior* to the PCA, then the supplementary variables to be projected would have needed a manual standardization as well. If a PCA was computed on Hellinger-transformed species abundance data (see Sect. 5.3.3) or on any data whose transformation involves some kind of standardization by rows, the *new species* to be projected into the biplot should be transformed using the same row sums and square-rooting as the data that have been submitted to the PCA.

5.3.2.5 Projecting Supplementary Objects into a PCA Biplot

If the PCA has been computed as above, using **vegan**'s function **rda()**, function **predict()** of package **vegan** cannot be used to project supplementary objects into a PCA biplot, at least if one wants the positions of the new sites to be estimated in the same way as those of the existing ones. The reason is that the output object of **rda()** has two classes: 'rda' and 'cca', and in this case **predict()** recognizes the class 'cca' and, accordingly, computes object scores as *weighted averages of the variables*. This is only correct for a correspondence analysis (CA) biplot (see Sect. 5.4.2.2).

However, another function **predict()**, of package **stats**, provides the correct site scores when the PCA is computed by some other **R** functions. As an example, let us compute a PCA of the Doubs environment data without sites 2, 8 and 22, using function **prcomp()**[3] of package **stats**, and then project the position of the missing sites in the ordination plot:

```
# PCA and projection of supplementary sites 2, 9 and 23 using
# prcomp() {stats} and predict( )
# Line numbers 8 and 22 are offset because of the deletion of
# empty site 8

# PCA using prcomp()
# Argument 'scale. = TRUE' calls for a PCA on a correlation matrix;
# it is equivalent to 'scale = TRUE' in function rda()
env.prcomp <- prcomp(env[-c(2, 8, 22), ], scale. = TRUE)

# Plot of PCA site scores using generic function plot()
plot(env.prcomp$x[ ,1], env.prcomp$x[ ,2], type = "n")
abline(h = 0, col = "gray")
abline(v = 0, col = "gray")
text(
  env.prcomp$x[ ,1],
  env.prcomp$x[ ,2],
  labels = rownames(env[-c(2, 8, 22), ])
)
```

[3]Do not use the alternative function **princomp()**, which computes variances with divisor n instead of $(n - 1)$.

```
# Prediction of new site scores
new.sit <- env[c(2, 8, 22), ]
pca.newsit <- predict(env.prcomp, new.sit)

# Projection of new site scores into the PCA plot
text(
   pca.newsit[, 1],
   pca.newsit[, 2],
   labels = rownames(pca.newsit),
   cex = 0.8,
   col = "blue"
)
```

When projecting new sites into a PCA, one must take the same care as with new variables. If the data submitted to the PCA have undergone a transformation prior to the analysis (not within it), the new site data must be transformed in the exact same way, i.e. with the same values of the transformation parameters. In our example this was not necessary, since the standardization of the variables was made within the PCA (argument `scale. = TRUE`) in **`prcomp()`**).

In contrast, if we want to compute and plot the scores of supplementary objects on the basis of a PCA computed by function **`rda()`**, we need to compute the new site scores by post-multiplying the new data (centred and, if necessary, scaled by the parameters of the original data set) by the matrix **U** of eigenvectors (Legendre and Legendre 2012, Eq. 9.18; see also the Code It Yourself corner at the end of this chapter). Furthermore, **`vegan`** applies a constant to the ordination scores, which must also be applied to the supplementary scores that we compute by hand. This constant can be retrieved from a **`scores()`** object. Furthermore, since the PCA was run on standardized data, the variables in the new site data have been standardized using the means and standard deviations of the *original* group of data, i.e. the ones that have been submitted to the PCA. The code is in the accompanying material of this book.

5.3.2.6 Combining Clustering and Ordination Results

Comparing a cluster analysis and an ordination can be fruitful to explain or confirm the differences between groups of sites. Here you will see two ways of combining these results. The first differentiates clusters of sites by colours on the ordination plot, the second overlays a dendrogram on the plot. Both will be done on a single PCA plot here (Fig. 5.3), but they can be drawn separately, of course.

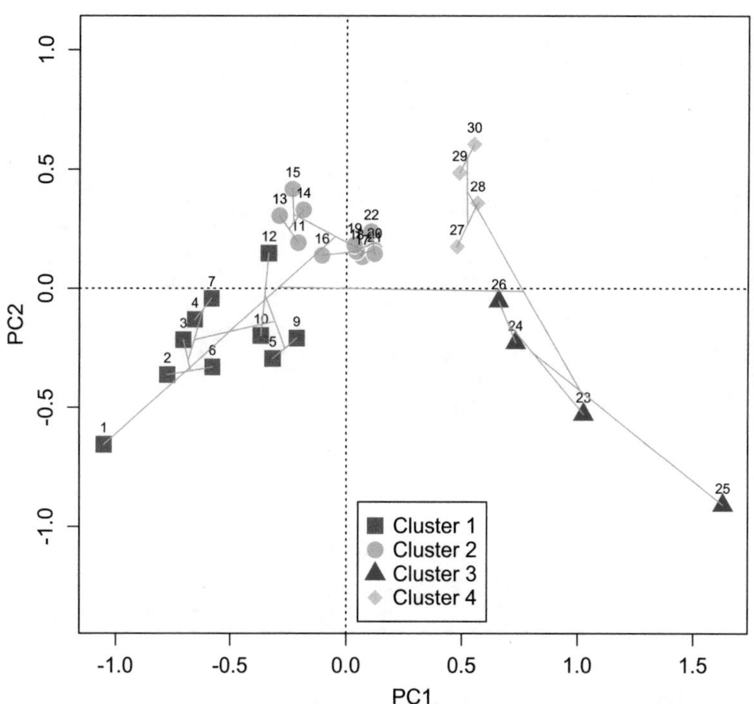

Fig. 5.3 PCA biplot (scaling 1) of the Doubs environmental data with overlaid clustering results

```
# Combining clustering and ordination results

# Clustering the objects using the environmental data: Euclidean
# distance after standardizing the variables, followed by Ward
# clustering
env.w <- hclust(dist(scale(env)), "ward.D")
# Cut the dendrogram to yield 4 groups
gr <- cutree(env.w, k = 4)
grl <- levels(factor(gr))

# Extract the site scores, scaling 1
sit.sc1 <- scores(env.pca, display = "wa", scaling = 1)

# Plot the sites with cluster symbols and colours (scaling 1)
p <- plot(
  env.pca,
  display = "wa",
  scaling = 1,
  type = "n",
  main = "PCA correlation + clusters"
)
```

```
abline(v = 0, lty = "dotted")
abline(h = 0, lty = "dotted")
for (i in 1:length(grl)) {
  points(sit.sc1[gr == i, ],
         pch = (14 + i),
         cex = 2,
         col = i + 1)
}
text(sit.sc1, row.names(env), cex = 0.7, pos = 3)
# Add the dendrogram
ordicluster(p, env.w, col = "dark grey")
# Add legend interactively
legend(
  locator(1),
  paste("Cluster", c(1:length(grl))),
  pch = 14 + c(1:length(grl)),
  col = 1 + c(1:length(grl)),
  pt.cex = 2
)
```

Hint *See how the coding of the symbols and colours is automatically conditioned on the number of groups: object* **`grl`** *has been set to contain numbers from 1 to the number of groups.*

5.3.3 PCA on Transformed Species Data

PCA being a linear method that preserves the Euclidean distance among sites, it is not naturally adapted to the analysis of species abundance data. However, transforming these after Legendre and Gallagher (2001) solves this problem (Sect. 3.5).

5.3.3.1 Application to the Hellinger-Transformed Fish Data

This application is based on functions **`decostand()`** and **`rda()`**. The graphical result is presented in Fig. 5.4. Readers are invited to experiment with other transformations, e.g. log-chord (see Sect. 3.3.1).

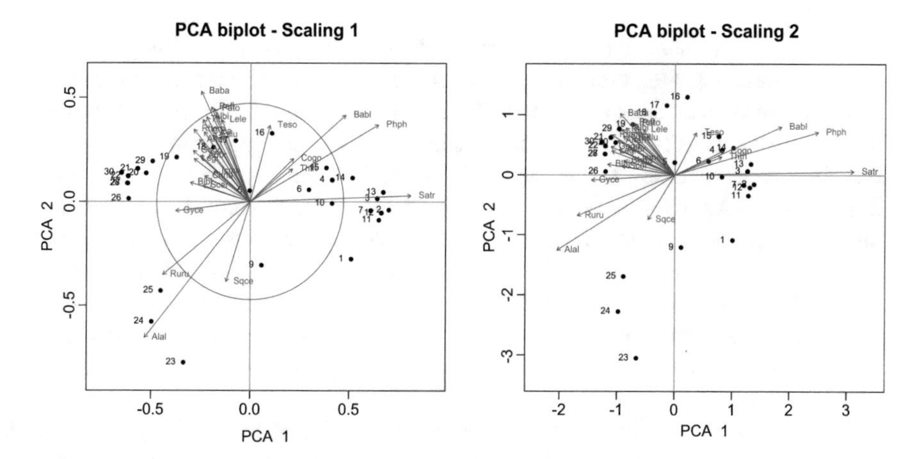

Fig. 5.4 PCA biplots of the Hellinger-transformed fish species data

```
# Hellinger pre-transformation of the species data
spe.h <- decostand(spe, "hellinger")
(spe.h.pca <- rda(spe.h))

# Scree plot and broken stick model
screeplot(
  spe.h.pca,
  bstick = TRUE,
  npcs = length(spe.h.pca$CA$eig)
)
# PCA biplots
spe.pca.sc1 <- scores(spe.h.pca, display = "species", scaling = 1)
spe.pca.sc2 <- scores(spe.h.pca, display = "species", scaling = 2)
par(mfrow = c(1, 2))
cleanplot.pca(spe.h.pca, scaling = 1, mar.percent = 0.06)
cleanplot.pca(spe.h.pca, scaling = 2, mar.percent = 0.06)
```

The species do not form clear groups like the environmental variables. However, see how the species replace one another along the site sequence.
In the scaling 1 biplot, observe that 8 species contribute strongly to axes 1 and 2. Are these species partly or completely the same as those identified as indicators of the groups in Sect. 4.11?
For comparison, repeat the PCA on the original object **spe** *without transformation. Which ordination shows the gradient of species contributions along the course of the river more clearly?*

Although PCA has a long history as a method devoted to tables of physical and chemical variables, the introduction of species data pre-transformations has opened up this powerful technique to the analysis of community data. Although PCA itself is not modified and remains a linear ordination model, the pre-transformations ensure that the species data are treated according to their specificity, i.e. without undue importance being given to double zeros. A scaling 1 PCA biplot thus reveals the underlying gradients structuring the community; the sites are ordered along the axes according to their positions along these gradients. The circle of equilibrium contribution allows the identification of the species contributing most to the plotted pair of axes. A scaling 2 biplot reveals the relationships among species in a correlation-like fashion; since the data have been transformed, the correlations are not equivalent to Pearson's *r* computed from the raw data.

Technical note: the chi-square transformation can also be applied to species data prior to PCA. In that case, the PCA solution is very similar, but not identical to a correspondence analysis (CA) of the species data (Sect. 5.4). Although the two methods preserve the chi-square distance among the sites, the calculation of the eigen-decomposition is not done in exactly the same way and leads to different sets of eigenvalues and eigenvectors.

5.3.3.2 Passive (*post hoc*) Explanation of Axes Using Environmental Variables

Although there are means of incorporating explanatory variables directly in the ordination process (canonical ordination, see Chap. 6), one may be interested in interpreting a simple ordination by means of external variables. This can be done in **`vegan`** by means of the function **`envfit()`**, which also works with CA (Sect. 5.4), PCoA (Sect. 5.5) and NMDS (Sect. 5.6). According to its author, Jari Oksanen, "**`envfit`** *finds vectors or factor averages of environmental variables.* [...] *The projections of points onto vectors have maximum correlation with corresponding environmental variables, and the factors show the averages of factor levels.*"

The result is an object containing coordinates of factor levels (points) or arrowheads (quantitative variables) that can be used to project these variables into the ordination diagram. Furthermore, **`envfit()`** computes a permutation test of the environmental variables and the **`plot()`** function allows users to draw only the variables with p-values equal to or smaller than a given level.

```
# A posteriori projection of environmental variables in a PCA
# A PCA scaling 2 plot is produced in a new graphic window.
biplot(spe.h.pca, main = "PCA fish abundances-scaling 2")
# Scaling 2 is default
(spe.h.pca.env <- envfit(spe.h.pca, env,  scaling = 2))
# Plot significant variables with a user-selected colour
plot(spe.h.pca.env, p.max = 0.05, col = 3)
# This has added the significant environmental variables to the
# last biplot drawn by R.
# BEWARE: envfit() must be given the same scaling  as the plot to
# which its result is added!
```

Hints *See how the plot of the environmental variables has been restricted to the significant variables by argument* `p.max = 0.05`.

This is a post hoc interpretation of ordination axes. Compare with Chap. 6.

Does this new information help interpret the biplot?

`envfit()` also proposes permutation tests to assess the significance of the R^2 of each explanatory variable regressed on the two axes of the biplot. But this is not, by far, the best way to test the effect of explanatory variables on a table of response variables. We will explore this topic in Chap. 6.

5.3.4 Domain of Application of PCA

Principal component analysis is a very powerful technique, but it has its limits. The main application of PCA in ecology is the ordination of sites on the basis of quantitative environmental variables or, after an appropriate transformation, of community composition data. PCA has originally been defined for data with multinormal distributions. In its applications in ecology, however, PCA is not very sensitive to departures from multinormality, as long as the distributions are not exaggeratedly skewed. The main computational step of PCA is the eigen-decomposition of a dispersion matrix (linear covariances or correlations). Covariances must in turn be computed on quantitative data — but see below for binary data. Here are, in more detail, the conditions of application of PCA:

- PCA must be computed on a table of dimensionally homogeneous variables. The reason is that it is the sum of the variances of the variables that is partitioned into eigenvalues. Variables must be in the same physical units to produce a meaningful sum of variances (the unit of a variance is the square of the unit of the variable from which it was computed), or they must be dimensionless, which is the case for standardized or log-transformed variables.

- PCA must not be computed on the transposed data matrix. The reason is that covariances or correlations among objects are meaningless.
- Covariances and correlations are defined for quantitative variables. However, PCA is very robust to variations in the precision of data. Since a Pearson correlation coefficient on semi-quantitative data is equivalent to a Spearman's correlation, a PCA on such variables yields an ordination where the relationship among variables is estimated using that measure.
- PCA can be applied to binary data. Gower (1966, *in* Legendre and Legendre 2012) has shown that with binary descriptors PCA positions the objects in the multidimensional space at distances that are the square roots of complements of simple matching coefficients S_1 (i.e., $\sqrt{1 - S_1}$) times a constant which is the square root of the number of binary variables.
- Species presence-absence data can be subjected to a chord, Hellinger or log-chord transformation prior to PCA. The justification is that the chord, Hellinger and log-chord distances computed on presence-absence data are equal to $\sqrt{2}\sqrt{1 - \text{Ochiai similarity}}$, so PCA after Hellinger or chord transformation preserves the Ochiai distance among objects in scaling type 1 plots. We also know that $\sqrt{1 - \text{Ochiai similarity}}$ is a metric and Euclidean distance (Legendre and Legendre 2012 Table 7.2), which is appropriate for ordination analysis of community composition presence-absence data.
- Avoid the mistake of interpreting the relationships among variables based on the proximities of the apices (tips) of the vector arrows instead of their angles in biplots.

5.3.5 *PCA Using Function* **PCA.newr()**

For someone who wants a quick assessment of the structure of his or her data, we provide functions **PCA.newr()** and **biplot.PCA.newr()** in file **PCA.newr.R**. Here is an example of how they work using the Doubs environmental data.

```
# PCA; scaling 1 is the default for biplots in this function
env.PCA.PL <- PCA.newr(env, stand = TRUE)
biplot.PCA.newr(env.PCA.PL)

# PCA; scaling 2 in the biplot
biplot.PCA.newr(env.PCA.PL, scaling = 2)
```

The graphs may be mirror images of those obtained with vegan. This is unimportant since the choice of the sign of the principal components, made within the **eigen()** *function used in PCA functions, is arbitrary.*

5.3.6 Imputation of Missing Values in PCA

In an ideal world, data would be perfect, with clear signal, as little noise as possible and, above all, without missing values. Unfortunately, it happens sometimes that for various reasons, in an otherwise good and interesting data set, values are missing. This poses problems. PCA cannot be computed on an incomplete data matrix. What should one do: delete the affected row(s) or column(s) and thereby loose precious existing data? Or, to avoid this, replace ("impute" in statistical language) the missing value(s) by some meaningful substitute, e.g. the mean of the corresponding variable (s) or an estimate obtained by regression analyses involving the other variables? Attractive as these two solutions appear, they do not consider the uncertainty attached to the estimation of the missing values and their use in further analyses. As a consequence, standard errors of estimators computed on the imputed data set are underestimated, leading to invalid tests if some are performed using these data.

To address these shortcomings, Josse and Husson (2012) proposed "*a regularized iterative PCA algorithm to provide point estimates of the principal axes and components* (. . .)". The PCA is computed iteratively, axis by axis, in a completely different way as the one used in **vegan**. When missing values are present, they are first replaced by the mean of the corresponding variable. Then a PCA is computed, and the missing values receive new estimates based on the result. The procedure is repeated until convergence, which corresponds to the point where the imputed values become stable. Complex procedures are implemented (1) to avoid overfitting of the imputation model (i.e., the problem of having too many parameters to adjust with respect to the amount of data, a situation occurring when many values are missing), and (2) to overcome the problem of the underestimation of the variance of the imputed data.

There is an important caveat here: the PCA solutions obtained with imputed values estimated using different numbers of PCA axes are not nested, i.e., a solution based on s axes is not identical to the first s axes of a solution with $(s + 1)$ or $(s + 2)$ axes.. Consequently, it is important to make an appropriate *a priori* decision about the desired number of axes. The decision may be empirical (for example, use the axes with sum of eigenvalues larger than the broken stick model prediction), or it can be informed by a cross-validation procedure on the data themselves, where data are deleted, reconstructed in PCAs with 1 to $n - 1$ dimensions, and the reconstruction error is computed. One retains the number of dimensions minimizing the reconstruction error.

Husson and Josse have grouped all necessary functions to perform iterative imputation of missing values in PCA in a package called **missMDA**. The function performing the imputation is **imputePCA()**. We shall now experiment it on the Doubs environmental data in two runs, one with only 3 missing values (i.e. 1% of the 319 values of the table) and the other with 32 (10%) missing values.

Our first experiment consists in removing three values selected to represent various situations. The first value will be close to the mean of a variable with a reasonably symmetrical distribution. pH has mean = 8.04. Let us delete the value in site 2 (8.0). The second will be close to the mean of a highly asymmetrical variable, pho, with mean 0.57. Let us delete the value in site 19 (row 18, 0.60). The third will also be removed from an asymmetrical variable, but far from its mean. bod has a mean of 5.01. Let us delete the value of site 23 (row 22, 16.4, the second highest value). How does **imputePCA()** perform?

```
# Imputation of missing values in PCA - 1: 3 missing values

# Replacement of 3 selected values by NA
env.miss3 <- env
env.miss3[2, 5] <- NA     # pH
env.miss3[18, 7] <- NA    # pho
env.miss3[22, 11] <- NA   # dbo

# New means of the involved variables (without missing values)
mean(env.miss3[, 5], na.rm = TRUE)
mean(env.miss3[, 7], na.rm = TRUE)
mean(env.miss3[, 11], na.rm = TRUE)

# Imputation
env.imp <- imputePCA(env.miss3)
# Imputed values
env.imp$completeObs[2, 5]     # Original value: 8.0
env.imp$completeObs[18, 7]    # Original value: 0.60
env.imp$completeObs[22, 11]   # Original value: 16.4

# PCA on the imputed data
env.imp3 <- env.imp$completeObs
env.imp3.pca <- rda(env.imp3, scale = TRUE)

# Procrustes comparison of original PCA and PCA on imputed data
pca.proc <- procrustes(env.pca, env.imp3.pca, scaling = 1)
```

The only good imputation is the one for pH, which falls right on the original value. Remember that pH is symmetrically distributed in the data. The two other imputations fall far away from the original values. In both cases the variables have strongly asymmetrical distributions. Obviously, this has more importance than the

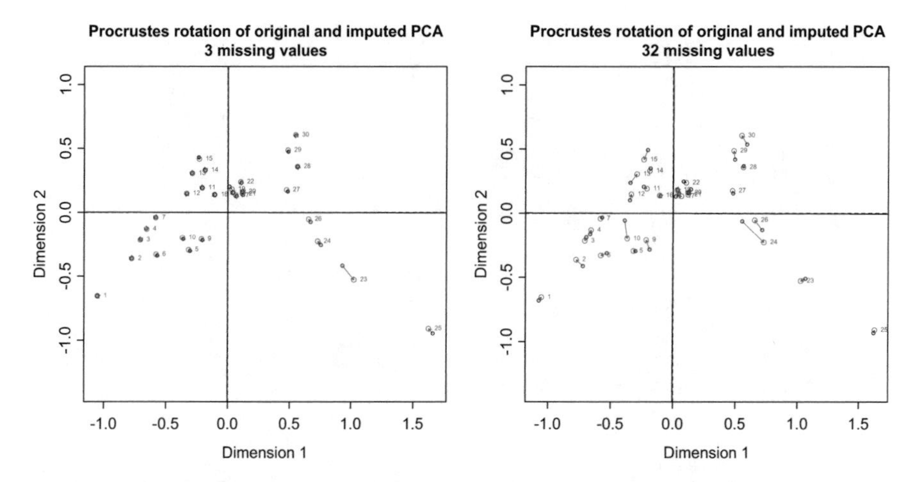

Fig. 5.5 Imputation of missing data in PCA. Procrustes rotation of the original PCA of environmental variables and the one performed on data where three missing values have been imputed. Scaling 1, only the sites are represented. Original sites: red; sites in imputed PCA: blue. Left: 3 missing values, or 1%; right: 32 missing values, or 10%

fact that the missing value is close or not to the mean of the variable (which was the case for `pho` but not `bod`).

Now, how does this affect the ordination result (2 axes)? To visualize this, let us compute a new PCA on the imputed matrix and compare it with the original PCA of environmental variables by means of a Procrustes rotation of the new PCA axes with respect to the original ones. Procrustes analysis finds the best superposition of two configurations (here the two ordination planes) of the same objects by rotation of one of the two sets; the result minimizes the squared distances between the corresponding objects (Legendre and Legendre 2012, p. 703). Procrustes rotation can be computed by means of function **`procrustes()`** of package **`vegan`**. The result is displayed in Fig. 5.5 left.

Obviously the differences between the real and reconstructed values had a very small impact on the ordination. The only visible difference is at site #23, where the difference between the real and imputed value is the largest.

Our second example is based on random removal of 32 values in the environmental matrix. In the particular example reported here, the affected sites were site 1 (4 NA), sites 2, 5 and 11 (3 NA), sites 19, 20, 22, 25 and 30 (2 NA) and sites 4, 12, 15, 16, 17, 18, 26, 27 and 28 (1 NA). Here the difference is larger (with the largest distance for site 1), but the overall shape of the ordination diagram, including the orientation of the variables (not shown here), is still reasonably well preserved (Fig. 5.5 right). This exercise shows that Josse and Husson's imputation technique can be useful to rescue a data set that would be almost useless if one removed all the rows or columns containing missing values (here 18 rows; all the columns are affected).

```
# Imputation of missing values in PCA - 2: 32 missing values

# Random replacement of 32 values (out of the 319) by NA
rnd <- matrix(sample(c(rep(1, 32), rep(0, 287))), 29, 11)
env.miss32 <- env
env.miss32[rnd == 1] <- NA
# How many NA in each site?
summary(t(env.miss32))

# Imputation
env.imp2 <- imputePCA(env.miss32)

# PCA on the imputed data
env.imp32 <- env.imp2$completeObs
env.imp32.pca <- rda(env.imp32, scale = TRUE)

# Procrustes comparison of original PCA and PCA on imputed data
pca.proc32 <- procrustes(env.pca, env.imp32.pca, scaling = 1)
par(mfrow = c(1, 2))
plot(pca.proc,
     main = "Procrustes rotation of original and imputed PCA\n
             3 missing values")
points(pca.proc, display = "target", col = "red")
text(
  pca.proc,
  display = "target",
  col = "red",
  pos = 4,
  cex = 0.6
)
plot(pca.proc32,
     main = "Procrustes rotation of original and imputed PCA\n
             32 missing values")
points(pca.proc32, display = "target", col = "red")
text(
  pca.proc32,
  display = "target",
  col = "red",
  pos = 4,
  cex = 0.6
)
```

Hint *In some of the plot titles above, the use of* `\n` *allows one to split a (long) title in two parts that are written on successive lines.*

In this second example, we removed 32 values randomly. Of course, in the real world, nobody does that. We did it for demonstration purposes only.

5.4 Correspondence Analysis (CA)

5.4.1 Introduction

For a long time, CA has been one of the favourite tools for the analysis of species presence-absence or abundance data. The raw data are first transformed into a matrix $\bar{\mathbf{Q}}$ of cell-by-cell contributions to the Pearson χ^2 statistic, and the resulting table is submitted to a singular value decomposition to compute its eigenvalues (which are the squares of the singular values) and eigenvectors. The result is an ordination in which the χ^2 distance (D_{16}) is preserved among sites instead of the Euclidean distance D_1. The χ^2 distance is not influenced by double zeros; it is an asymmetrical D function, as shown in Sect. 3.3.1. Therefore, CA is a method adapted to the analysis of species abundance data without pre-transformation. Note that the data submitted to CA must be frequencies or frequency-like, dimensionally homogeneous and non-negative; that is the case of species counts, biomasses, or presence-absence data.

For technical reasons due to the implicit centring of the frequencies in the calculation of the $\bar{\mathbf{Q}}$ matrix, CA ordination produces one axis fewer than min[n, p]. As in PCA, the orthogonal axes are ranked in decreasing order of the variation they represent, but instead of the total variance of the data, the variation is measured by a quantity called the total inertia (sum of squares of all values in matrix $\bar{\mathbf{Q}}$, see Legendre and Legendre 2012, under Eq. 9.25). Individual eigenvalues are always smaller than 1. To know the amount of variation represented along an axis, one divides the eigenvalue of this axis by the total inertia of the species data matrix.

In CA, both the objects and the species are generally represented as points in the same joint plot. As in PCA, two scalings of the results are available; they are most useful in ecology. In the explanation that follows, remember the objects (sites) of the data matrix are the rows and the species are the columns:

- **CA scaling 1**: rows (sites) are at the centroids of columns (species). This scaling is the most appropriate if one is primarily interested in the ordination of **objects (sites)**. In the multidimensional space, the χ^2 distance is preserved among objects. Interpretation: (1) the distances among objects in the reduced space approximate their χ^2 distances. Thus, object points that are close to one another are likely to be fairly similar in their species relative frequencies. (2) Any object found near the point representing a species is likely to contain a high contribution of that species. For presence-absence data, the object is more likely to possess the state "1" for that species.
- **CA scaling 2**: columns (species) are at the centroids of rows (sites). This scaling is the most appropriate if one is primarily interested in the ordination of **species**. In the multidimensional space, the χ^2 distance is preserved among species. Interpretation: (1) the distances among species in the reduced space approximate their χ^2 distances. Thus, species points that are close to one another are likely to have fairly similar relative frequencies along the objects. (2) Any species that lies

close to the point representing an object is more likely to be found in that object, or to have a higher frequency there than in objects that are further away in the joint plot.

The broken stick model, explained in Sect. 5.3.2.3, can be applied to CA axes for guidance as to the number of axes to retain. Our application below will concern the raw fish abundance data.

5.4.2 *CA Using Function* `cca()` *of Package* `vegan`

5.4.2.1 Running the Analysis and Drawing the Biplots

The calculations below closely resemble those used for PCA. First, let us run the analysis and draw its scree plot comparing the eigenvalues to the values of the broken stick model (Fig. 5.6):

```
# Compute CA
(spe.ca <- cca(spe))
summary(spe.ca)      # default scaling 2
summary(spe.ca, scaling = 1)
```

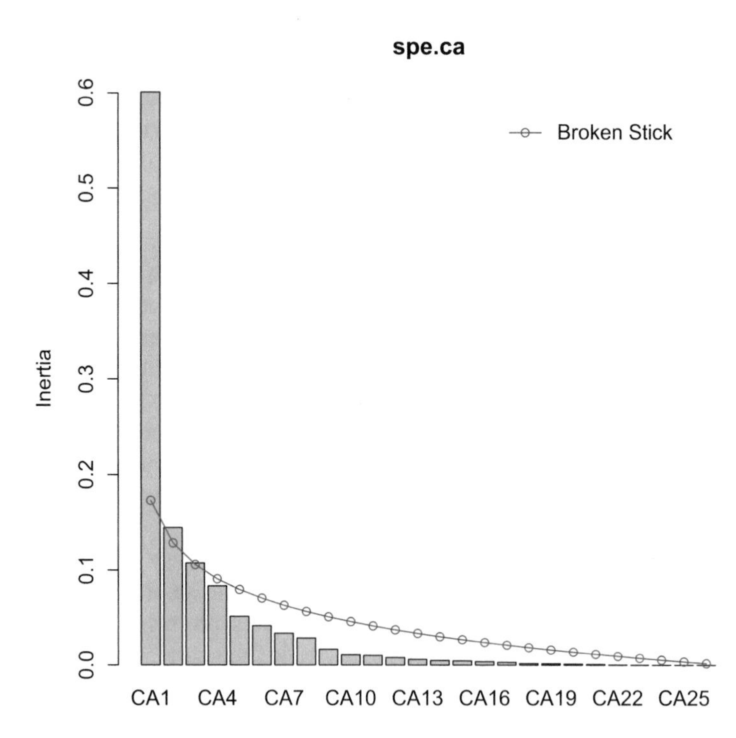

Fig. 5.6 Scree plot and broken stick model to help assess the number of interpretable axes in CA. Application to the Doubs fish raw abundance data

The first axis has a large eigenvalue. In CA, values over 0.6 indicate a very strong gradient in the data. What proportion of the total inertia does the first axis account for?
The eigenvalues are the same in both scalings. The scaling affects the eigenvectors to be drawn but not the eigenvalues.

```
# Scree plot and broken stick model using vegan's screeplot.cca()
screeplot(spe.ca, bstick = TRUE, npcs = length(spe.ca$CA$eig))
```

The first axis is extremely dominant, as can be seen from the bar plot as well as the numerical results.

It is time to draw the CA biplots of this analysis. Let us compare the two scalings (Fig. 5.7).

```
par(mfrow = c(1, 2))
# Scaling 1: sites are centroids of species
plot(spe.ca,
     scaling = 1,
     main = "CA fish abundances - biplot scaling 1"
)
# Scaling 2 (default): species are centroids of sites
plot(spe.ca, main = "CA fish abundances - biplot scaling 2")
```

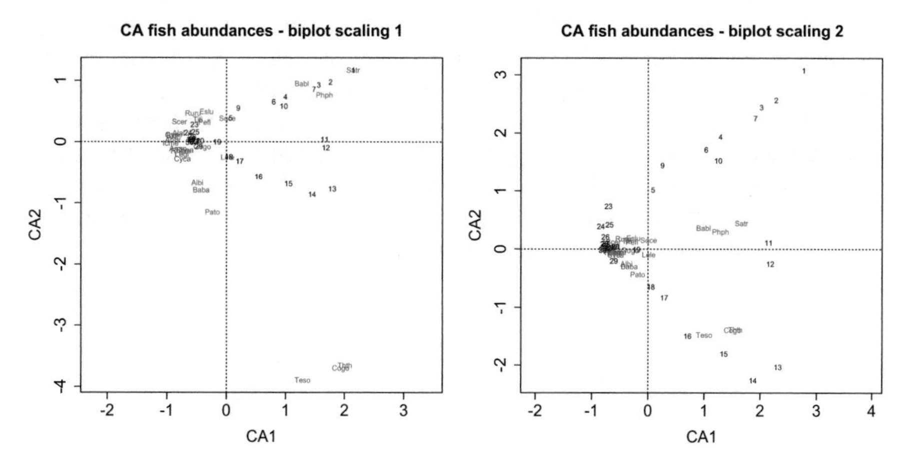

Fig. 5.7 CA biplots of the Doubs fish abundance data

Hint *Here you could also produce a clustering and overlay its result on the CA plot.*

The first axis opposes the lower section of the stream (sites 19–30) to the upper portion. This is clearly a strong contrast, which explains why the first eigenvalue is so high. Many species appear close to sites 19–30, indicating that they are more abundant downstream. Many of them are actually absent from the upper part of the river. The second axis contrasts the 10 upstream sites with the intermediate ones. Both groups of sites, which display short gradients on their own, are associated with characteristic species. The scaling 2 biplot shows how small groups of species are distributed among the sites. One can see that the grayling (`Thth`), the bullhead (`Cogo`) and the varione (`Teso`) are found in the intermediate group of sites (11–18), while the brown trout (`Satr`), the Eurasian minnow (`Phph`) and the stone loach (`Babl`) are found in a longer portion of the stream (approximately sites 1–18).

Observe how scalings 1 and 2 produce different plots. Scaling 1 shows the sites at the (weighted) centres of mass of the species. This is appropriate to interpret site proximities and find gradients or groups of sites. The converse is true for the scaling 2 biplot, where one can look for groups or replacement series of species. In both cases, interpretation of the species found near the origin of the graph should be done with care. This proximity could mean either that the species is at its optimum in the mid-range of the ecological gradients represented by the axes, or that it is present everywhere along the gradient.

5.4.2.2 Projection of Supplementary Sites or Species in a CA Biplot

Supplementary sites or species can be projected into a CA biplot using function **`predict()`** of **`stats`**. For a CA computed with **`vegan`**'s function **`cca()`**, this function computes the positions of supplementary sites in an appropriate manner (weighted averages); this is not the case for a PCA computed with **`rda()`** (Sect. 5.3.2.4). The data frame containing the supplementary items must have the exact same row names (for supplementary variables) or column names (for supplementary objects) as the data set that has been used to compute the CA. The examples below show (1) a CA with three sites removed, followed by the passive projection of these three sites, and (2) the same with three species.

```
# Projection of supplementary sites in a CA - scaling 1
sit.small <- spe[-c(7, 13, 22), ] # Data set with 3 sites removed
sitsmall.ca <- cca(sit.small)
plot(sitsmall.ca, display = "sites", scaling = 1)
# Project 3 sites
newsit3 <- spe[c(7, 13, 22), ]
ca.newsit <- predict(
                  sitsmall.ca,
                  newsit3,
                  type = "wa",
                  scaling = 1)
text(
  ca.newsit[, 1],
  ca.newsit[, 2],
  labels = rownames(ca.newsit),
  cex = 0.8,
  col = "blue"
)

# Projection of supplementary species in a CA - scaling 2
spe.small <- spe[, -c(1, 3, 10)]  # Data set with 3 species removed
spesmall.ca <- cca(spe.small)
plot(spesmall.ca, display = "species", scaling = 2)
# Project 3 species
newspe3 <- spe[, c(1, 3, 10)]
ca.newspe <- predict(
                  spesmall.ca,
                  newspe3,
                  type = "sp",
                  scaling = 2)
text(
  ca.newspe[, 1],
  ca.newspe[, 2],
  labels = rownames(ca.newspe),
  cex = 0.8,
  col = "blue"
)
```

5.4.2.3 *Post hoc* Curve Fitting of Environmental Variables

In Sect. 5.3.3.2 we used function **envfit()** to project environmental variables into a PCA biplot of the (transformed) fish data. However, the linear fitting and the projection of arrows only accounted for linear species-environment relationships. Sometimes one is interested to examine how selected environmental variables are connected to the ordination result, but on a broader, non-linear basis. This can be achieved by fitting surfaces of these environmental variables on the ordination plot.

Here we will again use function **envfit()** as a first step, but this time we propose to apply it with a formula interface, limiting the fitted model to two variables. The curve fitting itself is done by the **vegan** function **ordisurf()**, which fits smoothed two-dimensional splines by means of generalized additive models

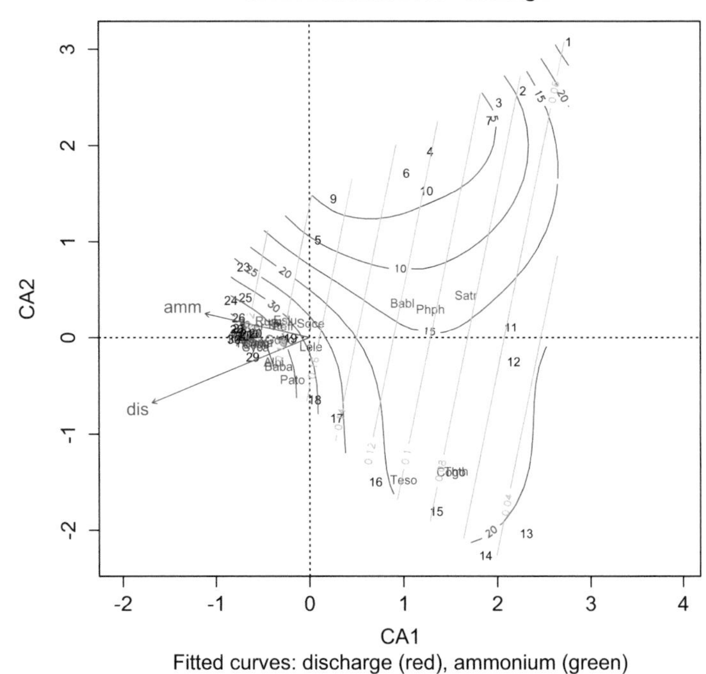

Fig. 5.8 CA biplot (scaling 2) of the Doubs fish abundance data with *a posteriori* curve fitting of two environmental variables: water discharge (red curves) and ammonium concentration (green curves)

(GAM). For this example, let us retain variables discharge (`dis`) and ammonium (`amm`), since both are important and are not overly correlated, as seen in Fig. 5.2. The resulting object will be used as in Sect. 5.3.3.2, but then we will add the fitted surfaces of the two selected environmental variables to the plot (Fig. 5.8):

```
plot(spe.ca, main = "CA fish abundances - scaling 2",
     sub = "Fitted curves: discharge (red), ammonium (green)")
spe.ca.env <- envfit(spe.ca ~ dis + amm, env)
plot(spe.ca.env)  # Two arrows
ordisurf(spe.ca, env$dis, add = TRUE)
ordisurf(spe.ca, env$amm, add = TRUE, col = "green")
```

On the figure, observe how the dis (red) fitted surface is strongly nonlinear, whereas the amm (green) surface is made of straight, parallel lines, indicating a linear fit.

5.4.2.4 Reordering the Data Table on the Basis of an Ordination Axis

A CA result is sometimes used to reorder the data table according to the first ordination axis. A compact form of ordered table is provided by the **vegan** function **vegemite()**, already used in Chap. 4, which can use the information provided by an ordination computed in **vegan**. One can also illustrate the ordered abundance table as a heat map by means of function **tabasco()**:

```
# Species data table ordered after the CA result
vegemite(spe, spe.ca)
# CA-ordered species table illustrated as a heat map
tabasco(spe, spe.ca)
```

The left-right and top-down orderings in this ordered table depend on the (arbitrary) orientation of the ordination axes. Observe that the ordering is not optimal since it is done only on the basis of the first CA axis. Therefore, sites 1 to 10 and 11 to 18 (separated along axis 2) and their corresponding characteristic species are interspersed.

5.4.3 *CA Using Function* CA.newr()

As in the case of PCA, we propose a simple CA function: **CA.newr()**. Here is how to use it on the fish data.

```
spe.CA.PL <- CA.newr(spe)
par(mfrow = c(1, 2))
biplot.CA(spe.CA.PL, scaling = 1, cex = 1)
biplot.CA(spe.CA.PL, scaling = 2, cex = 1)

# Ordering of the data table following the first CA axis
# The table is transposed, as in the vegemite() output
summary(spe.CA.PL)
t(spe[order(as.vector(spe.CA.PL$scaling1$sites[, 1])),
      order(as.vector(spe.CA.PL$scaling1$species[, 1]))])
```

Hints *Using* `$scaling2$sites` *and* `$scaling2$species` *(i.e. using the scaling 2 projection) would have produced the same ordered table.*

Argument `cex` *of the* **biplot()** *function is here to adapt the size of the symbols as well as the site and species names to the plot. The default is* `cex = 2`. *Smaller values produce smaller symbols and characters. They may be useful to draw plots containing many sites and species.*

An added bonus of **CA.newr()** is the possibility to display the cumulative fit of species and sites in terms of R^2. The maximum value is 1. These fits help identify the axes to which the species or sites contribute the most. For instance:

```
# Cumulative fit of species
spe.CA.PL$fit$cumulfit.spe
# Cumulative fit of sites
spe.CA.PL$fit$cumulfit.obj
```

The cumulative fit for species shows, for instance, that `Phph` is well fitted by axis 1 only (0.865) whereas `Cogo` is very well fitted by axes 1 and 2 together (0.914). On the other hand, `Lele` has no contribution to axes 1 and 2 but gets half of its fit (0.494) on axis 3. The same exercise can be done for the sites.

This information could be used for graphical purposes, e.g. to display only the sites or species whose cumulative fit is at least 0.5 on axis 2. Another application would involve the selection of several axes for a further analysis, retaining only enough to reach a cumulative fit of 0.8 for 80% of the sites (these figures are arbitrary and given as an example).

5.4.4 Arch Effect and Detrended Correspondence Analysis (DCA)

Long environmental gradients often support a succession of species. Since the species that are controlled by environmental factors tend to have unimodal distributions, a long gradient may encompass sites that, at opposite ends of the gradient, have no species in common; thus, their dissimilarity has the maximum possible value[4] (or their similarity is 0). But if one starts at any one point along the gradient and walks slowly towards an end, successive sites grow more and more different from the starting point until the maximum value of D is reached. Therefore, instead of a straight line, the gradient is represented as an arch on a pair of CA axes. Several detrending techniques have been proposed to counter this effect and straighten up gradients in ordination diagrams, leading to detrended correspondence analysis (DCA):

- Detrending by segments combined with nonlinear rescaling: axis I is divided into an arbitrary number of segments and, within each one, the mean of the object scores along axis 2 is made equal to zero. The arbitrarily selected number of

[4]The maximum value of the chi-square distance is $\sqrt{2y_{++}}$, where y_{++} is the sum of all frequencies in the table. This value is only obtained when there is a single species with abundance 1 in each of the two sites producing this maximum value, and these two species each have a total abundance of 1 in the data table. See Legendre and Legendre (2012, pp. 308–309) for details.

segments has a strong influence on the result. DCA results presented in the literature suggest that the scores along the second axis are essentially meaningless. The authors of this book strongly warn against the use of this form of DCA as an ordination technique; however, it may be used to estimate the "gradient length" of the first ordination axis, expressed in standard deviation units of species turnover. A gradient length larger than 4 indicates that some species have a unimodal response along the axis (ter Braak and Šmilauer 2002).

- Detrending by polynomials: another line of reasoning about the origin of the arch effect leads to the observation that when an arch occurs, the second axis can be seen as quadratically related to the first (i.e. it is the first axis to the power 2). This explains the parabolic shape of the scatter of points. Hence, a solution is to make the second axis not only linearly, but also quadratically independent from the first. Although intuitively attractive, this method of detrending should be applied with caution because it actually imposes a constraining model on the data.

DCA by segments is available in package **vegan** (function **decorana()**). In the output of this function, the gradient length of the axes is called "Axis lengths".

Given all its problems (see discussion in Legendre and Legendre 2012, p. 482–487), we will not describe this method further here. We now know that, even with long ecological gradients, a meaningful ordination can be obtained by using PCA of chord, Hellinger, or log-chord-transformed species data; see Sects. 3.5, 5.3.3, and 5.3.4.

An even more extreme effect of the same kind exists in PCA. It is called the horseshoe effect because, in the case of strong gradients, the sites at both ends bend inwards and appear closer than other pairs. This is due to the fact that PCA considers double zeros as resemblances. Consequently, sites located at opposite ends of an ecological gradient, having many double zeros, "resemble" each other in this respect. The Hellinger or chord transformations of the species data partly alleviate this problem.

5.4.5 Multiple Correspondence Analysis (MCA)

Multiple correspondence analysis (MCA) is the counterpart of PCA for the ordination of a table of categorical variables, i.e. a data frame in which all variables are factors. It is a special form of correspondence analysis where the variables are categorical. It has been designed primarily to analyse a series of individuals (e.g. persons in a survey, specimens in a taxonomic study) characterized by qualitative variables (e.g. questions with a choice of answers, or morphological characteristics). MCA can also be useful in environmental studies if the sites are described by qualitative variables.

In MCA, the variation of the data is expressed as inertia, as in CA. In most real cases, the inertia of the first few axes is relatively low when compared to the inertia of a CA axis or the variance of a PCA axis, because MCA computation involves the expansion of the data matrix, so that the levels of all *p* variables are represented by dummy binary variables. Such a matrix is called a complete disjunctive table.

MCA is implemented in the function **mca()** of package **MASS** and, with more options, in the function **MCA()** of package **FactoMineR**.

5.4.5.1 MCA on the Environmental Variables of the Oribatid Mite Data Set

As an example, let us compute an MCA on the environmental variables of our second data set. We have three qualitative variables: Substrate (7 classes), Shrubs (3 classes) and Microtopography (2 classes). Function **MCA()** offers the possibility of projecting supplementary variables into the MCA result. Note that *these supplementary variables are not involved in the computation of the MCA itself*; they are added to the result for heuristic purposes only. For more complex, simultaneous analyses of several data tables, see Chap. 6. Here we shall add two groups of supplementary variables: (1) the two quantitative environmental variables Substrate density and Water content, and (2) a 4-group classification of the Hellinger-transformed Oribatid mite data submitted to a Ward hierarchical clustering.

The numerical results of the analysis can be accessed by the name of the object followed by $..., e.g. `mite.env.MCA$ind$coord` for the site coordinates. The graphical results are grouped in Fig. 5.9.

```
# Preparation of supplementary data: mite classification
# into 4 groups
# Hellinger transformation of oribatid mite species data
mite.h <- decostand(mite, "hel")
# Ward clustering of mite data
mite.h.ward <- hclust(dist(mite.h), "ward.D2")
# Cut the dendrogram to 4 groups
mite.h.w.g <- cutree(mite.h.ward, 4)
# Assembly of the data set
mite.envplus <- data.frame(mite.env, mite.h.w.g)

# MCA of qualitative environmental data plus supplementary
# variables:
#   (1) quantitative environmental data,
#   (2) 4-group mite classification.
#   Default: graph=TRUE.

mite.env.MCA <- MCA(mite.envplus, quanti.sup = 1:2, quali.sup = 6)
mite.env.MCA
```

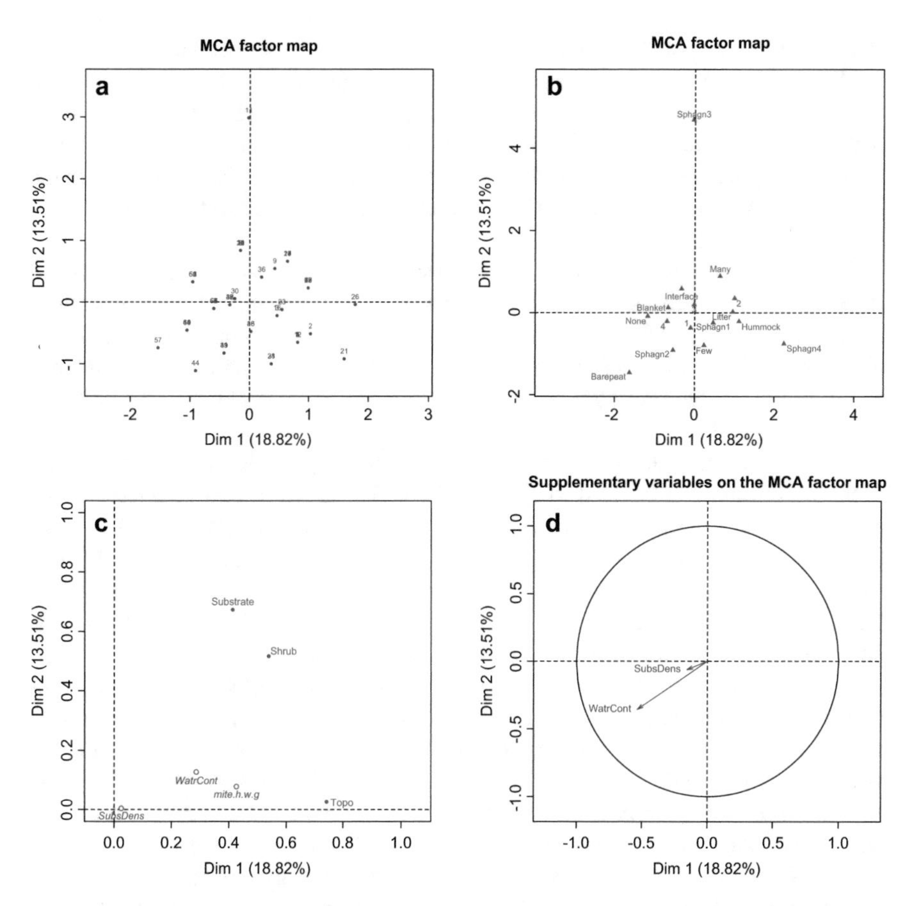

Fig. 5.9 Axes 1 and 2 of a multiple correspondence analysis (MCA) of the qualitative environmental variables of the oribatid mite data set with passive supplementary variables: two quantitative environmental variables and a 4-group typology of the oribatid mites. The two axes together represent 32.33% of the inertia of the dataset. (**a**) sites. (**b**) levels (modalities) of the variables involved in the analysis and of the qualitative supplementary variable (groups "1" to "4" marked by green triangles). (**c**) R^2 of the involved and supplementary variables with axes 1 and 2. (**d**) projection of the supplementary quantitative variables

Figure 5.9a represents the sites, each site being located at the weighted centroid of the dummy variables ("modalities" in MCA language) coded 1 for that site; in CA language it is a scaling 1 representation of the sites. Many site labels are blurred because several sites share the same profile. For instance, sites 1, 4, 5, 6, 7 and 12 share coordinates {0.811, −0.652} on the graph because their profile is the same: Sphagnum group 1, few shrubs and hummock. These modalities are indeed grouped in the same region of the graph as the groups of sites that share them (Fig. 5.9b). Two other features are worth mentioning. First, site 11 is isolated from all others at the top of the graph. This site is the only one that contains Sphagnum group 3. Second, pairs of sites 44 + 57 (left) and 21 + 26 (right) stand also relatively apart. The 44 + 57 pair shares the "Barepeat" modality and the pair 21 + 26 shares the "Sphagn4" modality. These two modalities have only two occurrences in the data set. These results show that MCA shares with CA the property of emphasizing rare events, which may be useful to identify rare characteristics or otherwise special features in a qualitative dataset. On the whole, Fig. 5.9a shows that the sites that are closest to the forest (small numbers) are on the right, and the sites that are closest to the free water (large numbers) are on the left.

Figure 5.9b shows the modalities at the weighted centroids of the sites (scaling 2 representation of the modalities). It allows the identification of rare modalities (e.g. "Sphagn3") and of groups of modalities (if any). Modalities projected in the same region of the graph are found in the same sites. Modalities projected close to the origin of the graph are common, so they do not contribute to discriminate groups of sites.

Figure 5.9c represents the squared correlation of the variables with the ordination axes. This graph can be used to identify the variables that are most related to the axes. For instance, variable "Topo" (blanket vs hummock) has an $R^2 = 0.7433$ on axis 1 but only 0.0262 on axis 2 (for this example, see `mite.env.MCA$var$eta2`). Variable "Shrub" has a similar R^2 with axes 1 and 2, and the R^2 of "Substrate" is larger on axis 2. Figure 5.9d represents the projections of the supplementary quantitative variables into the MCA.

Three of these graphs contain information about the supplementary variables added for heuristic purposes. In the graph of the modalities (Fig. 5.9b), one can also see the four groups of the mite community typology. The proximity of the groups with the environmental modalities allows an interpretation of the groups. The graph shows, for instance, that mite group 2 is found on sites with many shrubs, hummocks and forest litter. The two lower graphs (Fig. 5.9c, d) show that among the quantitative variables, Water content contributes more to the ordination plane (longer arrow) than Substrate density, but even Water content has a relatively low R^2. The mite typology has a much stronger relationship with axis 1 than with axis 2. Figure 5.9d shows the direction of largest variation of the two quantitative supplementary variables, mainly Water content, which is higher in the lower-left part of the graph. This confirms our observation on the graph of the sites: a gradient exists in the environmental variables, roughly in the direction of the Water content arrow.

Among the numerical results that are not graphically represented, the most important ones are the contributions of the modalities to the axes. Indeed, on the graphs, the rarest modalities stand out, but due to their rarity they don't contribute much to the axes. The contributions can be displayed (in our example) by typing `mite.env.MCA$var$contrib`. On axis 1, the highest contributions are those of modalities Hummock, Shrub-None and Blanket. On axis 2, Shrubs-Many, Sphagn3 and Shrubs-Few stand out. This points to the great importance of the two variables "Topo" and "Shrub". Are these linked together in some way? To find out, let us construct a contingency table between these two variables:

```
# Contingency table crossing variables "Shrub" and "Topo"
table(mite.env$Shrub, mite.env$Topo)
```

The table shows that sites where shrubs are present (modalities "`Few`" and "`Many`") are rather evenly represented on blankets and hummocks, but all sites devoid of shrubs have blanket-type soil coverage (or, conversely, no hummock is devoid of shrubs). Observe in Fig. 5.9b that modalities "`Blanket`" and "`None`" are close.

5.5 Principal Coordinate Analysis (PCoA)

5.5.1 Introduction

PCA as well as CA impose the distance that is preserved among objects: the Euclidean distance (and several others after pre-transformation of the species data) for PCA and the χ^2 distance for CA. If one wishes to ordinate objects on the basis of some other dissimilarity measure, more appropriate to the problem at hand, then PCoA is the method of choice. It provides a Euclidean representation of a set of objects whose relationships are measured by any dissimilarity measure chosen by the user. For example, if the coefficient is Gower's index $D = (1 - S_{15})$, which can combine descriptors of many mathematical types into a single measure of resemblance, then the ordination will represent the relationships among the objects based upon these variables and measured through the Gower index. This would not be possible with PCA or CA.

Like PCA and CA, PCoA produces a set of orthogonal axes whose importance is measured by eigenvalues. Since it is based on an association matrix, it can directly represent the relationships either among objects (if the association matrix was in Q mode) or among variables (if the association matrix was in R mode). If it is necessary to project variables, e.g. species, on a PCoA ordination of the objects (sites), the variables can be related *a posteriori* to the ordination axes using correlations or weighted averages and drawn on the ordination plot. In the case of association measures that have the Euclidean property (Sect. 3.3), PCoA behaves in a Euclidean manner. For instance, computing a Euclidean distance among sites and running a PCoA will yield the same results as running a PCA on a covariance matrix of the same data and looking at the scaling 1 ordination biplot. But if the association coefficient used is non-Euclidean, then PCoA may react by producing several negative eigenvalues in addition to the positive ones, plus a null eigenvalue in-between. The axes corresponding to negative eigenvalues cannot be represented on real ordination axes since they are complex. In most applications, this does not affect the representation of the objects on the several first principal axes, but it can lead to problems if the largest negative eigenvalues are of the same magnitude in absolute value as the first positive ones.

There are technical solutions to this problem, which consist in adding a constant to either the squared dissimilarities among objects (Lingoes correction) or to the dissimilarities themselves (Cailliez correction) (Gower and Legendre 1986). For hypothesis testing in two-way MANOVA (Sect. 6.3.3), the Lingoes correction is preferable because it produces a test with correct type I error, whereas the Cailliez correction produces a test with slightly inflated rate of type I error (Legendre and Anderson 1999). In the function **cmdscale()** presented below, the Cailliez correction is obtained with the argument `add = TRUE`. Note also that many similarity or dissimilarity coefficients are non-Euclidean, but the square root of the dissimilarity form ($\sqrt{1-S}$ or $\sqrt{D}$) of some of them is Euclidean. The simple matching (S_1), Jaccard (S_7), Sørensen (S_8) and percentage difference (aka Bray-Curtis, D_{14}) coefficients are examples. See Legendre and Legendre (2012) Tables 7.2 and 7.3. The function **dist.ldc()** of **adespatial** tells users whether a dissimilarity coefficient is Euclidean or not, and if its square root would be Euclidean.

Advanced note – One can avoid complex axes by keeping the eigenvectors with their original Euclidean norm (vector length = 1) instead of dividing each one by the square root of its eigenvalue, as is usual in the PCoA procedure. This workaround is used in the MEM spatial analysis presented in Chap. 7. It should not be used for routine ordination by PCoA since eigenvectors that have not been rescaled to $\sqrt{\text{eigenvalue}}$ cannot be used to produce plots that preserve the original dissimilarities among the objects.

The ordination axes of a PCoA can be interpreted like those of a PCA or CA: proximity of objects in the ordination represents their similarity in the sense of the association measure used.

Finally, **vegan** offers a weighted version of PCoA with function **wcmdscale()**. This function is based on **cmdscale()** with the added possibility to provide a vector of site weights. Weights are real positive numbers that can be larger than 1. A 0 weight returns NA for the corresponding object. Lingoes and Cailliez corrections are available through argument `add = "lingoes"` (default) or `"cailliez"`.

5.5.2 Application of PCoA to the Doubs Data Set Using cmdscale() and vegan

As an example, let us compute a matrix of percentage difference (aka Bray-Curtis) dissimilarities among sites, and subject this matrix to PCoA. In **vegan**, there is a way to project species abundances as weighted averages on a PCoA plot, by means of function **wascores()** (Fig. 5.10). Since species are projected as weighted averages of their contributions to the sites, their interpretation with respect to the sites is akin to that of a CA with scaling 2. It is also possible to project other (here, environmental) variables into the PCoA biplot using **envfit()**.

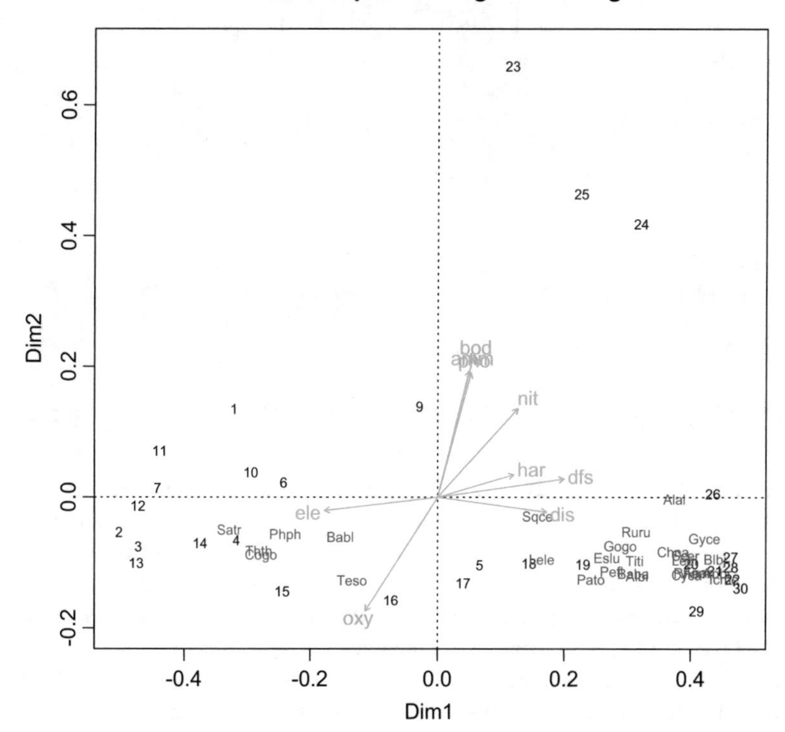

Fig. 5.10 PCoA biplot of a percentage difference dissimilarity matrix of the raw Doubs fish abundance data. *A posteriori* projection of the species as weighted averages using function **wascores()** (species abbreviations in red) and of environmental variables using function **envfit()** (green arrows). The relationships between species and sites are interpreted by proximities, as in CA

```
spe.bray <- vegdist(spe)
spe.b.pcoa <- cmdscale(spe.bray, k = (nrow(spe) - 1), eig = TRUE)
# Plot of the sites
ordiplot(scores(spe.b.pcoa, choices = c(1, 2)),
         type = "t",
         main = "PCoA with species weighted averages")
abline(h = 0, lty = 3)
abline(v = 0, lty = 3)
```

Don't worry about the warnings issued by **R** *and concerning the species scores. They are normal in this context, since species scores are not available at the first stage of a PCoA plot.*

```
# Add weighted average projection of species
spe.wa <- wascores(spe.b.pcoa$points[, 1:2], spe)
text(spe.wa, rownames(spe.wa), cex = 0.7, col = "red")
# A posteriori projection of environmental variables
(spe.b.pcoa.env <- envfit(spe.b.pcoa, env))
# Plot significant variables with a user-selected colour
plot(spe.b.pcoa.env, p.max = 0.05, col = 3)
```

Hint *Observe the use of two* **vegan** *functions,* **ordiplot()** *and* **scores()**, *to produce the ordination plot.* **vegan** *provides many special functions to handle outputs of its own analytical functions. Also, type* **?cca.object** *to learn about the internal structure of the* **vegan** *output objects.*

5.5.3 Application of PCoA to the Doubs Data Set Using pcoa()

There is another way to draw a double projection. It is based on correlations of the environmental variables with the PCoA ordination axes (see Legendre and Legendre 2012 p. 499). A PCoA computed on a matrix of Euclidean distances produces eigenvectors corresponding to what would be obtained in a scaling 1 PCA biplot of the same data. This representation can be drawn by using functions **pcoa()** and **biplot.pcoa()**, both available in the package **ape**.

Here is how these functions work. In our example, PCoA is run on a Euclidean distance matrix computed on a Hellinger-transformed species abundance matrix; the result of these two operations is a Hellinger distance matrix. In such a case, it is actually better (simpler and faster) to run a PCA directly on the transformed species data, but here the idea is to allow a comparison with the PCA run presented in Sect. 5.3.3. Two biplots are proposed, with projection of the raw and standardized species abundances. Compare the result below (Fig. 5.11) with the biplot of the PCA scaling 1 result in Fig. 5.4 (left).

```
spe.h.pcoa <- pcoa(dist(spe.h))
# Biplots
par(mfrow = c(1, 2))
# First biplot: Hellinger-transformed species data
biplot.pcoa(spe.h.pcoa, spe.h, dir.axis1 = -1)
abline(h = 0, lty = 3)
abline(v = 0, lty = 3)
# Second biplot: standardized Hellinger-transformed species data
spe.std <- scale(spe.h)
biplot.pcoa(spe.h.pcoa, spe.std, dir.axis1 = -1)
abline(h = 0, lty = 3)
abline(v = 0, lty = 3)
```

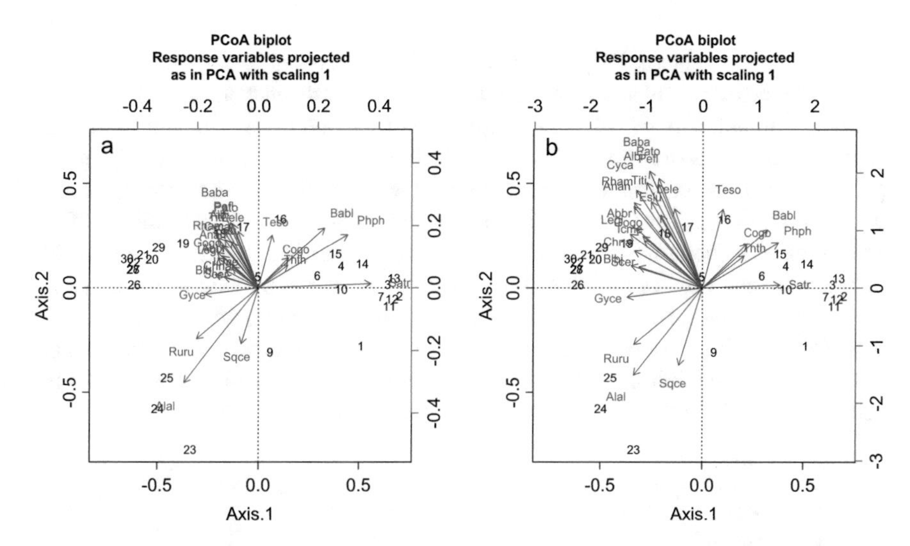

Fig. 5.11 PCoA biplots of the fish data obtained with functions **pcoa()** and **biplot.pcoa()**. The species are projected as in PCA with scaling 1. a: Hellinger-transformed raw species variables. b: standardized Hellinger-transformed species. The bottom and left-hand scales are for the objects, the top and right-hand scales are for the species

Hints *For projection of species data onto a PCoA plot, it is important to use the species data with the same transformation (if any) as the one used to compute the dissimilarity matrix. The standardization proposed here as an alternative may help in better visualizing the variables if they have very different variances.*

The argument `dir.axis1 = -1` *reverses axis 1 to make the result directly comparable with the PCA result in Fig. 5.4, scaling 1. Remember that the signs of ordination axes are arbitrary.*

How does this result compare with that of the PCA?

If one wishes to plot only a subset of the variables, one can provide a reduced data table as the second argument to the **biplot.pcoa()** function. For instance, in the second graph above, one could limit the plot to the four species that gave their names to Verneaux's zones (see Sect. 1.5.1):

```
# Third biplot: standardized Hellinger-transformed species data;
# only four species in the plot (figure not shown in the book)
biplot.pcoa(spe.h.pcoa, spe.h[, c(2, 5, 11, 21)], dir.axis1 = -1)
abline(h = 0, lty = 3)
abline(v = 0, lty = 3)
```

As mentioned above, PCoA should actually be reserved to situations where no Euclidean measure is available or selected. With Jaccard and Sørensen dissimilarity matrices computed by **ade4**, for example, the ordination is fully Euclidean because **ade4** takes the square root of the dissimilarities. In other cases, however, such as percentage difference dissimilarities computed with **vegan** (method="bray"), the dissimilarity matrices may not be Euclidean (see Sect. 3.3). This results in PCoA producing some negative eigenvalues. Lingoes and Cailliez corrections are available in the function **pcoa()**. This function provides the eigenvalues along with a broken stick comparison in its output. In the examples below, when a correction is requested for, the second column contains the corrected eigenvalues. However, the easiest way of obtaining a fully Euclidean ordination solution remains to take the square root of the dissimilarities before computing PCoA.

```
# PCoA on a Hellinger distance matrix
is.euclid(dist(spe.h))
summary(spe.h.pcoa)
spe.h.pcoa$values

# PCoA on a percentage difference dissimilarity matrix
is.euclid(spe.bray)
spe.bray.pcoa <- pcoa(spe.bray)
spe.bray.pcoa$values        # Observe eigenvalues 18 and following

# PCoA on the square root of a percentage difference
# dissimilarity matrix (aka Bray-Curtis)
is.euclid(sqrt(spe.bray))
spe.braysq.pcoa <- pcoa(sqrt(spe.bray))
spe.braysq.pcoa$values  # Observe the eigenvalues

# PCoA on a percentage difference dissimilarity matrix with
# Lingoes correction
spe.brayl.pcoa <- pcoa(spe.bray, correction = "lingoes")
spe.brayl.pcoa$values      # Observe the eigenvalues, col. 1 and 2

# PCoA on a percentage difference dissimilarity matrix with
# Cailliez correction
spe.brayc.pcoa <- pcoa(spe.bray, correction = "cailliez")
spe.brayc.pcoa$values      # Observe the eigenvalues, col. 1 and 2
```

If you want to choose the analysis displaying the highest proportion of variation on axes 1+2, which solution will you select among those above?

5.6 Nonmetric Multidimensional Scaling (NMDS)

5.6.1 Introduction

If the researcher's priority is not to preserve the exact dissimilarities among objects in an ordination plot, but rather to represent as well as possible the ordering relationships among objects in a small and specified number of axes, NMDS may be the solution. Like PCoA, NMDS can produce ordinations of objects from any dissimilarity matrix. The method can cope with missing values, as long as there are enough measures left to position each object with respect to a few others. NMDS is not an eigenvalue technique, and it does not maximise the variability associated with individual axes of the ordination. As a result, plots may arbitrarily be rotated or inverted. The procedure goes as follows (very schematically; for details see Legendre and Legendre 2012, p. 512 *et seq.*):

- Specify the desired number m of axes (dimensions) of the ordination.
- Construct an initial configuration of the objects in the m dimensions, to be used as a starting point of an iterative adjustment process. This is a tricky step, since the end-result may depend upon the starting configuration. A PCoA ordination may be a good starting point. Otherwise, try many independent runs with random initial configurations.
- An iterative procedure tries to position the objects in the requested number of dimensions in such a way as to minimize a stress function (scaled from 0 to 1), which measures how far the dissimilarities in the reduced-space configuration are from being monotonic to the original dissimilarities in the association matrix.
- The adjustment goes on until the stress value cannot be lowered, or until it reaches a predetermined low value (tolerated lack-of-fit).
- Most NMDS programs rotate the final solution using PCA, for easier interpretation.

For a given and small number of axes (e.g. $m = 2$ or 3), NMDS often achieves a less deformed representation of the dissimilarity relationships among objects than a PCoA in the same number of dimensions. But NMDS is a computer-intensive iterative technique exposed to the risk of suboptimal solutions. Indeed, the objective stress function to be minimized often reaches a local minimum larger than the true minimum.

5.6.2 Application to the Doubs Fish Data

NMDS can be performed in **R** with the elegant function **`metaMDS()`** of the **`vegan`** package. **`metaMDS()`** accepts raw data or dissimilarity matrices. Let us apply it to the fish abundances using the percentage difference index. **`metaMDS()`** uses random starts and iteratively tries to find the best possible solution. Species are added to

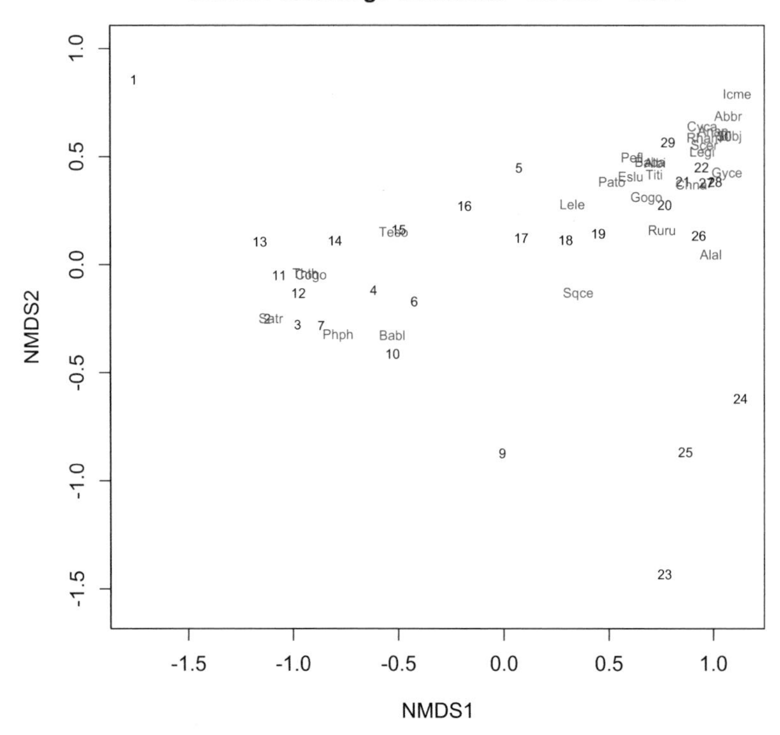

Fig. 5.12 NMDS biplot of a percentage difference dissimilarity matrix of the fish abundance data. Species were added using weighted averages. The relationships between species and sites are interpreted as in CA

the ordination plot as points using **wascores()**. See Fig. 5.12. Explanatory variables could also be added using **envfit()**.

```
spe.nmds <- metaMDS(spe, distance = "bray")
spe.nmds
spe.nmds$stress
plot(
  spe.nmds,
  type = "t",
  main = paste(
    "NMDS/Percentage difference - Stress =",
    round(spe.nmds$stress, 3)
  )
)
```

How does this result compare with those of PCA, CA and PCoA?

If one must use a dissimilarity matrix with missing values, NMDS can be computed with the function **isoMDS()**. An initial configuration must be provided in the form of a matrix positioning the sites (argument `y`) in the number of dimensions specified for the analysis (argument `k`). By default **isoMDS()** computes this initial configuration by PCoA, using function **cmdscale()**. To reduce the risk of reaching a local minimum, we suggest to use the function **bestnmds()** of the package **labdsv**. This function, which is a wrapper for **isoMDS()**, computes the analysis a user-specified number of times (argument `itr`) with internally produced random initial configurations. The solution with the smallest stress value is retained by the function.

A useful way to assess the appropriateness of an NMDS result is to compare, in a *Shepard diagram,* the dissimilarities among objects in the ordination plot with the original dissimilarities. In addition, the goodness-of-fit of the ordination can be measured as the R^2 of either a linear or a non-linear regression of the NMDS distances on the original dissimilarities. All this is possible in **R** using **vegan**'s functions **stressplot()** and **goodness()**(Fig. 5.13):

```
# Shepard plot and goodness of fit
par(mfrow = c(1, 2))
stressplot(spe.nmds, main = "Shepard plot")
gof <- goodness(spe.nmds)
plot(spe.nmds, type = "t", main = "Goodness of fit")
points(spe.nmds, display = "sites", cex = gof * 300)
```

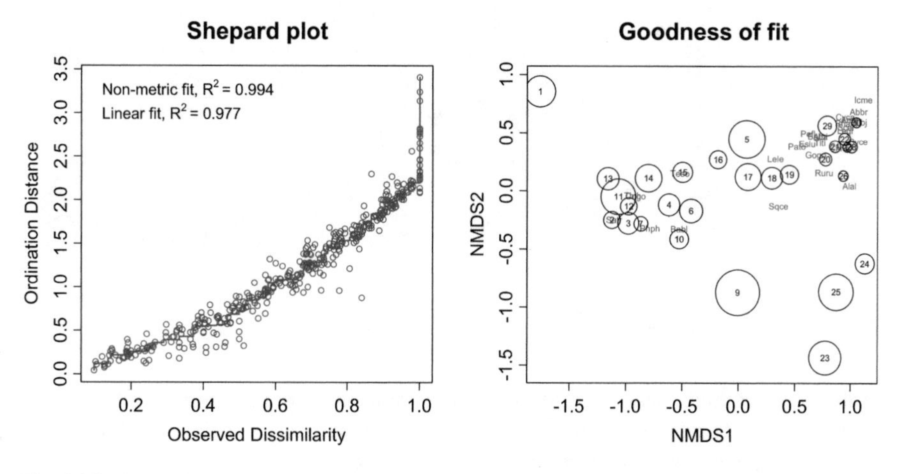

Fig. 5.13 Shepard and goodness of fit diagrams of the NMDS result presented in Fig. 5.12. Poorly fitted sites have larger bubbles

Hint *See how the goodness-of-fit of individual sites is represented using the results of the* **goodness()** *analysis by way of the* cex *argument of the* **points()** *function.*

As with the other ordination methods, it is possible to add information from a clustering result to an NMDS ordination plot. For instance, compute a Ward clustering of the percentage difference matrix, extract 4 groups and colorize the sites according to these groups:

```
# Ward clustering of percentage difference dissimilarity matrix
# and extraction of four groups
spe.bray.ward <-
 hclust(spe.bray, "ward.D") # Here better than ward.D2 for 4 groups
spe.bw.groups <- cutree(spe.bray.ward, k = 4)
grp.lev <- levels(factor(spe.bw.groups))

# Combination with NMDS result
sit.sc <- scores(spe.nmds)
p <-
  ordiplot(sit.sc, type = "n",
           main = "NMDS/% difference + clusters Ward/% difference")
for (i in 1:length(grp.lev))
{
  points(sit.sc[spe.bw.groups == i, ],
         pch = (14 + i),
         cex = 2,
         col = i + 1)
}
text(sit.sc, row.names(spe), pos = 4, cex = 0.7)
# Add the dendrogram
ordicluster(p, spe.bray.ward, col = "dark grey")
# Add a legend interactively
legend(
  locator(1),
  paste("Group", c(1:length(grp.lev))),
  pch = 14 + c(1:length(grp.lev)),
  col = 1 + c(1:length(grp.lev)),
  pt.cex = 2
)
```

5.6.3 PCoA or NMDS?

Principal coordinate analysis and nonmetric multidimensional scaling often pursue similar goals: obtain a meaningful ordination of objects on a small (usually 2 or 3) number of axes. As seen in the previous sections, in PCoA the solution is found using eigen-decomposition of the transformed dissimilarity matrix, whereas in NMDS the solution is found by an iterative approximation algorithm. The literature shows that some researchers favour the former and others the latter, generally without explicit justification. Users may wonder: are the two methods equally suitable in all situations? Do they produce identical or fairly similar solutions? If not, are there situations where one method is preferable over the other? How do they compare in their principles and applications? Some of their properties can be advocated in favour of the one or the other, depending on the application. Let us enumerate them.

NMDS

1. The iterative algorithm may find different solutions depending on the starting point of the calculation, which, in most instances, is a randomly chosen configuration.
2. The dissimilarities are distorted (stretched or squeezed) during NMDS calculation. That is an acknowledged property of the method. The distances in the ordination solution do not exactly correspond to the starting dissimilarities.
3. The first NMDS axis is not bound to maximize the variance of the observations. However, most if not all NMDS programs compute a PCA rotation of the NMDS result, so that the first axis maximizes the variance of the NMDS solution.
4. Different criteria are available in different NMDS functions to minimize the stress. The solution may differ depending on the criterion that one (or the R function) chooses to optimise.
5. **R** functions may contain several other arguments that may influence the solution. Except for highly experienced users, it is difficult to determine which combination of options is best for a particular data set.
6. Users must set k, the number of dimensions of the ordination solution. All ordination axes can be produced, i.e. min[p, $n-1$], but then the solution is not exact since the distances are distorted. It is recommended to set $k < (n - 1)/2$.
7. The stress statistic does not indicate the proportion of the variance of the data represented in the ordination solution. Instead, it indicates the amount of deformation of the original dissimilarities, which is very different.

Application to ecological analysis – When it is necessary to squeeze in two dimensions a PCA or PCoA solution that requires 3 or 4 dimensions, for instance in a figure drawn for publication, NMDS is useful to represent well the main dissimilarity relationships among the sites in 2-D. In such a case, it is preferable to use the PCA or PCoA ordination axes as input into NMDS to make sure that the NMDS solution will not diverge markedly from the metric ordination.

PCoA

1. PCoA finds the optimal solution by eigenvalue decomposition. The PCoA solution is unique.
2. In a PCoA, the dissimilarities are not distorted in the ordination solution.
3. PCoA can be used to find the first ordination axes for a given dissimilarity matrix. These axes are those that maximize the variance of the observations.
4. PCoA produces all ordination axes corresponding to a given dissimilarity matrix. The ordination solution is exact.
5. The full matrix of PCoA axes allows one to precisely reconstruct the distances among objects. *Reason:* the matrix of Euclidean distances among objects computed from the PCoA axes is strictly equal to the dissimilarity matrix subjected to PCoA, whatever the dissimilarity function that has been used. Proof of that property is found in Gower (1966) and in Legendre and Legendre (2012).
6. Hence, the set {dissimilarity function, ordination method} produces a unique transformation of the data, and the solution is reproduced exactly if one runs PCoA again on the same data, irrespective of the user or program, except for possible inversion of the signs along any one axis.
7. A pseudo-R^2 statistic is computed as the sum of the eigenvalues of the axes of interest (for example the first 2 axes) divided by the sum of all eigenvalues. It indicates the fraction of the variance of the data represented in the reduced-space ordination. This is a useful statistic to assess the ordination result.

Applications to ecological analysis – (1) PCoA produces ordinations of the objects in reduced 2-D or 3-D space. (2) PCoA can also act as a data transformation after computation of an appropriately chosen dissimilarity measure. The coordinates of the objects in full-dimensional PCoA space represent the transformed data. They can be used as starting point for new analyses, for example db-RDA (Chap. 6) or *k*-means partitioning (Chap. 4). *Recommendation:* use PCoA in most applications.

5.7 Hand-Written PCA Ordination Function

To conclude this chapter, let us dive into the bowels of an ordination method...

The Code It Yourself corner # 2

Legendre and Legendre (2012) provide the algebra necessary to program the ordination methods described above directly "from scratch", i.e., using the matrix algebra functions implemented in ***R****. While it is not the purpose of this book to do this for all methods, we provide an example that could stimulate the interest of users. After all, numerical ecology is a living science, and anybody could one day stumble upon a situation for which no ready-made function exists. The researcher may then be interested in developing his or her own method and write the functions to implement it.*

The example below is based on the algebraic development presented in Legendre and Legendre (2012), Sect. 9.1. It is presented in the form of a function, the kind that any user could write for her or his own use. The steps are the following for a PCA of a covariance matrix. To obtain a PCA of a correlation matrix, one has to standardize the data before using this function, or implement the standardization as an option in the function itself.

1 - Compute the covariance matrix ***S*** *of the original or centred data matrix.*

2 - Compute the eigenvectors and eigenvalues of ***S*** *(eqs. 9.1 and 9.2).*

3 - Extract matrix ***U*** *of the eigenvectors and compute* ***matrix F*** *of the principal components (eq. 9.4) for the scaling 1 biplot. This step involves the centred data.*

4 - Compute matrices ***U2*** *and* ***G*** *for the scaling 2 biplot.*

5 - Output of the results

```
# A simple function to perform PCA

myPCA <- function(Y) {
  Y.mat <- as.matrix(Y)
  object.names <- rownames(Y)
  var.names <- colnames(Y)

  # Centre the data (needed to compute matrix F)
  Y.cent <- scale(Y.mat, center = TRUE, scale = FALSE)

  # Covariance matrix S
  Y.cov <- cov(Y.cent)

  # Eigenvectors and eigenvalues of S (Legendre and Legendre 2012,
  # eq. 9.1 and 9.2)
  Y.eig <- eigen(Y.cov)

  # Copy the eigenvectors to matrix U (used to represent variables
  # in scaling 1 biplots)
  U <- Y.eig$vectors
  rownames(U) <- var.names

  # Compute matrix F (used to represent objects in scaling 1 plots)
  F <- Y.cent %*% U          # eq. 9.4
  rownames(F) <- object.names
```

```
  # Compute matrix U2 (to represent variables in scaling 2 plots)
  # eq. 9.8
  U2 <- U %*% diag(Y.eig$values ^ 0.5)
  rownames(U2) <- var.names

  # Compute matrix G (to represent objects in scaling 2 plots)
  # eq. 9.14
  G <- F %*% diag(Y.eig$values ^ 0.5)
  rownames(G) <- object.names

  # Output of a list containing all the results
  result <- list(Y.eig$values, U, F, U2, G)
  names(result) <- c("eigenvalues", "U", "F", "U2", "G")
  result
}
```

This function should give the exact same results as the function **PCA.newr()** used in Sect. 5.3.5. Now try it on the Hellinger-transformed fish species data and compare the results.

To make your function active, either **save it in a file** (called for instance **myPCA.R**) and **source it**, or (less elegant) copy the whole code directly into your **R** console.

```
# PCA on fish species using hand-written function
fish.PCA <- myPCA(spe.h)
summary(fish.PCA)
# Eigenvalues
fish.PCA$eigenvalues
# Eigenvalues expressed as percentages
(pv <-
  round(100 * fish.PCA$eigenvalues / sum(fish.PCA$eigenvalues),
  2))
# Alternate computation of total variation (denominator)
round(100 * fish.PCA$eigenvalues / sum(diag(cov(spe.h))), 2)
# Cumulative eigenvalues expressed as percentages
round(
  cumsum(100 * fish.PCA$eigenvalues / sum(fish.PCA$eigenvalues)),
  2)

# Biplots
par(mfrow = c(1, 2))
# Scaling 1 biplot
biplot(fish.PCA$F, fish.PCA$U)
# Scaling 2 biplot
biplot(fish.PCA$G, fish.PCA$U2)
```

Now you could plot other pairs of axes, for instance axes 1 and 3.

Compared to CA or PCoA, the code above is rather straightforward. But nothing prevents you from trying to program another method. You can also display the code of the `CA.newr()` and `pcoa()` functions and interpret it with the manual in hand.

Chapter 6
Canonical Ordination

6.1 Objectives

Simple (unconstrained) ordination analyses one data matrix and reveals its major structure in a graph constructed with a reduced set of orthogonal axes. It is therefore a passive form of analysis, and the user interprets the ordination results *a posteriori*, as described in Chap. 5. Canonical ordination, on the contrary, associates two or more data sets in the ordination process itself. Consequently, if one wishes to extract structures of a data set that are related to (or can be interpreted by) another data set, and/or formally test statistical hypotheses about the significance of these relationships, canonical ordination is the way to go.

Canonical ordination methods can be classified into two groups depending on the role played by the two matrices involved: symmetric and asymmetric.

Practically, you will:

- learn how to choose among various canonical ordination techniques: asymmetric [redundancy analysis (RDA), distance-based redundancy analysis (db-RDA), canonical correspondence analysis (CCA), linear discriminant analysis (LDA), principal response curves (PRC), co-correspondence analysis (CoCA)] and symmetric [canonical correlation analysis (CCorA), co-inertia analysis (CoIA) and multiple factor analysis (MFA)];
- explore methods devoted to the study of the relationships between species traits and environment;
- compute them using the correct options and properly interpret the results;
- apply these techniques to the Doubs River and other data sets;
- explore particular applications of some canonical ordination methods, for instance variation partitioning and multivariate analysis of variance by RDA;
- write your own RDA function.

D. Borcard et al., *Numerical Ecology with R*, Use R!,
https://doi.org/10.1007/978-3-319-71404-2_6

6.2 Canonical Ordination Overview

In the methods explored in Chap. 5, the ordination procedure itself is not influenced by external variables; these may only be considered after the computation of the ordination. One lets the data matrix express the relationships among objects and variables without constraint. This is an exploratory, descriptive approach. Canonical ordination, on the contrary, explicitly explores the relationships between two matrices: a response matrix and an explanatory matrix in some cases (asymmetric analysis), or two matrices with symmetric roles in other cases. Both matrices are used in the production of the ordination.

The way to combine the information of two (or, in some cases, more) data matrices depends on the method of analysis. We will first explore the two asymmetric methods that are mostly used in ecology nowadays, i.e., redundancy analysis (RDA) and canonical correspondence analysis (CCA). Both combine multiple regression with classical ordination (PCA or CA). Partial RDA will also be explored, as well as a procedure of variation partitioning. The significance of canonical ordinations will be tested by means of permutations. After that, we will devote sections to three other asymmetric methods: linear discriminant analysis (LDA), which looks for a linear combination of explanatory variables to explain a predefined grouping of the objects, principal response curves (PRC), developed for the analysis of multivariate results of designed experiments that involve repeated measurements, and co-correspondence analysis (CoCA), which is devoted to the simultaneous ordination of two communities sampled at the same sites. Then, we turn to three symmetric methods that compute eigenvectors describing the common structure of two or several data sets: canonical correlation analysis (CCorA), co-inertia analysis (CoIA) and multiple factor analysis (MFA). We conclude the chapter by visiting two methods devoted to the study of the relationship between species traits and the environment: the fourth-corner method and the RLQ analysis.

6.3 Redundancy Analysis (RDA)

6.3.1 Introduction

RDA is a method combining regression and principal component analysis (PCA). It is a direct extension of multiple regression analysis to model multivariate response data. RDA is an extremely powerful tool in the hands of ecologists, especially since the introduction of the Legendre and Gallagher (2001) transformations that opened RDA to the analysis of community composition data (transformation-based RDA, or tb-RDA).

Conceptually, RDA is a multivariate (meaning *multiresponse*) multiple linear regression followed by a PCA of the matrix of fitted values. It works as follows, using a matrix **Y** of **centred** response data and a matrix **X** of centred explanatory variables[1]:

- Regress each (centred) y variable on explanatory matrix **X** and compute the fitted ($\hat{y}$) (this is the only required matrix in most analyses) and residual (y_{res}) values (if needed). Assemble all vectors $\hat{y}$ into a matrix $\hat{\mathbf{Y}}$ of fitted values.
- Carry out a test of significance of the canonical relationship $\mathbf{Y} \sim \mathbf{X}$.
- If the test is significant, i.e. if the **X** variables explain significantly more of the variation of **Y** than random data would, compute a PCA of the matrix $\hat{\mathbf{Y}}$ of fitted values; this analysis produces a vector of canonical eigenvalues and a matrix **U** of **canonical eigenvectors**;
- Use matrix **U** to compute two types of ordination site scores: use either the matrix $\hat{\mathbf{Y}}$ of fitted values to obtain an ordination in the space of variables **X** (i.e., compute $\hat{\mathbf{Y}}\mathbf{U}$, which produces fitted site scores called "*Site constraints (linear combinations of constraining variables)*" in the summary of the RDA output, coded "lc" in `vegan`) or use the original centred matrix **Y** to obtain an ordination in the space of the original variables **Y** (i.e., compute **YU**, obtaining site scores called "*Site scores (weighted sums of site scores)*", coded "wa" in `vegan`).
- The residual values from the multiple regressions (i.e. $\mathbf{Y_{res}} = \mathbf{Y} - \hat{\mathbf{Y}}$) may also be submitted to a PCA to obtain an unconstrained ordination of the residuals. This partial PCA, which is not strictly speaking part of the RDA, is computed along with the constrained ordination by `vegan`'s `rda()` function.

Additional information on the algebra of RDA will be presented in the **Code it yourself** corner at the end of this section.

As can be seen from the conceptual steps presented above, RDA computes axes that are linear combinations of the explanatory variables. In other words, this method seeks, in successive order, a series of linear combinations of the explanatory variables that best explain the variation of the response data. The axes defined in the space of the explanatory variables are orthogonal to one another. RDA is therefore a constrained ordination procedure. The difference with unconstrained ordination is important: the matrix of explanatory variables conditions the "weights" (eigenvalues) and the directions of the ordination axes. In RDA, one can truly say that the axes *explain* or *model* (in the statistical sense) the variation of the dependent matrix. Furthermore, a global hypothesis (H_0) of absence of linear relationship between **Y** and **X** can be tested in RDA; this is not the case in PCA.

[1]For convenience, some programs and functions standardize the **X** variables at the beginning of the calculation. This does not change the RDA results because standardizing **X**, or not, does not change the matrix $\hat{\mathbf{Y}}$ of fitted values of the multivariate regression, upon which the RDA statistics, tests of significance, and PCA are computed.

An RDA produces min[p, m, $n-1$] canonical axes, where n is the number of objects, p is the number of response variables and m is the number of degrees of freedom of the model (rank of cov(**X**), which is $\leq$ number of numeric explanatory variables, including levels of factors if qualitative explanatory variables are included; a factor with k classes bears (k–1) degrees of freedom; if needed, it requires (k–1) dummy variables for coding). Each of the canonical axes is a linear combination (like a multiple regression model) of *all* explanatory variables. RDA may be computed, for convenience, using standardized explanatory variables; the fitted values of the regressions, as well as the canonical analysis results, are unchanged by standardization of the **X** variables. In **vegan**'s **rda()** function, the variation of the data matrix that cannot be explained by the environmental variables (i.e., the residuals of the regressions) is expressed by unconstrained PCA eigenvectors, which are given after the canonical eigenvectors.

For the reasons explained in Chap. 5 about PCA, an RDA can be computed on a covariance or a correlation response matrix. To obtain an analysis on the correlation response matrix, standardization of the response data is done by the option `scale = TRUE` in **vegan**'s **rda()**. This option should not be used with community composition data.

The statistical significance of an RDA (global model) and that of individual canonical axes can be tested by permutations. These tests will be introduced in due course.

6.3.2 *RDA of the Doubs River Data*

You will now explore various aspects of RDA. To achieve this, you will first prepare the data sets as usual, but also transform a variable and divide the explanatory variables into two subsets.

6.3.2.1 Preparation of the Data

```
# Load the required packages
library(ade4)
library(adegraphics)
library(adespatial)
library(vegan)
library(vegan3d)
library(MASS)
library(ellipse)
library(FactoMineR)
library(rrcov)
```

```
# Source additional functions that will be used later in this
# Chapter. Our scripts assume that files to be read are in
# the working directory.
source("hcoplot.R")
source("triplot.rda.R")
source("plot.lda.R")
source("polyvars.R")
source("screestick.R")

# Load the Doubs data. The file Doubs.Rdata is assumed to be in
# the working directory
load("Doubs.RData")
# Remove empty site 8
spe <- spe[-8, ]
env <- env[-8, ]
spa <- spa[-8, ]

# Set aside the variable 'dfs' (distance from the source) for
# later use
dfs <- env[, 1]

# Remove the 'dfs' variable from the env data frame
env2 <- env[, -1]
# Recode the slope variable (slo) into a factor (qualitative)
# variable to show how these are handled in the ordinations
slo2 <- rep(".very_steep", nrow(env))
slo2[env$slo <= quantile(env$slo)[4]] <- ".steep"
slo2[env$slo <= quantile(env$slo)[3]] <- ".moderate"
slo2[env$slo <= quantile(env$slo)[2]] <- ".low"
slo2 <- factor(slo2,
  levels = c(".low", ".moderate", ".steep", ".very_steep"))
table(slo2)
# Create an env3 data frame with slope as a qualitative variable
env3 <- env2
env3$slo <- slo2
# Create two subsets of explanatory variables
# Subset 1: Physiography (upstream-downstream gradient)
envtopo <- env2[, c(1 : 3)]
names(envtopo)
# Subset 2: Water quality
envchem <- env2[, c(4 : 10)]
names(envchem)

# Hellinger-transform the species data set
spe.hel <- decostand(spe, "hellinger")
```

> *Note that we have removed variable* `dfs` *from the environmental data frame and created a new object called* **env2**. *Apart from being a spatial rather than environmental variable,* `dfs`*, the distance from the source, is highly correlated with several other explanatory variables which are ecologically more explicit and therefore more interesting, like discharge, hardness and nitrogen content.*

6.3.2.2 RDA Using vegan

vegan allows the computation of an RDA in two different ways. The simplest syntax is to list the names of the data frames involved separated by commas:

```
simpleRDA <- rda(Y, X, W)
```

where **Y** is the response matrix, **X** is the matrix of explanatory variables, and **W** is an optional matrix of covariables (variables whose variation is to be controlled in a partial analysis, see Sect. 6.3.2.5).

This call, although simple, has some limitations. Its main drawback is that it does not allow qualitative variables of class "factor" to be included in the explanatory and covariable matrices. Therefore, in all but the simplest applications, it is better to use the formula interface:

```
formulaRDA <- rda(Y ~ var1 + factorA + var2*var3 + Condition(var4),
                  data = XWdata)
```

In this example, **Y** is the response matrix; the constraint includes a quantitative variable (var1), a factor (factorA), an interaction term between variables 2 and 3, whereas the effect of var4 is partialled out (so this is actually a partial RDA). The explanatory variables and the covariable are in object XWdata, which must have class data.frame.

This is the same kind of formula as used in **lm()** and other **R** functions devoted to regression. We will use it in the example below. For more information about this topic, consult the **rda()** documentation file.

Let us compute an RDA of the Hellinger-transformed fish species data, constrained by the environmental variables contained in **env3**, i.e. all environmental variables except `dfs`, and variable `slo` coded as a factor.

```
(spe.rda <- rda(spe.hel ~ ., env3))
summary(spe.rda)    # Scaling 2 (default)
```

Hints *Observe the shortcut (.) to tell the function to use all variables present in data frame env3, without having to enumerate them.*

If you don't want to see the site, species and constraint scores, add argument `axes = 0` *to the* **`summary()`** *call.*

Here the default choices are used, i.e. `scale = FALSE` *(RDA on a covariance matrix) and* `scaling = 2`.

Here is an excerpt of the output:

```
    Call:
rda(formula = spe.hel ~ ele + slo + dis + pH + har + pho + nit +
amm + oxy + bod, data = env3)

Partitioning of variance:
                  Inertia Proportion
Total              0.5025     1.0000
Constrained        0.3654     0.7271
Unconstrained      0.1371     0.2729

Eigenvalues, and their contribution to the variance

Importance of components:
                            RDA1     RDA2     RDA3     RDA4 ...
Eigenvalue                0.2281   0.0537 0.03212  0.02321 ...
Proportion Explained      0.4539   0.1069 0.06392  0.04618 ...
Cumulative Proportion     0.4539   0.5607 0.62466  0.67084 ...
                             PC1      PC2      PC3      PC4 ...
Eigenvalue               0.04581 0.02814  0.01528  0.01399 ...
Proportion Explained     0.09116 0.05601  0.03042  0.02784 ...
Cumulative Proportion    0.81825 0.87425  0.90467  0.93251 ...

Accumulated constrained eigenvalues
Importance of components:
                          RDA1   RDA2    RDA3    RDA4 ....
Eigenvalue              0.2281 0.0537 0.03212 0.02321 ...
Proportion Explained    0.6242 0.1470 0.08791 0.06351 ...
Cumulative Proportion   0.6242 0.7712 0.85913 0.92264 ...
```

```
Scaling 2 for species and site scores
* Species are scaled proportional to eigenvalues
* Sites are unscaled: weighted dispersion equal on all
  dimensions
* General scaling constant of scores:  1.93676

Species scores

          RDA1     RDA2      RDA3      RDA4      RDA5     RDA6
Cogo  0.13386  0.11619 -0.238205  0.018531  0.043161 -0.02973
Satr  0.64240  0.06654  0.123649  0.181606 -0.009584  0.02976
(...)
Anan -0.19440  0.14152  0.033624  0.017384  0.008122  0.01764

Site scores (weighted sums of species scores)

        RDA1      RDA2     RDA3      RDA4      RDA5      RDA6
1    0.40149 -0.154133  0.55506  1.601005  0.193044  0.916850
2    0.53522 -0.025131  0.43393  0.294832 -0.518997  0.458849
(...)
30 -0.48931  0.321574  0.31409  0.278210  0.488026 -0.150951

Site constraints (linear combinations of constraining variables)

        RDA1      RDA2     RDA3      RDA4      RDA5     RDA6
1    0.55130  0.002681  0.47744  0.626961 -0.210684  0.31503
2    0.29736  0.105880  0.64854  0.261364 -0.057127  0.09312
(...)
30 -0.42566  0.338206  0.24941  0.345838  0.404633 -0.13761

Biplot scores for constraining variables

                  RDA1      RDA2     RDA3     RDA4      RDA5
ele             0.8239 -0.203027  0.46599 -0.16932  0.003151
slo.moderate   -0.3592 -0.008707 -0.21727 -0.18287  0.158087
(...)
bod            -0.5171 -0.791730 -0.15644  0.22067  0.075935 -

Centroids for factor constraints

                  RDA1      RDA2     RDA3     RDA4     RDA5
slo.low        -0.2800  0.005549 -0.09029  0.07610 -0.07882
(...)
slo.very_steep 0.3908 -0.094698  0.28941  0.02321 -0.12175
```

As in the case of PCA, this output requires some explanations. Some of the results are similar to those of a PCA, but additional elements are provided.

- **Partitioning of variance**: the overall variance is partitioned into constrained and unconstrained fractions. The constrained fraction is the amount of variance of the **Y** matrix explained by the explanatory variables. Expressed as a proportion, it is equivalent to an R^2 in multiple regression; in RDA this quantity is also called the bimultivariate redundancy statistic. However, **this R^2 is biased**, like the unadjusted R^2 of multiple regression, as shown by Peres-Neto et al. (2006). We will present the computation of an adjusted, unbiased R^2 below.
- **Eigenvalues and their contribution to the variance**: this analysis yielded 12 canonical axes (with eigenvalues labelled RDA1 to RDA12) and 16 additional, unconstrained axes for the residuals (with eigenvalues labelled PC1 to PC16). The results give the eigenvalues themselves, as well as the cumulative proportion of variance explained (for the RDA axes) or represented (for the residual axes). The last cumulative value is therefore 1. The cumulative contribution to the variance obtained by the 12 canonical axes is the (biased) proportion of the total variance of the response data explained by the RDA. It is the same value as the "Proportion constrained" presented above; it is 0.7271 in this example.
- One feature of the eigenvalues is worth mentioning. Observe that the canonical eigenvalues RDA1 to RDA12 are (of course) decreasing in values; however, the first residual eigenvalue (PC1) is *larger* than the last canonical eigenvalue (in this example it is actually larger than most RDA eigenvalues). This means that the first residual structure (axis) of the data has more variance than some of the structures that can be explained by the explanatory variables in **X**. It is up to the user to exploit this information, for example by plotting the first pair of residual axes and devising hypotheses about the causes of the features revealed. These causes should not, however, involve variables that have already been used in the model, except if one suspects that products (interactions) or higher-order combinations (e.g. squared variables) may be required.
- An important distinction must be made: the canonical (RDAx) eigenvalues measure amounts of variance **explained** by the RDA model, whereas the residual (PCx) eigenvalues measures amounts of variance **represented** by the residual axes, but not explained by any model.
- **Accumulated constrained eigenvalues**: these are cumulative amounts of variance expressed as proportions of the total **explained** variance, as opposed to their contributions to the *total* variance described above.
- **Species scores** are the coordinates of the tips of the vectors representing the response variables in the bi- or triplots. As in PCA, they depend on the scaling chosen: scaling 1 or scaling 2.
- **Site scores (weighted sums of species scores)**: coordinates of the sites as expressed in the space of the response variables **Y**.
- **Site constraints (linear combinations of constraining variables)**: coordinates of the sites in the space of the explanatory variables **X**. These are the fitted site scores.

- **Biplot scores for constraining variables**: coordinates of the tips of the vectors representing the explanatory variables. These coordinates are obtained by computing correlations between the explanatory variables and the fitted site scores (Legendre and Legendre 2012 Eq. 11.21); in scaling 2 these correlations are the biplot scores; in scaling 1 these correlations are transformed to produce the biplot scores (Legendre and Legendre 2012 Eq. 11.20). All variables, including $k - 1$ levels of factors with k levels, are represented in this table. For factors, however, a representation of the centroids of the levels is preferable. See below.
- **Centroids for factor constraints**: coordinates of centroids of levels of factor variables, i.e., means of the scores of the sites possessing state "1" for a given level.

In the **`rda()`** output, an interesting information is missing: the canonical coefficients, i.e., the equivalent of regression coefficients for each explanatory variable on each canonical axis. These coefficients (Legendre and Legendre 2012 Eq. 11.19) can be retrieved by typing **`coef()`**:

```
# Canonical coefficients from the rda object
coef(spe.rda)
```

Hint *Type* **`?coef.cca`** *and see how to obtain fitted and residual values. There is also a* **`calibrate()`** *function allowing the projection of new sites into a canonical ordination result for bioindication purposes, although with some conditions. See Sect. 6.3.2.7.*

6.3.2.3 Retrieving, Interpreting and Plotting Results From a **`vegan`** RDA Output Object

The various elements making up the **`rda()`** output object can be retrieved in the same way as for a PCA. This is useful when you want to use the results outside the functions provided by **`vegan`** to handle them.

As mentioned above, the **R^2 of a RDA is biased** like the ordinary R^2 of multiple regression, and for the same reason (Peres-Neto et al. 2006). On the one hand, any variable included in an explanatory matrix **X** increases the R^2, irrespective of it being related, or not, to the response data. On the other hand, the accumulation of explanatory variables inflates the apparent amount of explained variance because of random correlations. This problem can be cured by **adjusting the R^2** using Ezekiel's formula (Ezekiel 1930), which is also valid in the multivariate case:

$$R^2_{adj} = 1 - \frac{n-1}{n-m-1}\left(1 - R^2\right) \tag{6.1}$$

where n is the number of objects and m is the number of degrees of freedom of the model (i.e., the rank of the explanatory matrix, which is in many cases the number of quantitative explanatory variables plus, if present, the degrees of freedom associated with each factor: $k - 1$ d. f. for a factor with k levels). Ezekiel's adjustment can be used as long as the number of degrees of freedom of the model is not overly large with respect to the number of observations. As a rule of thumb, this adjustment may be overly conservative when $m > n/2$. An adjusted $\boldsymbol{R}^2$ near 0 indicates that **X** does not explain more of the variation of **Y** than random normal deviates would do. Adjusted $\boldsymbol{R}^2$ values can be negative, indicating that the explanatory variables **X** do worse than a set of m random normal deviates would.

In our example, $n = 29$ and $m = 12$ (remember that one of the 10 variables is a factor with $k = 4$ levels, so it takes up 3 degrees of freedom). The R^2 and adjusted R^2 can be computed using **vegan**'s function **RsquareAdj()**.

```
# Unadjusted R^2 retrieved from the rda object
(R2 <- RsquareAdj(spe.rda)$r.squared)
# Adjusted R^2 retrieved from the rda object
(R2adj <- RsquareAdj(spe.rda)$adj.r.squared)
```

As one can see, the adjustment has substantially reduced the value of the R^2. The adjusted R^2 measures the unbiased amount of explained variation and will be used later for variation partitioning.

Let us now plot the results of our RDA, using the fitted ("lc") site scores for the objects (Fig. 6.1). We can call this a triplot since there are three different entities in the plot: sites, response variables and explanatory variables. To differentiate the latter two, we will draw arrowheads only on the vectors of the quantitative explanatory variables, not on the response variable vectors.

```
# Scaling 1
plot(spe.rda,
  scaling = 1,
  display = c("sp", "lc", "cn"),
  main = "Triplot RDA spe.hel ~ env3 - scaling 1 - lc scores"
)
```

This plot displays all our entities: sites, species, explanatory variables as arrows (with heads), or centroids, depending on their types. We represent the species by lines without arrowheads to make them appear different from the explanatory variables, after retrieval from the output object:

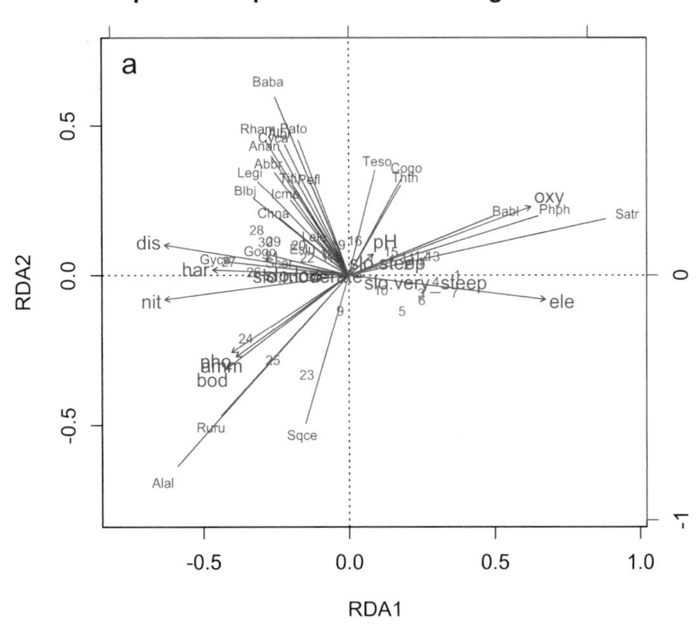

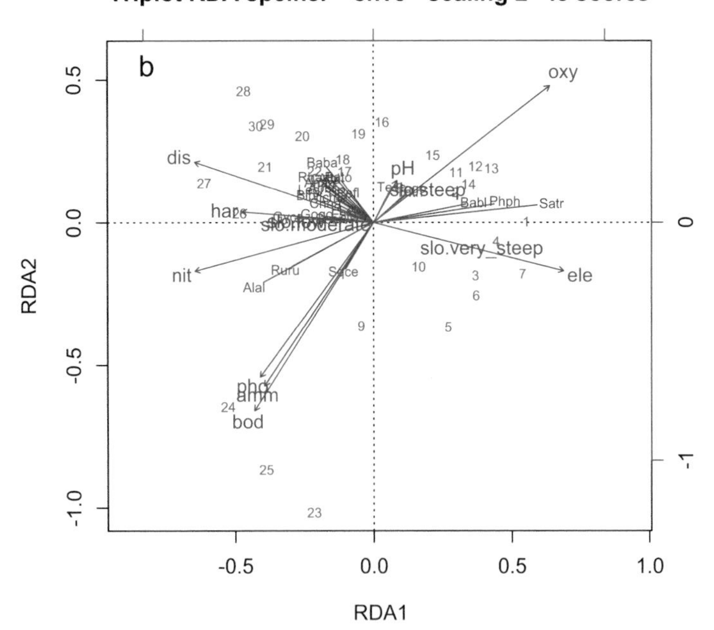

Fig. 6.1 RDA triplots of the Hellinger-transformed Doubs fish abundance data constrained by all environmental variables except `dfs`. a: scaling 1; b: scaling 2. Fitted site scores. The bottom and left-hand scales are for the objects and the response variables, the top and right-hand scales are for the explanatory variables.

```
spe.sc1 <-
  scores(spe.rda,
         choices = 1:2,
         scaling = 1,
         display = "sp"
)
arrows(0, 0,
  spe.sc1[, 1] * 0.92,
  spe.sc1[, 2] * 0.92,
  length = 0,
  lty = 1,
  col = "red"
)

# Scaling 2
plot(spe.rda,
  display = c("sp", "lc", "cn"),
  main = "Triplot RDA spe.hel ~ env3 - scaling 2 - lc scores"
)
spe.sc2 <-
  scores(spe.rda,
         choices = 1:2,
         display = "sp"
)
arrows(0, 0,
  spe.sc2[, 1] * 0.92,
  spe.sc2[, 2] * 0.92,
  length = 0,
  lty = 1,
  col = "red"
)
```

Hints *In the* **`scores()`** *function argument,* `choices=` *indicates which axes are to be selected. Also, be careful to specify the scaling if it is different from 2.*

The **`plot()`** *function used above has an argument* `type` *that determines how the sites must be displayed. For small data sets (as here), the default is* `type="text"` *and so the site labels are printed, but for large data sets it switches to* `"points"`*. If you want to force site labels instead of points with a large data set, write* `type="text"`*, but be prepared to see a crowded plot.*

See how to choose the elements to be plotted, using the argument `display=c(...)`*. In this argument,* `"sp"` *stands for* ***sp****ecies,* `"lc"` *for fitted site scores* (**l**inear **c**ombinations of explanatory variables), `"wa"` *for site scores in the species space* (**w**eighted **a**verages in CCA or weighted sums in RDA), *and* `"cn"` *for* ***con****straints (i.e., the explanatory variables).*

In the **`arrows()`** *call, the scores are multiplied by 0.92 so that the arrows do not cover the names of the variables. Adjust this factor by trial and error.*

The two RDA triplots use the fitted site scores (called '`lc`' in **`vegan`**). The choice between these and the site scores that are weighted sums of species scores (called '`wa`' in **`vegan`** because in CA/CCA they are weighted averages) for the triplots depends on the context and the purpose of the plot. On the one hand, the fitted site scores are strictly orthogonal linear combinations of the explanatory variables; they represent clearly and exclusively what can be modelled using the explanatory variables at hand. We advocate the use of these lc scores in most situations because the "true" ordination diagram of RDA is the ordination of the $\hat{\mathbf{Y}}$ matrix of fitted values. On the other hand, the site scores that are weighted sums of species appear more robust to noise in the environmental variables: McCune (1997) showed that if the latter contain much random error, the resulting lc plots may be completely scrambled. However, weighted sums of species ("wa") scores are "contaminated" scores, halfway between the model fitted by the RDA procedure and a PCA of the original data, and as such it is not clear how they should be interpreted.

The weighted sums of species ("wa") triplot is preferable in one case: when RDA is used as a form of analysis of variance (Sect. 6.3.2.9), because in that situation all replicate sites with the same combination of factor levels are represented on top of one another in the fitted site scores (lc) triplot. In such an analysis, one can even plot *both* types of scores, with the lc scores showing the centroid of the sites sharing a given combination of factor levels and the wa scores showing their dispersion.

Of course, the triplots can also be drawn using the wa scores. In the **`plot()`** function, replace argument `display = "lc"` by `"wa"`.

Independently of the choice of site scores, the interpretation of the constrained triplots must be preceded by a test of statistical significance of the global canonical relationship; see below. As in multiple regression, a nonsignificant result must not be plotted and interpreted; it must be discarded.

For the species and sites, the interpretation of the two scalings is the same as in PCA. However, the presence of vectors and centroids of explanatory variables calls for additional interpretation rules. Here are the essential ones (see Legendre and Legendre 2012 p. 640–641):

- *Scaling 1* – distance triplot: (1) The angles **between response and explanatory variables** in the triplot reflect their correlations (*but **not** the angles among the response variables*). (2) The relationship between the **centroid** of a qualitative explanatory variable and a response variable (species) is found by projecting the centroid at right angle on the response variable, as for individual objects, since we are projecting the centroid of a group of objects. (3) Distances among **centroids**, and between **centroids** and individual objects, approximate their Euclidean distances.
- *Scaling 2* – correlation triplot: (1) Projecting an object at right angle on a response or a **quantitative explanatory variable** approximates the value of the object along that variable. (2) The angles in the triplot between response and **explanatory variables**, and between response variables themselves or **explanatory variables themselves**, reflect their correlations. (3) The relationship between the **centroid** of a qualitative explanatory variable and a response variable

(species) is found by projecting the centroid at right angle on the response variable (as for individual objects). (4) *Distances among* ***centroids****, and between centroids and individual objects, do not approximate their Euclidean distances.*

On these bases, it is now possible to interpret the triplots. Let us take as examples the pair of plots representing the fitted site scores (Fig. 6.1). The numerical output shows that the first two canonical axes explain together 56.1% of the total variance of the response data, the first axis alone explaining 45.4%. These are unadjusted values, however. Since the adjusted R^2 of the RDA[2] is $R^2_{\text{adj}} = 0.5224$, the proportions of accumulated **constrained** eigenvalues (i.e., proportions with respect to the *explained* variance, not the total variance) show that the first axis alone explains $0.5224 \times 0.6242 = 0.326$ or 32.6% variance, and the first two axes together $0.5224 \times 0.7712 = 0.4029$ or 40.3% variance. We can be confident that the major trends have been well modelled in this analysis. Because ecological data are generally quite noisy, one should not expect very high values of R^2_{adj}. Furthermore, the first unconstrained eigenvalue (PC1) is comparatively small (0.04581), compared to the first constrained eigenvalue (0.2281), which means that it does not display any dominant residual structure of the response data.

These triplots show that oxygen (`oxy`), elevation (`ele`), nitrates (`nit`) and discharge (`dis`), as well as slope (mainly the level `slo.very_steep`) play an important role in the dispersion of the sites along the first axis. Both triplots oppose the upper and lower parts of the river along the first axis. The scaling 2 triplot shows three groups of fish species correlated with different sets of explanatory variables: the brown trout (`Satr`), Eurasian minnow (`Phph`) and stone loach (`Babl`) are found in the first half of the sites, and are correlated with high oxygen content and slope as well as high elevation. The bleak (`Alal`), roach (`Ruru`) and European chub (`Sqce`), on the opposite, are related to sites 23, 24 and 25 characterized by high phosphates (`pho`), ammonium (`amm`) and biological oxygen demand (`bod`) levels. Most other species are bunched together away from these extremes. They show mostly shorter projections, indicating that they are either present over most portions of the river or related to intermediate ecological conditions.

To avoid the need to manually retrieve the biplot scores and ask for the arrows, we provide a homemade function called **`triplot.rda()`** to produce an RDA triplot in one command. The function offers many choices: lc or wa scores, scaling 1 or 2, choice of a subgroup of variables (species) as well as several graphical parameters. In the example below, we propose to draw only the species that reach a (cumulative) goodness-of-fit of at least 0.6 (an arbitrary value) in the ordination plane formed by axes 1 and 2. The higher the goodness-of-fit, the better the species is fitted on the corresponding axis. Goodness-of-fit can be retrieved from an **`rda()`** object by function **`goodness()`** of package **`vegan`**.

The function works as follows. Scaling 1: Fig. 6.2a; scaling 2: Fig. 6.2b.

[2]Obtained with `RsquareAdj(spe.rda)`.

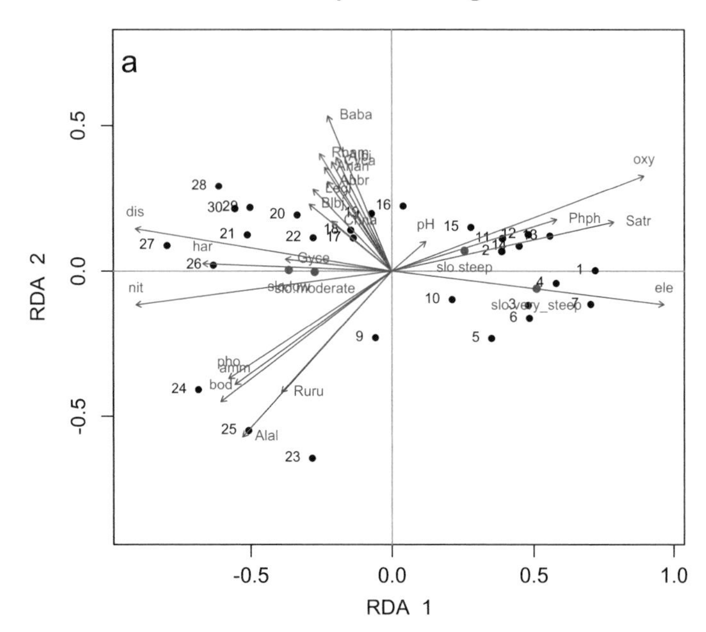

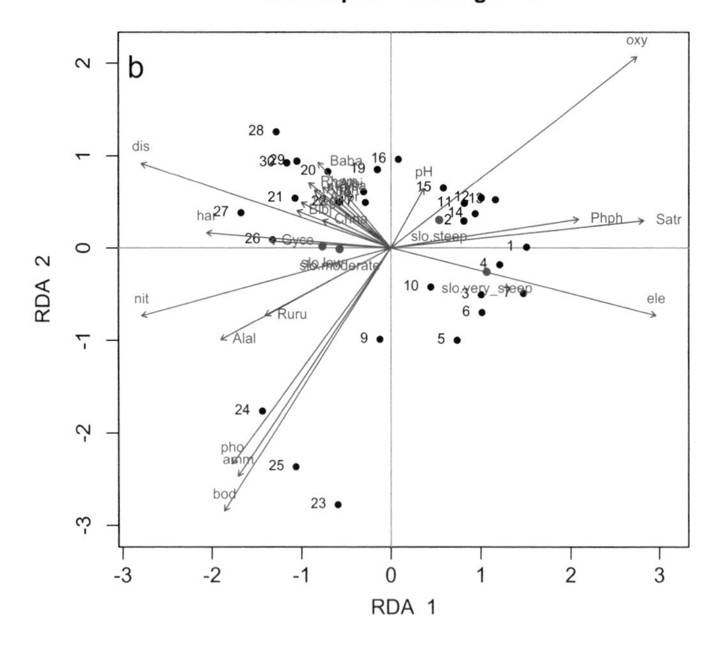

Fig. 6.2 Triplots of the RDA drawn with the **`triplot.rda()`** function with fitted site scores, a subset of species and all environmental variables except `dfs`. (**a**) scaling 1; (**b**) scaling 2

```
# Select species with goodness-of-fit at least 0.6 in the
# ordination plane formed by axes 1 and 2
spe.good <- goodness(spe.rda)
sel.sp <- which(spe.good[, 2] >= 0.6)
# Triplots with homemade function triplot.rda(), scalings 1 and 2
triplot.rda(spe.rda,
  site.sc = "lc",
  scaling = 1,
  cex.char2 = 0.7,
  pos.env = 3,
  pos.centr = 1,
  mult.arrow = 1.1,
  mar.percent = 0.05,
  select.spe = sel.sp
)
triplot.rda(spe.rda,
  site.sc = "lc",
  scaling = 2,
  cex.char2 = 0.7,
  pos.env = 3,
  pos.centr = 1,
  mult.arrow = 1.1,
  mar.percent = 0.05,
  select.spe = sel.sp
)
```

Hint *Argument* `mar.percent` *can receive a negative value, e.g.* `-0.1`*, which shrinks the margins of the graph. In some graphs, the arrows (species and/or environmental variables) must be shortened to fit the size of the cluster of site points. Shrinking the margin compensates for that, allowing all elements of the graph to occupy the whole plotting area.*

> *Display the documentation in the upper part of the file of function* **`triplot.rda()`** *and explore the various possibilities to improve the look of your triplots.*

6.3.2.4 Permutation Tests of RDA Results

Due to widespread problems of non-normal distributions in ecological data, classical parametric tests are often not appropriate in this field. This is why most methods of ecological data analysis nowadays resort to permutation tests whenever possible. In RDA, the use of parametric tests is possible only when the response variables are standardized and the error distribution is normal (Miller 1975, Legendre et al. 2011); this is clearly not the case for community composition data, for example. So all RDA programs for ecologists implement permutation tests. The principle of a permutation test is to generate a reference distribution of the chosen statistic under the null hypothesis H_0 by randomly permuting appropriate elements of the data a large number of times and recomputing the statistic each time. Then one compares the true value of the statistic to this reference distribution. The *p*-value is computed as the proportion of the permuted values equal to or larger than the true (unpermuted) value of the statistic for a one-tailed test in the upper tail, like the *F*-test used in RDA. The true value is included in this count. The null hypothesis is rejected if this *p*-value is equal to or smaller than the predefined significance level α.

Three elements are critical in the construction of a permutation test: (1) the choice of the permutable units, (2) the choice of the statistic, and (3) the permutation scheme.

The **permutable units** are often the response data (random permutation of the rows of the **Y** data matrix), but sometimes other permutable units must be defined. In partial canonical analysis (Sect. 6.3.2.5), for example, it is the residuals of some regression model that are permuted. See Legendre and Legendre (2012, pp. 651 *et sq*. and especially Table 11.6 p. 653). In the simple case presented here, the null hypothesis H_0 states (loosely) that no (linear) relationship exists between the response data **Y** and the explanatory variables **X**. In terms of permutation tests, this means that the sites in matrix **Y** can be permuted randomly to produce realizations of this H_0, thereby destroying the possible relationship between a given fish assemblage and the ecological condition of its site. A permutation test repeats the calculation with permuted datas 100, 1000 or 10,000 times (including the true, unpermuted value) to produce a large sample of test statistics to which the true value is compared.

The **test statistic** (often called *pseudo-F*) is defined as follows:

$$F = \frac{\mathrm{SS}\left(\widehat{\mathbf{Y}}/m\right)}{\mathrm{RSS}/(n-m-1)} \tag{6.2}$$

where m is the number of canonical eigenvalues (or degrees of freedom of the model), $\mathrm{SS}(\hat{\mathbf{Y}})$ (explained variation) is the sum-of-squares of the table of fitted values, and RSS (residual sum of squares) is the total sum-of-squares of **Y**, SS(**Y**), minus the explained variation $\mathrm{SS}(\hat{\mathbf{Y}})$.

The test of significance of individual axes is based on the same principle. The first canonical eigenvalue is tested as follows: that eigenvalue is the numerator of the *F*-statistic, whereas the SS(**Y**) minus that eigenvalue divided by $(n-1\ -\ 1)$ is the denominator (since $m\ =\ 1$ eigenvalue in this case). The test of the subsequent canonical axes is more complicated: the previously tested canonical axes have to be included as covariables in the analysis, as in Sect. 6.3.2.5, and the RDA is recomputed; see Legendre et al. (2011) for details.

The **permutation scheme** describes how the units are permuted. In most cases the permutations are free, i.e. all units are considered equivalent and fully exchangeable, and the data rows of **Y** are permuted. However, in some situations permutations may be restricted within subgroups of data, for instance when a multistate covariable or a multi-level experimental factor is included in the analysis.

With this in mind, we can now test our RDA results. Let us run a global test first, followed by a test of the canonical axes. The test function is called **anova()** because it relies on a ratio between two measures of variation. Do not confuse it with the classical analysis-of-variance (ANOVA) test.

```
## Global test of the RDA result
anova(spe.rda, permutations = how(nperm = 999))
## Tests of all canonical axes
anova(spe.rda, by = "axis", permutations = how(nperm = 999))
```

The test of the axes can only be run with the formula interface. How many canonical axes are significant at the α = 0.05 level?

Of course, given that these tests are available, it is useless to apply other criteria like the broken-stick or Kaiser-Guttman's criterion to determine the interpretability of canonical axes. These could be applied to the residual, unconstrained axes, however. Let us apply the (generally quite liberal) Kaiser-Guttman criterion:

```
# Apply Kaiser-Guttman criterion to residual axes
spe.rda$CA$eig[spe.rda$CA$eig > mean(spe.rda$CA$eig)]
```

There may still some interesting variation in these data that has not been explained by our set of environmental variables.

6.3.2.5 Partial RDA

Partial canonical ordination is the multivariate equivalent of partial linear regression. For example, it is possible to run an RDA of a (transformed) plant species data matrix **Y**, explained by a matrix of climatic variables **X**, in the presence of soil covariables **W**. Such an analysis would allow the user to display the patterns of the species data uniquely explained by a linear model of the climatic variables when the effect of the soil constraints is held constant.

We will now run an example using the Doubs data. At the beginning of this chapter, we created two objects containing subsets of environmental variables. One subset contains physiographic variables (`envtopo`), i.e. elevation, slope (the original, quantitative variable, not the 4-level factor used in the previous RDA) and discharge; the other contains variables describing water chemistry (`envchem`), i.e. pH, hardness, phosphates, nitrates, ammonium, oxygen content as well as biological oxygen demand. The analysis that follows will determine whether water chemistry significantly explains the fish species patterns when the effect of the topographic gradient is held constant.

In **vegan**, partial RDA can be run in two ways, depending on whether one uses the simple syntax or the formula interface. In the first case, explanatory variables and covariables may or may not be assembled in separate objects, which can be vectors, matrices or data frames; factor variables cannot be used with that notation, however. With the formula interface, the **X** and **W** variables must be in the same object, which must be a data frame and may contain factor variables.

```
# Simple syntax; X and W may be in separate tables of quantitative
# variables
(spechem.physio <- rda(spe.hel, envchem, envtopo))
summary(spechem.physio)
# Formula interface; the X and W variables must be in the same
# data frame
(spechem.physio2 <-
  rda(spe.hel ~ pH + har + pho + nit + amm + oxy + bod
        + Condition(ele + slo + dis), data = env2))
```

Hint *The formula interface may seem cumbersome, but it allows a better control of the model and the use of factors and interactions among the constraints* (**X**) *or the conditions* (**W**). *For example, one could have used the factor-transformed* `slo` *variable in the second analysis, but not in the first one.*

The results of the two analyses are identical.

Here again, some additional explanations are needed about the summary output.

- **Partitioning of variance**: This item now shows four components. The first one (`Total`) is, as usual, the total inertia (variance in this case) of the response data. The second line (`Conditioned`) gives the amount of variance that has been explained by the covariables and removed[3]. The third line (`Constrained`) gives the amount of variance uniquely explained by the explanatory variables. The fourth line (`Unconstrained`) gives the residual variance. **Beware**: the values given as proportions (right-hand column) are **unadjusted** and are therefore biased. For a proper computation of unbiased, adjusted R^2 and partial R^2, see below (variation partitioning).
- **Eigenvalues, and their contribution to the variance after removing the contributions of conditioning variables**: these values and proportions are *partial* in the sense that the effects of the covariables have been removed. The sum of all these eigenvalues corresponds therefore to the sum of the constrained and

[3]Mathematically, to partial out the effect of a matrix **W** from a canonical ordination of **Y** by **X**, one computes the residuals of a multivariate multiple regression of **X** on **W** and uses these residuals as the explanatory variables.

unconstrained (residual) variances, excluding the conditioned (i.e., removed) variance.

We can now test the partial RDA and, if it is significant, draw triplots of the first pair of axes (Fig. 6.3).

```
# Test of the partial RDA, using the results with the formula
# interface to allow the tests of the axes to be run
anova(spechem.physio2, permutations = how(nperm = 999))
anova(spechem.physio2,
      permutations = how(nperm = 999),
      by = "axis"
)

# Partial RDA triplots (with fitted site scores)
# Scaling 1
triplot.rda(spechem.physio,
  site.sc = "lc",
  scaling = 1,
  cex.char2 = 0.8,
  pos.env = 3,
  mar.percent = 0
)
# Scaling 2
triplot.rda(spechem.physio,
  site.sc = "lc",
  scaling = 2,
  cex.char2 = 0.8,
  pos.env = 3,
  mult.spe = 1.1,
  mar.percent = 0.04
)
```

Hint *See the hints in Sect. 6.3.3 about the two ways of coding formulas containing covariables.*

As could be expected, the results of this partial analysis differ somewhat from the previous results, but not fundamentally. Although there are interesting features to discuss about the amounts of variance explained, we will postpone this aspect until we have seen a quick and elegant manner to compute the adjusted R^2 of partial analyses, by means of variation partitioning (Sect. 6.3.2.8). On the triplots, the explanatory variables show the same relationships to one another, but some of them [hardness (`har`) and nitrates (`nit`)] are less important to explain the fish community structure, as shown by their shorter vectors. This may be due to the fact that these two variables are well correlated with the positions of the sites along the

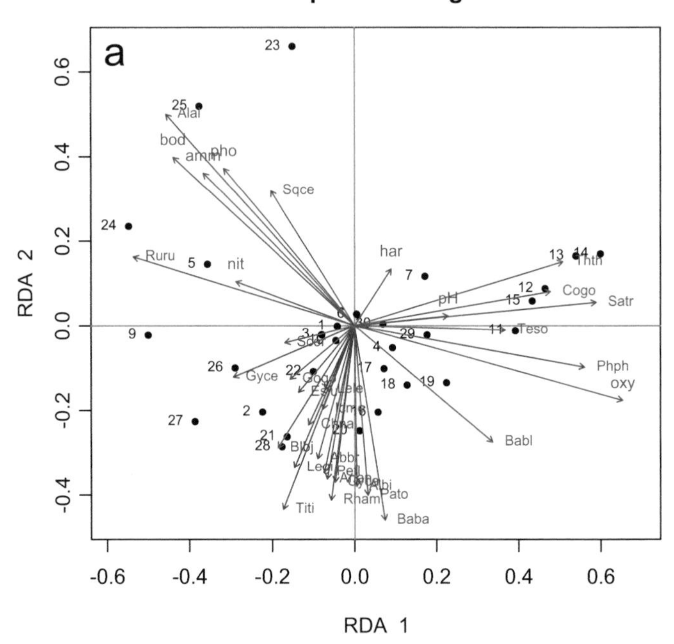

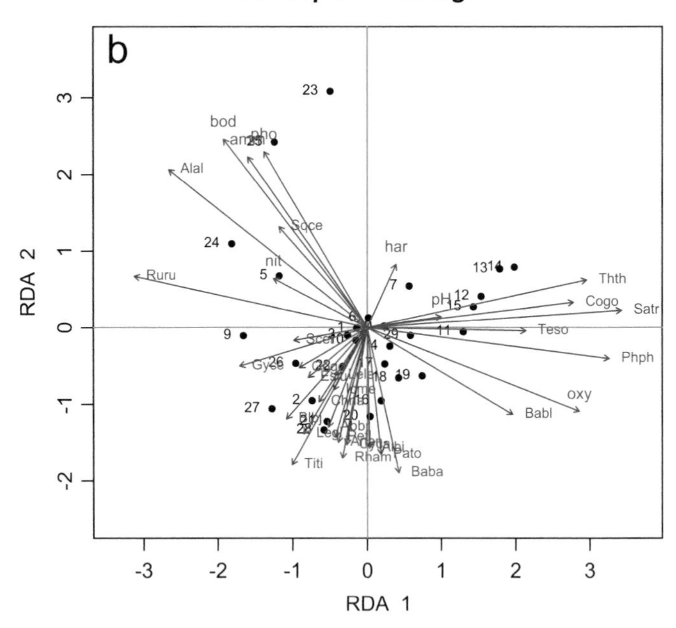

Fig. 6.3 RDA triplots, Doubs Hellinger-transformed fish data explained by chemistry, controlling for physiography. Fitted site scores. a: scaling 1; b: scaling 2

river, and therefore their apparent effect on the fish community may have been spurious and has been "removed" by the analysis, which controlled for the effect of the physiographic variables. The scaling 1 triplot shows that the sites are not as cleanly ordered by their succession along the river. This indicates that the chemical variables that are important for the fishes do not necessarily follow that order and that the fish community responds significantly to these chemical constraints irrespective of their locations along the river.

6.3.2.6 Selection of Explanatory Variables

It happens sometimes that one wishes to reduce the number of explanatory variables. The reasons vary: search for parsimony, rich data set but poor *a priori* hypotheses, very small *n*, or a method producing a large set of explanatory variables which must be reduced afterwards (as in eigenvector-based spatial analysis, see Chap. 7). In the Doubs data, there could be two reasons (albeit not compelling) to reduce the number of explanatory variables: search for parsimony, and possible strong linear dependencies (correlations) among the explanatory variables in the RDA model, which could render the regression coefficients of the explanatory variables in the model unstable.

Linear dependencies can be explored by computing the X variables' variance inflation factors (VIF), which measure to what extent each variable in a data set X is collinear with the others. High VIFs are found when two variables are highly intercorrelated or when one variable is a strong linear combination of several others. VIF values above 20 indicate strong collinearity. Ideally, VIFs above 10 should be at least examined, and avoided if possible. High VIFs may indicate variables that are functionally related to one another. In that case, one can remove a variable from such a group on the basis of ecological reasoning. For example, if two variables measure the same basic environmental property but in different manners, e.g. total N vs NO_3^-, one of them can be removed without harm.

Contrary to what one sometimes reads, variables X with high VIFs should generally not be manually removed before the application of a procedure of selection of variables. Indeed, two highly correlated variables that are both strong predictors of one or several of the response variables Y may both contribute significantly, in complementary manners, to the linear model of these response variables. A variable selection procedure is the appropriate way of determining if that is the case.

For a matrix **X** containing quantitative variables only, VIF indices can be easily computed with the following R code:

```
vif <- diag(solve(cor(X)))
```

VIFs can also be computed in **vegan** after RDA or CCA. The algorithm in the **vif.cca()** function allows users to include factors in the RDA; the function will compute VIF after breaking down each factor into dummy variables. If **X** contains quantitative variables only, the **vif.cca()** function produces the same results as the equation above. Here is an example where VIFs are computed after RDA for a matrix **X** containing quantitative variables plus a factor (the variable "slo"). Several VIF values are above 10 or even 20 in these analyses, so that a reduction of the number of explanatory variables is justified.

```
# Variance inflation factors (VIF) in two RDAs
# First RDA of this Chapter: all environmental variables
# except dfs
vif.cca(spe.rda)
# Partial RDA – physiographic variables only
vif.cca(spechem.physio)
```

No single, perfect method exists to reduce the number of variables, besides the examination of all possible subsets of explanatory variables, which is time-prohibitive for real data sets. In multiple regression, the three usual methods are forward selection, backward elimination and stepwise selection of explanatory variables, the latter being a combination of the first two. In RDA, forward selection is the method most often applied because it works even in cases where the number of explanatory variables is larger than $(n - 1)$. This method works as follows:

- Compute m RDAs of the response data with one of the m explanatory variables in turn.
- Select the "best" explanatory variable on the basis of criteria that are developed below. If it is significant…
- … the next task is to look for a 2nd (3rd, 4th, etc.) variable to include in the explanatory model. Compute all models containing the previously selected variable(s) plus one of the remaining explanatory variables. Identify the "best" new variable; include it in the model if its partial contribution is significant at the pre-selected significance level.
- The process continues until no more significant variable can enter the model.

Several criteria exist to decide when to stop variable selection. The traditional one is a pre-selected significance level α (selection is stopped when no additional variable has a partial permutational p-value smaller than or equal to α). However, this criterion is known to be overly liberal, either by sometimes selecting a "significant" model when none should have been identified (hence inflating type I error), or by including too many explanatory variables into the model (hence inflating the amount of explained variance). Blanchet et al. (2008a) addressed this double problem and proposed solutions to improve this technique:

- To prevent the problem of inflation of the overall type I error, a global test using all explanatory variables is first run. If, and only if, that test is significant, the forward selection is performed;
- To reduce the risk of incorporating too many variables into the model, the adjusted coefficient of multiple determination R^2_{adj} of the global model (containing all the potential explanatory variables) is computed, and used as a second stopping criterion. Forward selection is stopped if one of the following criteria is reached: the α significance level or the global R^2_{adj}; in other words, if a candidate variable is deemed nonsignificant or if it brings the R^2_{adj} of the current model over the value of the R^2_{adj} of the global model.

Another criterion applied in model selection is Akaike's information criterion (AIC). In multivariate applications like RDA, an AIC-like criterion can be computed, but, according to Oksanen (2017, documentation file of function **ordistep()** in **vegan**) it may not be completely trustworthy. Furthermore, experience shows that it tends to be very liberal.

Three functions are mostly used in ecology for variable selection in the RDA context: **forward.sel()** of package **adespatial** and **ordistep()** and its offspring **ordiR2step()** of package **vegan**. Let us explore these functions by presenting them and applying them each in turn.

Forward selection with function forward.sel()

Function **forward.sel()** requires a response data matrix and an explanatory data matrix which, unfortunately, must contain quantitative variables only. Factors must be recoded in the form of dummy variables.

To identify the "best" explanatory variables in turn, and decide when to stop the selection, **forward.sel()** finds the explanatory variable with the highest R^2 (first variable) or partial (additional) R^2 (for the following variables). The decision to retain a variable or to stop the procedure can be made on several criteria: a pre-selected significance level α (argument `alpha`), and the adjusted coefficient of multiple determination R^2_{adj} of the global model containing all the potential explanatory variables (argument `adjR2thresh`). Note that **forward.sel()** allows for additional, more arbitrary stopping criteria such as a maximum number of variables entered, a maximum R^2 and a minimum additional contribution to the R^2. These criteria are rarely used.

Let us apply **forward.sel()** to our fish and environmental data. Since this procedure does not allow for factor variables, we will use the `env2` data set, which contains only quantitative variables, in the analysis.

```
# RDA with all explanatory variables except dfs
spe.rda.all <- rda(spe.hel ~ ., data = env2)
# Global adjusted R^2
(R2a.all <- RsquareAdj(spe.rda.all)$adj.r.squared)
# Forward selection using forward.sel() {adespatial}
forward.sel(spe.hel, env2, adjR2thresh = R2a.all)
```

The result shows that one can devise an explanatory model of the fish community that is much more parsimonious than the model involving all explanatory variables. Three variables (`ele`, `oxy` and `bod`) suffice to obtain an R^2_{adj} close (0.5401) to that of the global model (0.5864).

Forward selection with function `ordistep()`

The working of the **`ordistep()`** function of **`vegan`** is inspired from the **`step()`** function of **`stats`** for model selection. **`ordistep()`** accepts quantitative as well as factor explanatory variables, allows forward, backward and stepwise (a combination) selection. It can be applied to RDA, CCA (Sect. 6.4) and db-RDA (Sect. 6.3.3). In this function, one first provides an "empty" model, i.e., a model with intercept only, and a "scope" model containing all candidate variables. In forward selection, variables are added in order of decreasing *F*-values, each addition being tested by permutation, and stops when the permutational probability exceeds the predefined α significance level (here argument `Pin`). In case of equality of the *F*-statistic of two variables during selection, the variable that has the lowest value of the Akaike Information Criterion (AIC) is selected for inclusion in the model. **`ordistep()`** also allows backward selection (variables are removed from the complete model until only the significant ones remain) and a "stepwise" selection (argument `direction = "both"`) where variables enter the model if their permutational probability is smaller than or equal to the predefined `Pin` value, and removed if, within the model under construction, their probability (which changes after each inclusion of a new variable), becomes larger than `Pout`.

```
# Forward selection using vegan's ordistep()
# This function allows the use of factors.
mod0 <- rda(spe.hel ~ 1, data = env2)
step.forward <-
  ordistep(mod0,
           scope = formula(spe.rda.all),
           direction = "forward",
           permutations = how(nperm = 499)
)
RsquareAdj(step.forward)
```

Hint *If you don't want the screen display of the intermediate results of* **ordistep()** *and* **ordiR2step()** *during the computation, add argument* `trace = FALSE` *to the* **ordistep()** *call. At the end, display the table of results by typing* `name-of-the-ordistep-result-object$anova`.

The selected variables are the same four in this example as found with **forward.sel()** without the R^2_{adj} stopping criterion of Blanchet et al. (2008a) (not shown): `ele, oxy, bod` and `slo`. Its R^2_{adj} exceeds the R^2_{adj} of the model containing all variables, although not by much (0.5947).

Backward elimination with function ordistep()

ordistep() allows backward elimination. In this case, only the model containing all candidate explanatory variables must be provided to the function. Backward elimination returns the same result as forward selection with our data; this will not necessarily be the case for other data:

```
# Backward elimination using vegan's ordistep()
step.backward <-
  ordistep(spe.rda.all,
           permutations = how(nperm = 499)
)
RsquareAdj(step.backward)
```

Hint *When the* `'scope'` *argument is missing,* `direction = "backward"` *becomes the default.*

Forward selection with function ordiR2step()

Function **ordiR2step()**, which is limited to RDA and db-RDA and does not allow backward elimination, uses the same two criteria as **forward.sel()** (α level and R^2_{adj} of the global model) with the added bonus that it accepts factor variables. An application with the same quantitative environmental data matrix as the one used with **forward.sel()** returns, of course, the same result:

```
# Forward selection using vegan's ordiR2step()
step2.forward <-
  ordiR2step(mod0,
             scope = formula(spe.rda.all),
             direction = "forward",
             R2scope = TRUE,
             permutations = how(nperm = 199)
)
RsquareAdj(step2.forward)
```

Hint *See argument* `R2scope = TRUE`, *which asks* `ordiR2step()` *to compute the* R^2_{adj} *of the global model to be used as a second stopping criterion. Actually* `TRUE` *is the default; we added it to emphasize its importance.*

A forward selection on object `env3`, which contains a factor variable instead of the quantitative `slo`, would return a different result because the transformation of our quantitative `slo` variable into a four-level factor produced a loss of information, which resulted in the R^2_{adj} of the global RDA being smaller. Note also that a factor represents a single variable in this procedure. Its levels are not selected separately.

Forward selection within a partial RDA with functions `ordistep()` and `ordiR2step()`

It could happen that a researcher would like to forward-select explanatory variables while keeping others constant. This is possible with **`ordistep()`** and **`ordiR2step()`**. The conditioning variables (i.e., variables to be held constant) must be specified in the two models provided to the functions. As an illustration, let us run a forward selection of environmental variables while holding the slope constant, using **`ordiR2step()`**.

```
# Partial forward selection with variable slo held constant
mod0p <- rda(spe.hel ~ Condition(slo), data = env2)
mod1p <- rda(spe.hel ~ . + Condition(slo), data = env2)
step.p.forward <-
  ordiR2step(mod0p,
             scope = formula(mod1p),
             direction = "forward",
             permutations = how(nperm = 199)
)
```

As you can see, the result is different from those obtained above, because the influence of the slope is taken into account when starting the forward selection.

Parsimonious RDA

These analyses show that the most parsimonious attitude would be to settle for a model containing only three explanatory variables: elevation, oxygen and biological oxygen demand. What would such an analysis look like? Although it is already made and stored in object `step2.forward`, let us run it again with an explicit formula:

```
(spe.rda.pars <- rda(spe.hel ~ ele + oxy + bod, data = env2))
anova(spe.rda.pars, permutations = how(nperm = 999))
anova(spe.rda.pars, permutations = how(nperm = 999), by = "axis")
(R2a.pars <- RsquareAdj(spe.rda.pars)$adj.r.squared)
# Compare the variance inflation factors
vif.cca(spe.rda.all)
vif.cca(spe.rda.pars)
```

These results are fascinating in that they demonstrate how a parsimonious approach can help improve the quality of a model. With a moderate cost in explanatory power, we produced a model that is as highly significant, has no harmful collinearity (all VIFs are now well below 10), and can be decomposed into *three significant* canonical axes, whereas the global model produced only two significant axes.

It is now time to produce a triplot of this result (Fig. 6.4). We present the scaling 1 triplot. We will compare it to the triplot of the global analysis (Fig. 6.1a).

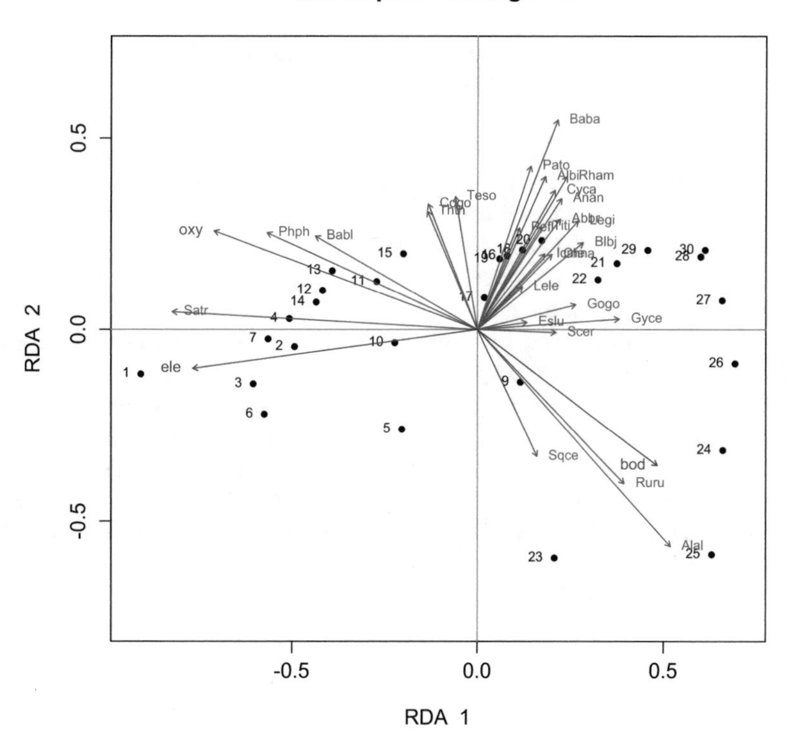

Fig. 6.4 RDA triplot, Hellinger-transformed Doubs fish data constrained by three environmental variables. Fitted site scores. Scaling 1

```
# Triplots of the parsimonious RDA (with fitted site scores)
par(mfrow = c(1, 2))
# Scaling 1
triplot.rda(spe.rda.pars,
  site.sc = "lc",
  scaling = 1,
  cex.char2 = 0.8,
  pos.env = 2,
  mult.spe = 0.9,
  mult.arrow = 0.92,
  mar.percent = 0.01
)
# Scaling 2 : see accompanying material
```

Since there is now a third significant canonical axis, you could plot other combinations: axes 1 and 3, and axes 2 and 3.

This triplot indeed presents the same structures as the one produced with all explanatory variables (Fig. 6.1a). The sites and species show the same relationships. The three selected explanatory variables are sufficient to reveal the major features of the data.

6.3.2.7 Environmental Reconstruction: Projecting New Sites in an RDA to Estimate the Values of Explanatory Variables

An RDA model is generally used to interpret the structure of the response data as explained by the set of independent variables. But if the model is built from species that act as bioindicators of environmental conditions represented by the explanatory variables and if the model explains a fairly large amount of variance, then RDA can be applied in an opposite way, i.e. to estimate the values of (quantitative) explanatory variables on the basis of the abundances of the species. This is sometimes also called calibration or bioindication. It can be computed with function **`calibrate()`** of package **`vegan`**.

As an illustration, let us imagine two new sites along the Doubs River where fishes were captured and counted, but no environmental variable was measured. In this example, the new sites are made of the (rounded) mean abundances of the 27 fish species in the first 15 sites (new site 1) and of the last 14 sites (new site 2). We now want to use our parsimonious RDA result to estimate the values of variables `ele`, `oxy` and `bod` on the basis of the fish abundances of the new sites.

```
# New (fictitious) objects with fish abundances
# Variables(species) must match those in the original data set in
# name, number and order
site1.new <- round(apply(spe[1:15, ], 2, mean))
site2.new <- round(apply(spe[16:29, ], 2, mean))
obj.new <- t(cbind(site1.new, site2.new))
# Hellinger transformation of the new sites
obj.new.hel <- decostand(obj.new, "hel")
# Calibration
calibrate(spe.rda.pars, obj.new.hel)
# Compare with real values at sites 7 to 9 and 22 to 24 :
env2[7:9, c(1, 9, 10)]
env2[22:24, c(1, 9, 10)]
```

Are the calibrated values realistic? Which ones are good, which fall far from the true values?

Note that this is but a very small incursion into the world of environmental reconstruction. Many more functions are available to that effect, especially in package **rioja**, which is devoted to the analysis of stratigraphic data.

6.3.2.8 Variation Partitioning

A common occurrence in ecology is that one has two or more sets of explanatory variables pertaining to different classes. In the Doubs data, we have already split the environmental variables into a first subset of physiographic and a second subset of chemical variables. For various reasons, one might be interested not only in a partial analysis like the one that we conducted above, but in quantifying the variation explained by all subsets of the variables when controlling for the effect of the other subsets. In multivariate ecological analysis, a procedure of variation partitioning has been proposed to that effect by Borcard et al. (1992) and improved by the use of adjusted R^2 by Peres-Neto et al. (2006). When two explanatory data sets are used, the total variation of **Y** is partitioned as in Fig. 6.5 (left). The figure also shows the fractions resulting from partitioning by three and four sets of explanatory variables.

```
# Explanation of fraction labels (two, three and four explanatory
# matrices) with optional colours
par(mfrow = c(1, 3), mar = c(1, 1, 1, 1))
showvarparts(2, bg = c("red", "blue"))
showvarparts(3, bg = c("red", "blue", "yellow"))
showvarparts(4, bg = c("red", "blue", "yellow", "green"))
```

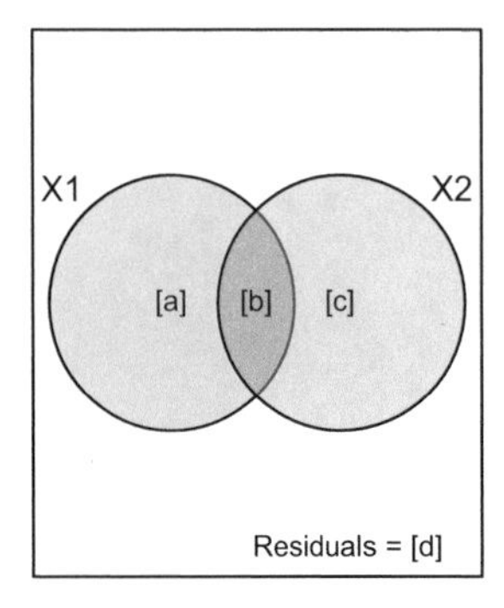

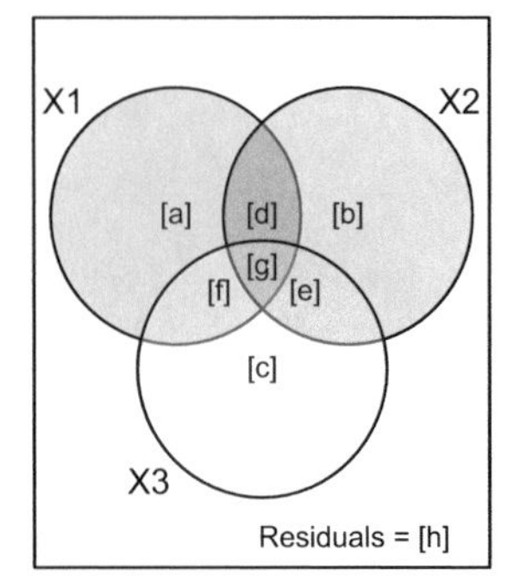

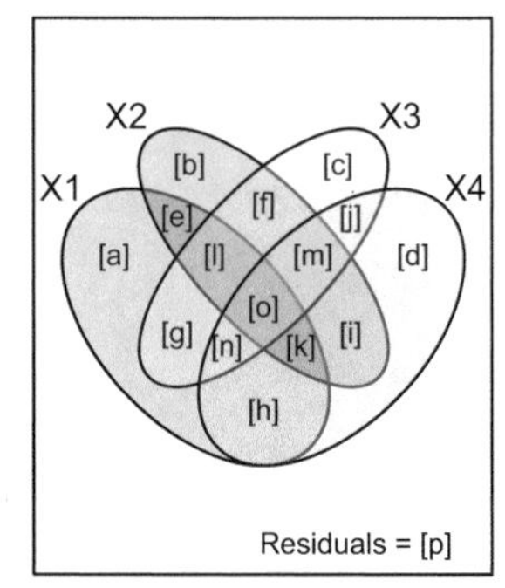

Fig. 6.5 Venn diagrams of the variation partitioning of a response data set **Y** explained by two (left), three (centre) and four (right) data sets **X1** to **X4**. The enclosing rectangles represent the total sum-of-squares of **Y**

In a partitioning by two explanatory matrices **X** and **W**, both explain some variation of the response data. Since the explanatory data sets are generally not orthogonal to one another (except in some cases addressed later), some amount of variation is explained jointly by the two sets (fraction [b] of Fig. 6.5 left). Consequently, the variation explained by all variables together is less than the sum of the variations explained by the various subsets. This is why one can express the variation explained by **X** as fraction [a + b], the variation explained by **W** (**X2** in Fig. 6.5 left) as fraction [b + c], and the unexplained variation as fraction [d]. In the case of two explanatory data sets, three RDAs are needed to partition the variation into the four individual fractions [a], [b], [c] and [d].

The conceptual steps are the following:

- If necessary, forward-select the explanatory variables *separately* in each subset.
- Run an RDA of the response data **Y** by **X**. This yields fraction [a + b].
- Run an RDA of the response data **Y** by **W**. This yields fraction [b + c].
- Run an RDA of the response data **Y** by **X** and **W** together. This yields fraction [a + b + c].
- **Compute the adjusted R^2 (R^2_{adj}) of the three RDAs above.**
- Compute the fractions of adjusted explained variation by subtraction:
  - fraction $[a]_{adj} = [a + b + c]_{adj} - [b + c]_{adj}$
  - fraction $[c]_{adj} = [a + b + c]_{adj} - [a + b]_{adj}$
  - fraction $[b]_{adj} = [a + b]_{adj} - [a]_{adj} = [b + c]_{adj} - [c]_{adj}$
  - fraction $[d]_{adj} = 1 - [a + b + c]_{adj}$

The three RDAs can be tested as usual, and fractions [a] and [c] can be computed and tested by means of partial RDA. Fraction [b], however, is not an adjusted component of variance and cannot be estimated and tested by regression methods. It has zero degree of freedom. Note also that there is no equation for computing an adjusted R^2 directly for a partial RDA. The subtractive procedure described above goes around this difficulty. It has been shown by Peres-Neto et al. (2006) to produce unbiased estimates of the fractions of variation. Remember also that adjusted R^2 can be negative (Sect. 6.3.2.3). Negative R^2_{adj} can be ignored (considered as null) for the ecological interpretation of the results.

The whole procedure of variation partitioning (except for the preliminary step of forward selection) can be run in one **R** command with up to four explanatory matrices. The function to use, available in **vegan**, is **varpart()**. Let us apply it to the Doubs data and follow up with tests of all testable fractions.

```
## 1. Variation partitioning with all explanatory variables
##    (except dfs)
(spe.part.all <- varpart(spe.hel, envchem, envtopo))
plot(spe.part.all, digits = 2, bg = c("red", "blue"))
```

The plot gives correct values of the adjusted R squares, but the sizes of the circles in the Venn diagram are not to scale.

This first partitioning shows that both sets of explanatory variables contribute to the explanation of the species data. The unique contribution of the chemical variables (fraction [a], $R^2_{adj} = 0.241$) is more than twice as large as that of physiography (fraction [c], $R^2_{adj} = 0.112$). The variation explained jointly by the two sets (fraction [b], $R^2_{adj} = 0.233$) is also large. This indicates that the chemical and physiographic variables are intercorrelated. This is a good reason to make an effort towards parsimony, and to combine variation partitioning with forward selection.

```
## 2. Variation partitioning after forward selection of explanatory
##    variables
# Separate forward selection in each subset of environmental
# variables
spe.chem <- rda(spe.hel, envchem)
R2a.all.chem <- RsquareAdj(spe.chem)$adj.r.squared
forward.sel(spe.hel,
            envchem,
            adjR2thresh = R2a.all.chem,
            nperm = 9999
)
spe.topo <- rda(spe.hel, envtopo)
R2a.all.topo <- RsquareAdj(spe.topo)$adj.r.squared
forward.sel(spe.hel,
            envtopo,
            adjR2thresh = R2a.all.topo,
            nperm = 9999
)
# Parsimonious subsets of explanatory variables, based on forward
# selections
names(envchem)
envchem.pars <- envchem[, c(4, 6, 7)]
names(envtopo)
envtopo.pars <- envtopo[, c(1, 2)]
```

```
# Variation partitioning
(spe.part <- varpart(spe.hel, envchem.pars, envtopo.pars))
plot(spe.part,
     digits = 2,
     bg = c("red", "blue"),
     Xnames = c("Chemistry", "Physiography"),
     id.size = 0.7
)
# Tests of all testable fractions
# Test of fraction [a+b]
anova(rda(spe.hel, envchem.pars), permutations = how(nperm = 999))
# Test of fraction [b+c]
anova(rda(spe.hel, envtopo.pars), permutations = how(nperm = 999))
# Test of fraction [a+b+c]
env.pars <- cbind(envchem.pars, envtopo.pars)
anova(rda(spe.hel, env.pars), permutations = how(nperm = 999))
# Test of fraction [a]
anova(rda(spe.hel, envchem.pars, envtopo.pars),
      permutations = how(nperm = 999)
)
# Test of fraction [c]
anova(rda(spe.hel, envtopo.pars, envchem.pars),
      permutations = how(nperm = 999)
)
```

Are any of these components non-significant?

As expected, forward-selecting the explanatory variables *independently* in each subset (chemistry and physiography) does nothing to prevent inter-set correlations; some of the variables retained in each set are correlated with those of the other set. Therefore, fraction [b] remains important. Beware: conducting variable selection on the union of the two explanatory data sets would make fraction [b] very small or empty because pairs or groups of collinear variables would be less likely to be retained in the model. In this type of analysis, we do not want to eliminate the common fraction, we want to estimate its magnitude and interpret it. If one wants to estimate how much of the variation of **Y** is explained jointly by two or more explanatory data sets (usually because they represent different categories of constraints), it is important to carry out forward selection separately on the sets of explanatory variables.

In this example, the two independent forward selections run above have retained the same variables as when the whole set had been submitted to forward selection (i.e., `ele`, `oxy`, `bod`) **plus** variables `slo` and `nit`. The latter is strongly correlated to `ele` ($r = -0.75$). This means that in the RDA with the chemical variables, `nit` has explained some of the same structures as `ele` was explaining in the physiography RDA. Removing `nit` from the chemical variables and running the partitioning again changes the result:

```
## 3. Variation partitioning without the 'nit' variable
envchem.pars2 <- envchem[, c(6, 7)]
(spe.part2 <- varpart(spe.hel, envchem.pars2, envtopo.pars))
plot(spe.part2, digits = 2)
```

This last partitioning has been run to demonstrate the effect of correlation between the explanatory variables of different sets. However, one generally applies variation partitioning to *assess* the magnitude of the various fractions, including the common ones. If the aim is to minimize the correlation among variables, other approaches are preferable (examination of the VIFs, global forward selection).

With this warning in mind, we can examine what the comparison between the two latter partitionings tells us. Interestingly enough, the overall amount of variation explained is about the same (0.595 instead of 0.590). The [b] fraction has dropped from 0.196 to 0.088 and the fraction (elevation + slope) explained uniquely by physiography has absorbed the difference, rising from 0.142 to 0.249. This does not mean that elevation is a better *causal* candidate than nitrates to explain fish communities. Comparison of the two analyses rather indicates that nitrate content is related to elevation, *just as the fish communities are*, and that the interpretation of variables elevation and nitrates must be done with caution since their causal link to the fish communities cannot be untangled. On the other hand, elevation is certainly related to other, unmeasured environmental variables that have an effect on the communities, making it a good proxy for them.

The comparisons above tell us that:

1. Forward selection provides a parsimonious solution without sacrificing real explanatory power: the R^2_{adj} of the three partitionings are approximately equal.
2. The common fractions, which are one of the reasons why partitioning is computed, must be interpreted with caution, even when the variables responsible for them are biologically legitimate.
3. Forward-selecting *all* explanatory variables before attributing the remaining ones to subsets is in contradiction with the aim of variation partitioning, except to help identify the magnitude of the effect of some variables responsible for the common fractions.
4. Forward selection and variation partitioning are powerful statistical tools. As such they can help support sound ecological reasoning but they cannot replace it.

Final note about the [b] fraction This common fraction should **never** be mistaken for an interaction term in the analysis of variance sense. One more time, let us stress that the common fractions arise because *explanatory variables in different sets* are correlated. In contrast, in a replicated two-way ANOVA, an interaction measures the influence of the levels of one factor on the effect of the other factor on the response variable. In other words, an interaction is present when the effect of one factor changes across the levels of the other factor. An interaction is most easily measured when the two factors are linearly independent (uncorrelated, scalar product of 0) as in balanced ANOVA, a case where the [b] fraction is equal to 0. This completes the demonstration that a [b] fraction is not an interaction.

6.3.2.9 RDA as a Tool for Multivariate ANOVA

In its classical, parametric form, multivariate analysis of variance (MANOVA) has stringent conditions of application and restrictions (e.g. multivariate normality of each group of data, homogeneity of the variance-covariance matrices, number of response variables smaller than the number of objects minus the number of degrees of freedom of the MANOVA model). It is practically never well-adapted to ecological data, despite its obvious interest for the analysis of the results of ecological experiments.

Fortunately, for MANOVA, RDA offers an elegant alternative to parametric analysis, while adding the versatility of permutation tests and the possibility of representation of the results in triplots. The trick is to use factor variables and their interactions as explanatory variables in RDA. In the example below, the factors are coded as orthogonal Helmert contrasts to allow the testing of the factors and interaction in a way that provides the correct F values. The interaction is represented by the products of the variables coding for the main factors. The properties of Helmert contrasts are the following for a balanced design: (1) the sum of each coding variable is zero; (2) all variables coding for a factor or the interaction are orthogonal (their scalar products are all zero); (3) the groups of variables coding for the main factors and their interaction are all orthogonal to one another.

To illustrate this application, let us use a part of the Doubs data to construct a fictitious balanced two-way ANOVA design. We will use the first 27 sites (site 8 has already been excluded in Sect. 6.3.2.1), leaving the two last out.

We create a first factor representing elevation. This factor will have 3 levels, one for each group of sites 1–10, 11–19 and 20–28. Remember that the empty site 8 has been removed from the data, so the design is balanced with 9 sites per group.

The second factor will mimic the pH variable as closely as possible. In the real data, pH is fairly independent from elevation ($r = -0.05$), but no direct codification allows the creation of a 3-level factor orthogonal to our first factor. Therefore, we will simply create an artificial, 3-level factor orthogonal to elevation and *approximately* representing pH. Be aware that we do this for illustration purposes only, and that such a manipulation would not be tolerable in the analysis of real data.

The end result is a balanced two-way crossed design with two factors of three levels each. After having tested for the homogeneity of variance-covariance matrices for the two factors, for demonstration purposes, we will test the two main factors and their interaction, each time using the appropriate terms as covariables. Using Helmert contrasts instead of factors allows an explicit control of all the terms of the model specification. Note that coding the factors with binary variables would not make the coding variables orthogonal among the groups (factors and interaction). At the end of this section, however, we show how to compute the whole permutational MANOVA in one single command by means of the function **adonis2()**.

The condition of homogeneity of the variance-covariance matrices still applies even with this permutational approach. It can be tested with function **betadisper()** of package **vegan** which applies the Anderson (2006) testing method. It is possible to test the homogeneity of variances with respect to each factor separately. However, there is a risk in the case of crossed designs. When an interaction exists between the two factors, then the hypothesis of no heterogeneity of variances with respect to one factor may be rejected because the within-group dispersions differ due to the varying effect of the levels of the other factor. To avoid this, one can create an artificial factor crossing the two real ones, i.e., defining the cell-by-cell attribution of the data. The test becomes a test of homogeneity of within-cell dispersions.

```
# Creation of a factor 'elevation' (3 levels, 9 sites each)
ele.fac <- gl(3, 9, labels = c("high", "mid", "low"))
# Creation of a factor mimicking 'pH'
pH.fac <-
  as.factor(c(1, 2, 3, 2, 3, 1, 3, 2, 1, 2, 1, 3, 3, 2,
              1, 1, 2, 3, 2, 1, 2, 3, 2, 1, 1, 3, 3))
# Is the two-way factorial design balanced?
table(ele.fac, pH.fac)
# Creation of Helmert contrasts for the factors and the interaction
ele.pH.helm <-
  model.matrix(~ ele.fac * pH.fac,
               contrasts = list(ele.fac = "contr.helmert",
                                pH.fac = "contr.helmert"))[, -1]
ele.pH.helm
```

We removed the first column of the matrix of Helmert contrasts (...[,-1]) because it contains a trivial column of 1 (intercept).

Examine the matrix of contrasts. Which columns represent ele.fac? pH.fac? The interaction?

```
# Check property 1 of Helmert contrasts: all variables sum to 0
apply(ele.pH.helm, 2, sum)
# Check property 2 of Helmert contrasts: their crossproducts
# must be 0 within and between groups (factors and interaction)
crossprod(ele.pH.helm)

# Verify multivariate homogeneity of within-group covariance
# matrices using the betadisper() function (vegan package)
# implementing Marti Anderson's testing method (Anderson 2006)
cell.fac <- gl(9, 3)
spe.hel.d1 <- dist(spe.hel[1:27, ])

# Test of homogeneity of within-cell dispersions
(spe.hel.cell.MHV <- betadisper(spe.hel.d1, cell.fac))
anova(spe.hel.cell.MHV)     # Parametric test, not recommended
permutest(spe.hel.cell.MHV)

# Alternatively, test homogeneity of dispersions within each
# factor.
# These tests ore more robust with this small example because
# there are now 9 observations per group instead of 3.

# Factor "elevation"
(spe.hel.ele.MHV <- betadisper(spe.hel.d1, ele.fac))
permutest(spe.hel.ele.MHV) # Permutation test
# Factor "pH"
(spe.hel.pH.MHV <- betadisper(spe.hel.d1, pH.fac))
anova(spe.hel.pH.MHV)
permutest(spe.hel.pH.MHV) # Permutation test
```

The within-group dispersions are homogeneous. We can proceed with anova.

```
# Test the interaction first. Factors ele and pH (columns 1-4)
# are assembled to form the matrix of covariables for the test.
interaction.rda <-
  rda(spe.hel[1:27, ],
      ele.pH.helm[, 5:8],
      ele.pH.helm[, 1:4])
anova(interaction.rda, permutations = how(nperm = 999))
```

Is the interaction significant? A significant interaction would preclude the global interpretation of the tests of the main factors since it would indicate that the effect of a factor depends on the level of the other factor.

```
# Test the main factor ele. The factor pH and the interaction
# are assembled to form the matrix of covariables.
factor.ele.rda <-
  rda(spe.hel[1:27, ],
      ele.pH.helm[, 1:2],
      ele.pH.helm[, 3:8])
anova(factor.ele.rda,
      permutations = how(nperm = 999),
      strata = pH.fac
)
```

Is the factor ele significant?

```
# Test the main factor pH. The factor ele and the interaction
# are assembled to form the matrix of covariables.
factor.pH.rda <-
  rda(spe.hel[1:27, ],
      ele.pH.helm[, 3:4],
      ele.pH.helm[, c(1:2, 5:8)])
anova(factor.pH.rda,
  permutations = how(nperm = 999),
  strata = ele.fac
)
```

Is the factor pH significant?

```
# RDA with the significant factor ele
ele.rda.out <- rda(spe.hel[1:27, ] ~ ., as.data.frame(ele.fac))
# Triplot with "wa" sites related to factor centroids, and species
# arrows
plot(ele.rda.out,
  scaling = 1,
  display = "wa",
  main = "Multivariate ANOVA, factor elevation - scaling 1 -
      wa scores")
```

```
ordispider(ele.rda.out, ele.fac,
  scaling = 1,
  label = TRUE,
  col = "blue"
)
spe.sc1 <-
  scores(ele.rda.out,
  scaling = 1,
  display = "species")
arrows(0, 0,
  spe.sc1[, 1] * 0.3,
  spe.sc1[, 2] * 0.3,
  length = 0.1,
  angle = 10,
  col = "red"
)
text(
  spe.sc1[, 1] * 0.3,
  spe.sc1[, 2] * 0.3,
  labels = rownames(spe.sc1),
  pos = 4,
  cex = 0.8,
  col = "red"
)
```

Hints *To convince yourself that this procedure is the exact equivalent of an ANOVA, apply it to species 1 only, and then run a traditional ANOVA on species 1 with factors* `ele.fac` *and* `pH.fac`*, and compare the F-values. The probabilities may differ slightly since they are permutational in RDA.*

In the permutational tests of each main effect, the argument `strata` *restricts the permutations within the levels of the other factor. This ensures that the proper* H_0 *is produced by the permutation scheme. Note that the use of blocks allows one to apply these analyses to nested (e.g. split-plot) designs.*

If you want to create a matrix of Helmert contrasts with the crossed main factors only, without the interaction terms, replace '' by '+' in the formula: ...*`~ele.fac + pH.fac`*... Examine the documentation files of functions* **`model.matrix()`**, **`contrasts()`** *and* **`contr.helmert()`**.

In the triplot section, observe how we use function **`ordispider()`** *to relate the "wa" site scores to the centroids of the factor levels.*

Figure 6.6 represents the "wa" scores of the sites (to show their dispersion) related to the centroids of the factor levels by means of straight lines, in scaling 1. The species arrows have been added (in red) to expand the interpretation.

Figure 6.6 cleanly shows the relationships of selected groups of species with elevation. This factor can be seen as a proxy for several important environmental variables, so that this analysis is quite informative. The figure also allows one to

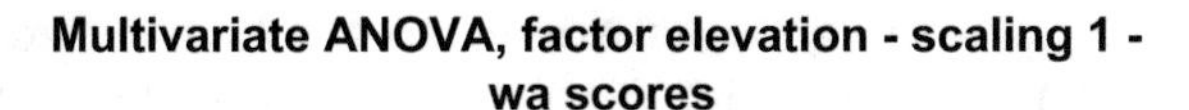

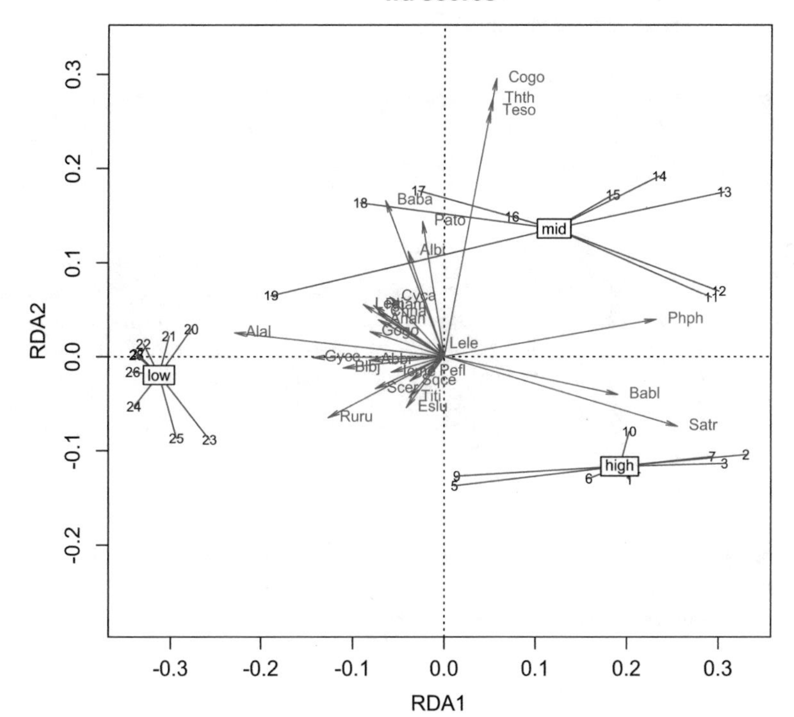

Fig. 6.6 Triplot of a multivariate ANOVA by RDA; 27 sites of the Doubs fish species explained by factor "elevation" (3 levels: low, mid and high). Scaling 1

compare the dispersions of the levels on the ordination axes. Here, the first axis discriminates between the low-elevation sites one the left, and the mid- and high-elevation sites on the right. The dispersion of the "wa" scores of the low-elevation sites is much smaller on the first axis than the dispersion of the two other groups. The second axis contrasts the mid- and high elevation sites, with the low-elevation sites in-between. The within-group dispersions of the three groups of sites on this axis are about the same.

If the design is unbalanced, this procedure can still be applied, but the contrasts will not be orthogonal and therefore the tests of significance of the factors and interaction will have reduced power to detect significant effects. Actually, non-orthogonal contrasts produce a common fraction (fraction [b] in variation partitioning with two explanatory matrices) that cannot be attributed to one or the other source of variation, thereby making the tests of the main factors and interaction more difficult to interpret.

Finally, **vegan** proposes yet another path to run an analysis of variance using a site by species or dissimilarity response matrix: functions **adonis()** and

adonis2(). Let us briefly present **adonis2()**, which is based on McArdle and Anderson (2001). See also Sect. 6.3.3. In this context, this is how the overall MANOVA would be computed:

```
adonis2(spe.hel[1:27, ] ~ ele.fac * pH.fac,
        method = "euc",
        by = "term"
)
```

Hint *The argument* `by = term`*, which is the default option, asks the function to compute the sums of squares and assess the significance of the terms in the order of their mention in the code. In balanced designs this is not important: reversing the order of the two factors will produce the exact same result. In the case of unbalanced designs, however, changing the order changes the results because of the non-orthogonality of the factors, which in fact introduces a [b] fraction in the sense of Sect. 6.3.2.8. This is another good reason to avoid unbalanced designs whenever possible.*

Finally, Laliberté et al. (2009) offer a function called **manovRDa.R** to compute a two-way MANOVA by RDA for fixed and random factors. It is available on the Web page http://www.elaliberte.info/code.

6.3.2.10 Nonlinear Relationships in RDA

Another point is worth mentioning. RDA carries out a multivariate linear regression analysis followed by a PCA of the fitted values. Consequently, other methods based upon the multiple regression equation can be used in RDA as well. Consider the fact that all RDA models presented above only used explanatory variables to the power 1. However, it is frequent that raw species responses actually have unimodal distributions along an environmental gradient, showing an ecological optimum and some tolerance to the variation of a given environmental constraint. A strictly linear model would strongly suffer from lack-of-fit in such a case. Plotting all possible pairs of response versus explanatory variables to detect such nonlinearities would be too cumbersome. An interesting shortcut to identify and model unimodal responses is to provide second-degree explanatory variables along with the first-degree terms (i.e., provide the terms for a quadratic model), then run forward selection. This procedure will retain the relevant terms (called monomials), be they of the first or second degree. Of course, the interpretation of such results is more complex, so that it should be applied only when one has serious reasons to suspect unimodal distributions causing nonlinear relationships. Third-order monomials of explanatory variables may even be useful in the case of unimodal but highly skewed response variable distributions (Borcard et al. 1992). Let us compute two examples of polynomial

RDA. The first one involves a single explanatory variable and shows the effect of incorporating a second-degree term into the model. The second example proposes a full RDA with all explanatory variables (except `dfs`) and their second-degree terms, followed by forward selection.

A single explanatory variable

In the Doubs data, there are several species that are found mostly in the central section of the river. Their relationship with the variable "distance from the source" (`dfs`) is therefore unimodal: absence first, then presence, then absence again. Such a simple case could be a good candidate to experiment with a second-degree polynomial, i.e. a sum of terms encompassing a variable to exponents 1 and 2. When interpreting the result (Fig. 6.7), note that species having their optimum around mid-river will point to the *opposite* direction from the `dfs`-squared variable `dfs2`, because the quadratic term of a unimodal model, which is concave down, is negative. Species with arrows pointing in the same direction as the `dfs2` variable may be more present at both ends of the river than in the middle (no species shows this

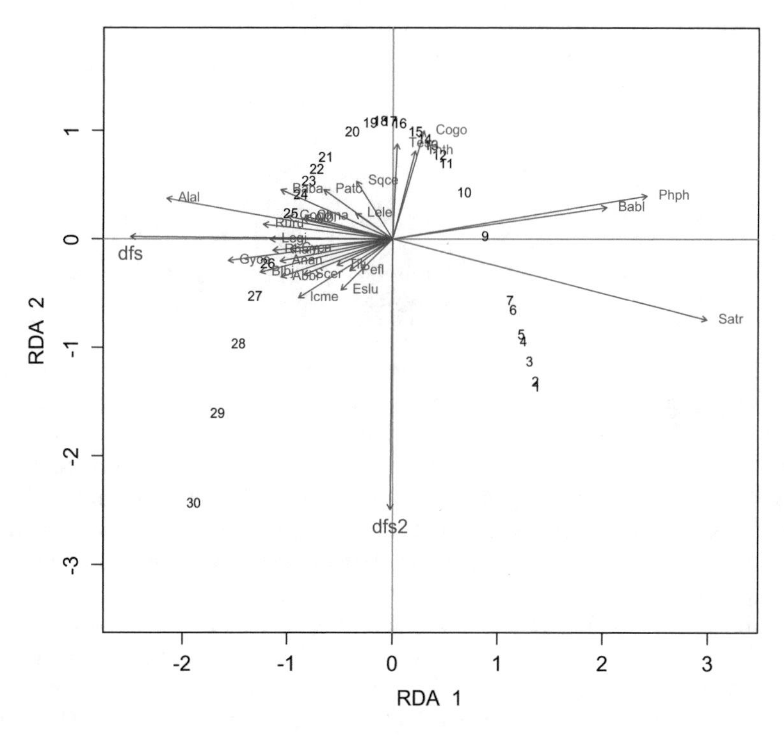

Fig. 6.7 Scaling 2 triplot of the Hellinger-transformed fish species explained by an orthogonal second-degree polynomial of the variable "distance from the source" (`dfs`)

particular distribution in the Doubs data set). Arrows of species that are more present at one end of the river point in the direction of the `dfs` variable or opposite to it.

```
# Create a matrix of dfs and its orthogonal second degree term
# using function poly()
dfs.df <- poly(dfs, 2)
colnames(dfs.df) <- c("dfs", "dfs2")
# Verify that the polynomial terms are orthogonal
cor(dfs.df)
# Find out if both variables are significant
forward.sel(spe.hel, dfs.df)

# RDA and test
spe.dfs.rda <- rda(spe.hel ~ ., as.data.frame(dfs.df))
anova(spe.dfs.rda)

# Triplot using "lc" (model) site scores and scaling 2
triplot.rda(spe.dfs.rda,
  site.sc = "lc",
  scaling = 2,
  plot.sites = FALSE,
  pos.env = 1,
  mult.arrow = 0.9,
  move.origin = c(-0.25, 0),
  mar.percent = 0
)
```

Hint *If one computes raw second-degree variables by hand (i.e., by squaring first-degree variables) or by applying argument* `raw = TRUE` *to function* **`poly()`**, *it is better to* ***centre*** *the first-degree variables before computing the second-degree terms, otherwise the latter will be strongly linearly related to the former. This is not necessary when using* **`poly()`** with `raw = FALSE` *(default).*

We drew a scaling 2 triplot because we were interested primarily in the relationships among the species. At first glance, this triplot looks like some mistake has been made, but in fact it displays exactly what has been asked for. The surprising feature is, of course, the curved distribution of the sites in the plot. Since we modelled the data by means of a second-degree function (i.e., a parabolic function) of the distance from the source, the modelled data (option `"lc"`) form a parabola. If you want a triplot showing the sites in a configuration closer to the data, replace `"lc"` by `"wa"` in the plotting function above.

To illustrate the interpretation of the first-order `dfs` and the second-order `dfs2` variables, let us take some fish species as examples. We will map them along the river to make the comparison easier (Fig. 6.8). The four examples are the brown trout (`Satr`), the grayling (`Thth`), the bleak (`Alal`) and the tench (`Titi`). The code is the one used in Sect. 2.2.3 to produce Fig. 2.3.

Comparing the ordination triplot with these four maps shows how to interpret the fish vectors in combination with the two variables `dfs` and `dfs2`. Among all

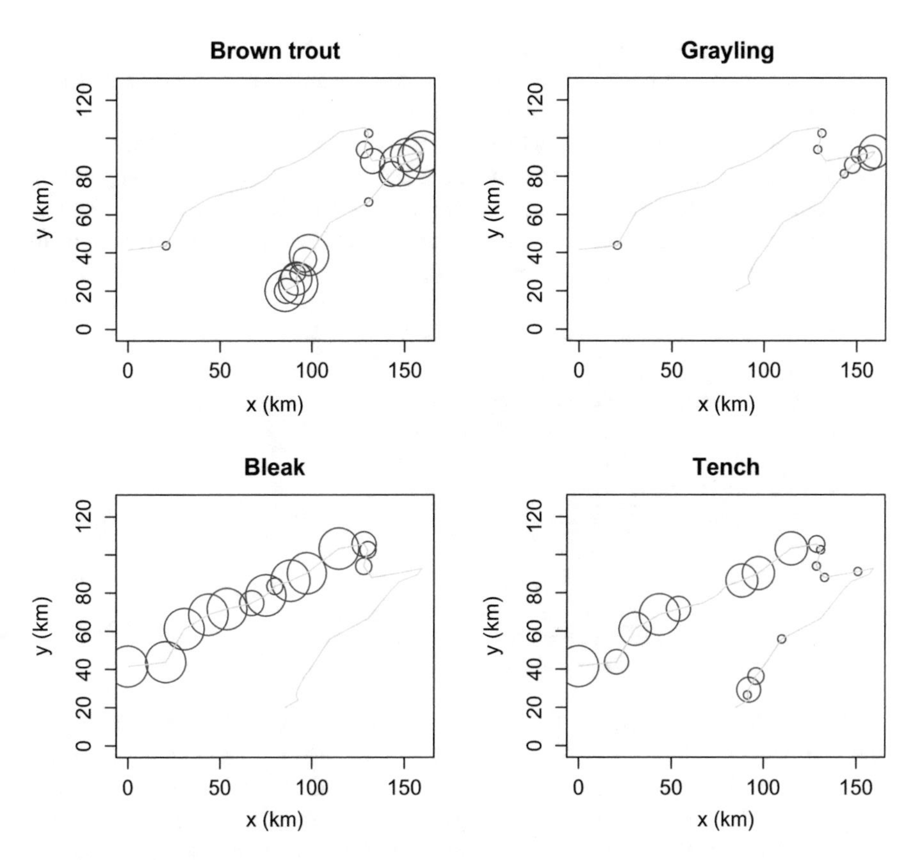

Fig. 6.8 Bubble plots of the abundances of four fish species, to explain the interpretation of the second-degree RDA presented above

species, the brown trout (`Satr`) is most strongly linked to the upper half of the river; its vector is opposed to `dfs` and orthogonal to (i.e. independent of) `dfs2`; the grayling (`Thth`) is characteristic of the central part of the river. Note that its vector on the triplot is *opposed* to that of variable `dfs2`. The bleak (`Alal`) is abundant in the lower half of the river, as confirmed by its vector pointing in the direction of `dfs`, orthogonal to `dfs2` and directly opposite to `Satr`. Finally, the tench (`Titi`) is present in three different zones along the river, which results in vector pointing halfway between the `dfs` and `dfs2` vectors.

Of course, since we have other, more explicit environmental variables at our disposal, this exercise may seem unnecessary. But it shows that in other cases, adding a second-degree term may indeed add explanatory power to the model and improve its fit; the forward selection result confirmed that `dfs2` added a significant contribution in this example, the R^2_{adj} being raised from 0.428 to 0.490.

RDA with all explanatory variables and their second-degree terms

To run this RDA we need to compute the second-degree terms of all explanatory variables except `dfs`. Raw polynomial terms are highly correlated, so it is preferable to use orthogonal polynomials, which can be computed by function **`poly()`** of package **`stats`**. However, when a matrix is provided to it, **`poly()`** computes all polynomial terms including the ones combining the variables, e.g. xy, x^2y^2, and so on. What we need here are only the variables and their respective (orthogonal) quadratic terms. Our function **`polyvars()`** takes care of this.

```
env.square <- polyvars(env2, degr = 2)
names(env.square)
spe.envsq.rda <- rda(spe.h ~ ., env.square)
R2ad <- RsquareAdj(spe.envsq.rda)$adj.r.squared
spe.envsq.fwd <-
   forward.sel(spe.h,
               env.square,
               adjR2thresh = R2ad)
spe.envsq.fwd
envsquare.red <- env.square[, sort(spe.envsq.fwd$order)]
(spe.envsq.fwd.rda <- rda(spe.h ~., envsquare.red))
RsquareAdj(spe.envsq.fwd.rda)
summary(spe.envsq.fwd.rda)
```

The result is quite different from the one obtained with the RDA based only on the first-degree variables (see Sect. 6.3.2.6). In that analysis, three variables (`ele`, `oxy` and `bod`) were selected, yielding a model with an $R^2_{\text{adj}} = 0.5401$. The polynomial RDA is far less parsimonious, having retained nine explanatory variables. As a bonus, the adjusted R^2_{adj} is now 0.7530, an important increase. Actually, the nine terms retained belong to five different variables. Indeed, the linear and quadratic terms of four variables have been retained, those of `ele`, `oxy`, `slo` and `amm`. This means that among the species that respond to these variables, some of them do it in a linear way (modelled by the first-degree term) and others by showing a unimodal response (modelled by the second-degree term). A triplot of this analysis allows a finer interpretation Fig. 6.9.

```
# Triplot using lc (model) site scores and scaling 2
triplot.rda(spe.envsq.fwd.rda,
  site.sc = "lc",
  scaling = 2,
  plot.sites = FALSE,
  pos.env = 1,
  mult.arrow = 0.9,
  mult.spe = 0.9,
  mar.percent = 0
)
```

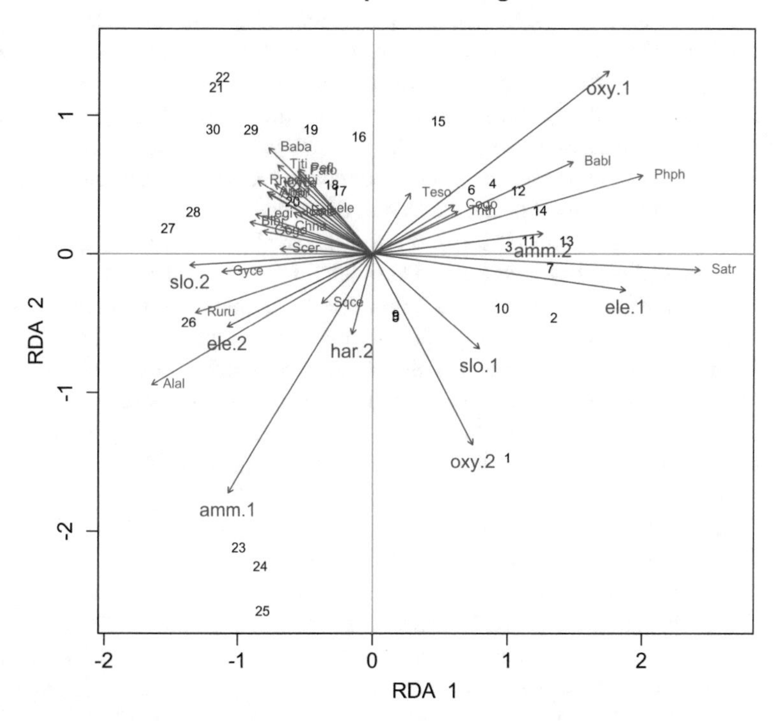

Fig. 6.9 Triplot of a second-degree polynomial RDA of the Doubs fish data, after forward selection of environmental variables. Scaling 2

Many fine features can be observed in this triplot. As an example, note that the arrows of the tench (`Titi`) and squared oxygen content (`oxy.2`) point in opposite directions. Examination of the raw data shows that the tench tends to be more abundant in places where oxygen concentration is intermediate. This is also roughly true for many of the species pointing in the same direction as `Titi`. Remember from the simpler example above that the mode of the species distribution points towards the *opposite* direction of the arrow of the corresponding squared explanatory variable.

Higher-degree polynomials of explanatory variables are also used in spatial analysis (see Chap. 7), where the explanatory variables are spatial coordinates and the higher-degree variables form trend-surface analysis models. This does not, preclude the use of polynomials with other types of explanatory variables.

6.3.3 *Distance-Based Redundancy Analysis (db-RDA)*

Ecologists have long needed methods to analyse community composition data in a multivariate framework. The need was particularly acute in ecological experiments designed to be analysed by multifactorial analysis of variance. We have seen that community composition data with large numbers of zeros must be transformed before they are used in MANOVA and other Euclidean-based models of analysis. The Legendre and Gallagher (2001) transformations (Sect. 3.5) are one way to solve that problem: transformed species data were used in the ANOVA by RDA described in Sect. 6.3.2.9. These transformations cover only the chord, Hellinger, chi-square, and Ochiai distance cases, however. Ecologists may want to compute RDA based on other dissimilarity indices that cannot be computed by a data transformation followed by the calculation of the Euclidean distance.

Legendre and Anderson (1999) proposed the method of distance-based redundancy analysis (db-RDA) to solve that problem. They showed that RDA could be used as a form of ANOVA that was applicable to community composition data if these were transformed in some appropriate way, which went through the calculation of a dissimilarity matrix of the user's choice. This approach remains fully valid and useful for all dissimilarity measures that cannot be obtained by a data transformation followed by the calculation of the Euclidean distance. Among the dissimilarities devoted to communities of living organisms, let us mention some measures for binary data (e.g. Jaccard ($\sqrt{1 - S_7}$), Sørensen ($\sqrt{1 - S_8}$), and quantitative dissimilarity measures like percentage difference (aka Bray-Curtis, D_{14}), asymmetric Gower ($\sqrt{1 - S_{19}}$), Whittaker (D_9) and Canberra (D_{10}). Dissimilarities intended for other types of data, e.g. symmetric Gower ($\sqrt{1 - S_{15}}$), Estabrook-Rogers ($\sqrt{1 - S_{16}}$), and the generalized Mahalanobis distance for groups of observations, can also be used in canonical ordination through db-RDA when the analysis concerns response variables describing the physical environment. Examples of papers involving db-RDA are Anderson (1999), Geffen et al. (2004) and Lear et al. (2008). The method goes as follows:

- Compute a Q-mode dissimilarity matrix for the response data.
- Compute a principal coordinate analysis (PCoA) of the dissimilarity matrix, correcting for negative eigenvalues if necessary, using the Lingoes correction. Keep *all* principal coordinates in a file; they express all the variance of the data as seen through the dissimilarity measure.
- Run and test an RDA of the principal coordinates created above (which act as the response data) constrained by the explanatory variables available in the study. The explanatory variables may, for example, represent the factors of a manipulative or mensurative experiment.

These steps are quite simple; they can be run one by one in R with a few lines of code only, but more directly **vegan** proposes the function **capscale()** to that effect. This function allows the direct plotting of the weighted average species scores if the user provides the matrix of species data in argument `comm`.

There is another way of computing db-RDA. McArdle and Anderson (2001) proposed an alternative method that runs the analysis directly on a dissimilarity response matrix without having to go through PCoA. This alternative way is provided in **vegan** by function **dbrda()**. Unfortunately, species scores cannot be added directly to the plot with this function. But the main interest in this alternative procedure is that, according to the authors' simulations, the permutation test computed on their method based on a dissimilarity matrix has correct type I error when applied to multifactorial ANOVA designs. The method based on PCoA data obtained with the Lingoes correction shares the same property, whereas tests on data obtained with the Cailliez correction produce inflated type I error rates (McArdle and Anderson 2001).

For the sake of example, we will compute a db-RDA on the fish data (reduced to 27 sites) constrained by the factor `ele` created in Sect. 6.3.2.9. Let us do it on the basis of a percentage difference dissimilarity matrix. Now, we remember from Chap. 3 that this measure is not Euclidean, which will result in the production of one or several negative eigenvalues if no correction is applied. The corrections may consist in square-rooting the dissimilarity matrix or applying the Lingoes correction (Sect. 5.5).

Our aim is double: we want a test with a correct type I error rate, and we want to plot the results with the species scores as well. Function **`dbrda()`** applied to a square-rooted dissimilarity matrix will provide the test, and function **`capscale()`** with the Lingoes correction will provide the material for the plot (shown in Fig. 6.10).

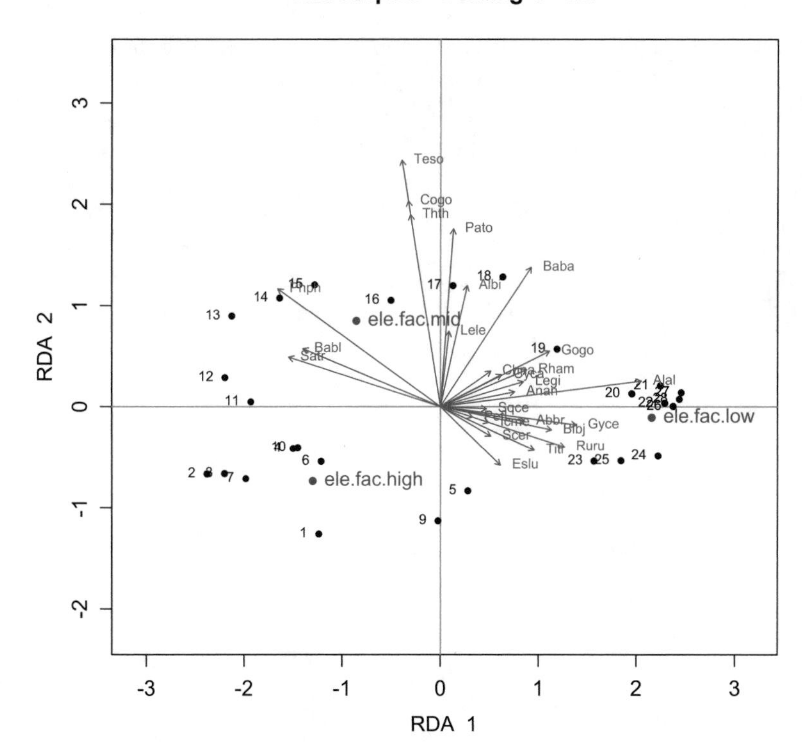

Fig. 6.10 Triplot of the db-RDA analysis of 27 sites of the Doubs data constrained by factor `ele`; computation with **`capscale()`**. Percentage difference dissimilarity, Lingoes correction. Scaling 1. Sites are represented by their "wa" scores to show their dispersion around the centroids of the levels of factor ele

```
# Rename columns of matrix of Helmert contrasts (for convenience)
colnames(ele.pH.helm) <-
  c("ele1", "ele2", "pH1", "pH2", "ele1pH1", "ele1pH2",
    "ele2pH1", "ele2pH2" )
# Create the matrix of covariables. MUST be of class matrix,
# NOT data.frame
covariables <- ele.pH.helm[, 3:8]
# Compute the dissimilarity response matrix with vegan's vegdist()
spe.bray27 <- vegdist(spe[1:27, ], "bray")
# … or with function dist.ldc() of adespatial
spe.bray27 <- dist.ldc(spe[1:27, ], "percentdiff")

# 1. dbrda() on the square-rooted dissimilarity matrix
bray.env.dbrda <-
  dbrda(sqrt(spe.bray27) ~ ele1 + ele2 + Condition(covariables),
        data = as.data.frame(ele.pH.helm),
        add = FALSE)
anova(bray.env.dbrda, permutations = how(nperm = 999))

# 2. capscale() with raw (site by species) data
bray.env.cap <-
  capscale(spe[1:27, ] ~ ele1 + ele2 + Condition(covariables),
           data = as.data.frame(ele.pH.helm),
           distance = "bray",
           add = "lingoes",
           comm = spe[1:27, ])
anova(bray.env.cap, permutations = how(nperm = 999))

# Plot with "wa" scores to see dispersion of sites around the
# factor levels
triplot.rda(bray.env.cap, site.sc = "wa", scaling = 1)
```

Hints *In* **cmdscale()**, *the argument* add = TRUE *adds a constant to the distances to avoid negative eigenvalues. This is the Cailliez correction. In* **capscale()**, **dbrda()** *and* **adonis2()**, add = "cailliez" *also produces the Cailliez correction. Another option is* add = "lingoes", *which produces the Lingoes correction. The Lingoes and Cailliez corrections are also available in functions* **pcoa() {ape}** *and* **wcmdscale() {vegan}**. *As mentioned above, the Lingoes correction is preferable before tests of significance in MANOVA by db-RDA.*

In **capscale()** *and* **dbrda()**,*covariables can be provided in two ways:*

(1) Constraining variables and covariables can be found in the same object, which ***must*** *be a data frame, and the formula must be complete with all variables and covariables stated explicitly; covariables can be quantitative or factors;*

(2) If all covariables are quantitative (or all factors coded as dummy variables perceived by R as quantitative, as in our object covariables *above), one can simplify the coding: the constraining variables are in a* ***data frame****, and the covariables are in a separate object that* ***must*** *be of class* ***matrix****; in this case the object can be called globally without having to detail the covariables by names.*

When using raw response data, the association coefficient is determined by the argument distance.

The results of the two analyses are slightly different because (1) the test is not performed in the same manner and (2) the correction to make the response matrix Euclidean is not the same.

6.3.4 A Hand-Written RDA Function

The code below is another exercise in coding matrix algebra in R; just follow the equations.

The Code It Yourself corner #3

```
myRDA <- function(Y, X)
{
    ## 1. Preparation of the data
    Y.mat <- as.matrix(Y)
    Yc <- scale(Y.mat, scale = FALSE)
    X.mat <- as.matrix(X)
    Xcr <- scale(X.mat)

    # Dimensions
    n <- nrow(Y)
    p <- ncol(Y)
    m <- ncol(X)
```

```
## 2. Computation of the multivariate linear regression
# Matrix of regression coefficients (eq. 11.9)
B <- solve(t(Xcr) %*% Xcr) %*% t(Xcr) %*% Yc

# Matrix of fitted values (eq. 11.10)
Yhat <- Xcr %*% B

# Matrix of residuals
Yres <- Yc - Yhat

## 3. PCA on fitted values
# Covariance matrix (eq. 11.12)
S <- cov(Yhat)

# Eigenvalue decomposition
eigenS <- eigen(S)

# How many canonical axes?
kc <- length(which(eigenS$values > 0.00000001))

# Eigenvalues of canonical axes
ev <- eigenS$values[1 : kc]
# Total variance (inertia) of the centred matrix Yc
trace = sum(diag(cov(Yc)))

# Orthonormal eigenvectors (contributions of response
# variables, scaling 1)
U <- eigenS$vectors[, 1 : kc]
row.names(U) <- colnames(Y)

# Site scores (vegan's wa scores, scaling 1; eq.11.17)
F <- Yc %*% U
row.names(F) <- row.names(Y)

# Site constraints (vegan's 'lc' scores, scaling 1;
# eq. 11.18)
Z <- Yhat %*% U
row.names(Z) <- row.names(Y)

# Canonical coefficients (eq. 11.19)
CC <- B %*% U
row.names(CC) <- colnames(X)

# Species-environment correlations
corXZ <- cor(X, Z)

# Diagonal matrix of weights
D <- diag(sqrt(ev / trace))
```

```
    # Biplot scores of explanatory variables
    coordX <- corXZ %*% D     # Scaling 1
    coordX2 <- corXZ          # Scaling 2
    row.names(coordX) <- colnames(X)
    row.names(coordX2) <- colnames(X)

    # Scaling to sqrt of the relative eigenvalue
    # (for scaling 2)
    U2 <- U %*% diag(sqrt(ev))
    row.names(U2) <- colnames(Y)
    F2 <- F %*% diag(1/sqrt(ev))
    row.names(F2) <- row.names(Y)
    Z2 <- Z %*% diag(1/sqrt(ev))
    row.names(Z2) <- row.names(Y)

    # Unadjusted R2
    R2 <- sum(ev/trace)
    # Adjusted R2
    R2a <- 1 - ((n - 1)/(n - m - 1)) * (1 - R2)

    # 4. PCA on residuals
    # Write your own code as in Chapter 5. It could begin
    # with :
    #     eigenSres <- eigen(cov(Yres))
    #     evr <- eigenSres$values

    # 5. Output
    result <-
      list(trace, R2, R2a, ev, CC, U, F, Z, coordX,
           U2, F2, Z2, coordX2)
    names(result) <-
      c("Total_variance", "R2", "R2adj", "Can_ev",
        "Can_coeff", "Species_sc1", "wa_sc1", "lc_sc1",
        "Biplot_sc1", "Species_sc2", "wa_sc2", "lc_sc2",
        "Biplot_sc2")
    result
}
```

Apply your homemade function to the Doubs fish and environmental data:

```
doubs.myRDA <- myRDA(spe.hel, env2)
summary(doubs.myRDA)
```

6.4 Canonical Correspondence Analysis (CCA)

6.4.1 Introduction

The canonical counterpart of CA, canonical correspondence analysis, has been acclaimed by ecologists ever since its introduction (ter Braak, 1986, 1987, 1988). It shares many characteristics with RDA, so that a detailed description is not necessary here. Basically, it is a weighted form of RDA applied to the same matrix $\bar{Q}$ of contributions to the χ^2 statistic as used in CA (Legendre and Legendre 2012 Sect. 11.2). CCA shares the basic properties of CA, combined with those of a constrained ordination. It preserves the χ^2 distance among sites, and species are represented as points in the triplots. ter Braak (1986) has shown that, provided that some conditions are fulfilled[4], CCA is a good approximation of a multivariate Gaussian regression. One particularly attractive feature of a CCA triplot is that species are ordered along the canonical axes following their ecological optima. This allows a relatively easy ecological interpretation of species assemblages. Also, species scores can be used as synthetic descriptors in a clustering procedure (for instance *k*-means partitioning) to produce a typology of the species in assemblages.

CCA does have some drawbacks, however, related to the mathematical properties of the χ^2 distance. Legendre and Gallagher (2001) state that "*a difference between abundance values for a common species contributes less to the distance than the same difference for a rare species, so that rare species may have an unduly large influence on the analysis*". Despite its widespread use, "*the χ^2 distance is not unanimously accepted among ecologists; using simulations, Faith* et al. *(1987) concluded that it was one of the worst distances for community composition data*" (Legendre and Gallagher 2001). Its use should be limited to situations where rare species are well sampled and are seen as potential indicators of particular characteristics of an ecosystem; the alternative is to eliminate rare species from the data table before CCA. These problems, among other points, have led to the development of the species pre-transformations to open these data to the realm of RDA, ANOVA and other linear methods. The proportion of total inertia represented by explained inertia (inertia is the measure of variation of the data in CCA), which can be interpreted as an R^2, is also biased, but Ezekiel's adjustment cannot be used. However, a bootstrap procedure has been developed for its estimation (Peres-Neto et al. 2006).

Despite these shortcomings, CCA is still widely used and deserves an illustration.

---

[4]Two important conditions are that the species must have been sampled along their whole ecological range and that they display unimodal responses towards their main ecological constraints. These conditions are difficult to test formally, but graphs of species abundances in sites arranged along their scores on the first few CA ordination axes may help visualize their distributions along the main ecological gradients.

6.4.2 CCA of the Doubs River Data

6.4.2.1 CCA Using vegan

Let us run a CCA of the Doubs data using **vegan**'s **cca()** function and the formula interface. The species data are the raw, untransformed abundances and the explanatory variables are all the ones in object **env3**. Do **not** use Hellinger, log-chord, or chord-transformed data, which are meant to be used with RDA; the preserved distance would no longer be the χ^2 distance and the results could not be interpreted. Furthermore, the row sums of the data table, which are used as weights in the CCA regressions, have no identifiable meanings for such transformed data.

```
(spe.cca <- cca(spe ~ ., env3))
summary(spe.cca)    # Scaling 2 (default)
# Unadjusted and adjusted R^2 - like statistics
RsquareAdj(spe.cca)
```

Hint *In CCA the measure of explained variation is not the "true" R^2 but a ratio of inertias. Furthermore, the "adjusted R^2" is estimated by a bootstrap procedure (Peres-Neto et al. 2006). For a CCA, the function* **RsquareAdj()** *does it with 1000 permutations by default. Consequently, the values obtained can vary from one run to another.*

The differences with an RDA output are the following:

- The variation is now expressed as **Mean squared contingency coefficient**.
- The maximum number of canonical axes in CCA is min[($p-1$), m, $n-1$]. The minimum number of residual axes is min[($p-1$), $n-1$]. In our example these numbers are the same as in RDA.
- The species scores are represented by **points** in the triplot.
- Site scores are weighted **averages** (instead of weighted sums) of species scores.

6.4.2.2 CCA Triplot

The **vegan**-based code to produce CCA triplots is similar to the one used for RDA, except that the response variables (species) are represented by points and thus arrows are not available for them (Fig. 6.11).

```
par(mfrow = c(1, 2))
# Scaling 1: species scoresscaled to the relative eigenvalues,
# sites are weighted averages of the species
plot(spe.cca,
  scaling = 1,
  display = c("sp", "lc", "cn"),
  main = "Triplot CCA spe ~ env3 - scaling 1"
)
# Default scaling 2: site scores scaled to the relative
# eigenvalues, species are weighted averages of the sites
plot(spe.cca,
  display = c("sp", "lc", "cn"),
  main = "Triplot CCA spe ~ env3 - scaling 2"
)
```

In CCA as in RDA, the introduction of explanatory variables calls for additional interpretation rules for the triplots. Here are the essential ones:

- *Scaling 1* – (1) Projecting an object at right angle on a **quantitative explanatory variable** approximates the position of the object along that variable. (2) An object found near the point representing the **centroid of a class of a qualitative explanatory variable** is more likely to possess that class of the variable. (3) Distances among **centroids** of qualitative explanatory variables, and between **centroids** and individual objects, approximate χ^2 distances.
- *Scaling 2* – (1) The optimum of a species along a **quantitative environmental variable** can be obtained by projecting the species at right angle on the variable. (2) A species found near the **centroid of a class of a qualitative environmental variable** is likely to be found frequently (or in larger abundances) in the sites possessing that class of the variable. (3) *Distances among* ***centroids****, and between centroids and individual objects, do not approximate* χ^2 *distances.*

The scaling 1 triplot focuses on the distance relationships among sites, but the presence of species with extreme scores renders the plot difficult to interpret beyond trivialities (Fig. 6.11a). Therefore, it may be useful to redraw it without the species (Fig. 6.12 left):

```
# CCA scaling 1 biplot without species (using lc site scores)
plot(spe.cca,
  scaling = 1,
  display = c("lc", "cn"),
  main = "Biplot CCA spe ~ env3 - scaling 1"
)
```

Here the response of the fish communities to their environmental constraints is more apparent. One can see two well-defined groups of sites, one linked to high elevation and very steep slope (sites 1–7 and 10) and another with the highest oxygen contents (sites 11–15). The remaining sites are distributed among various conditions towards more eutrophic waters. Remember that this is a constrained ordination of the *fish* community data, not a PCA of the site environmental variables.

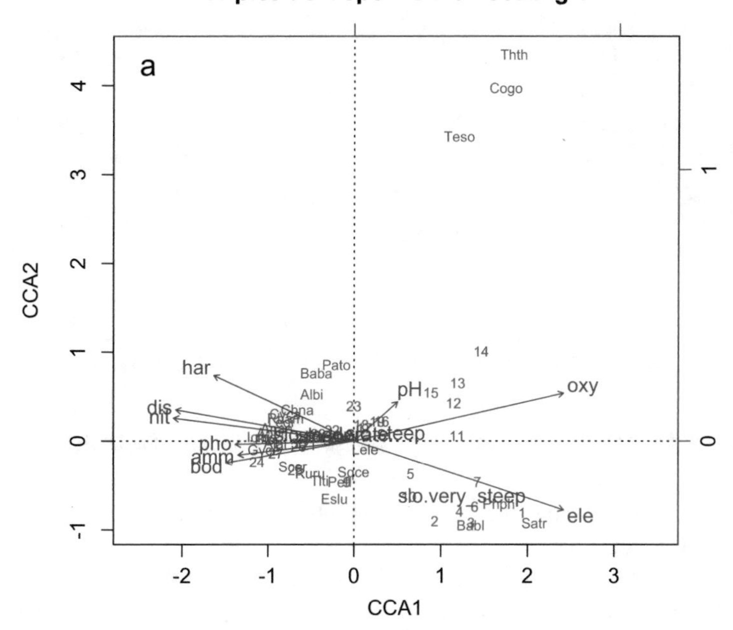

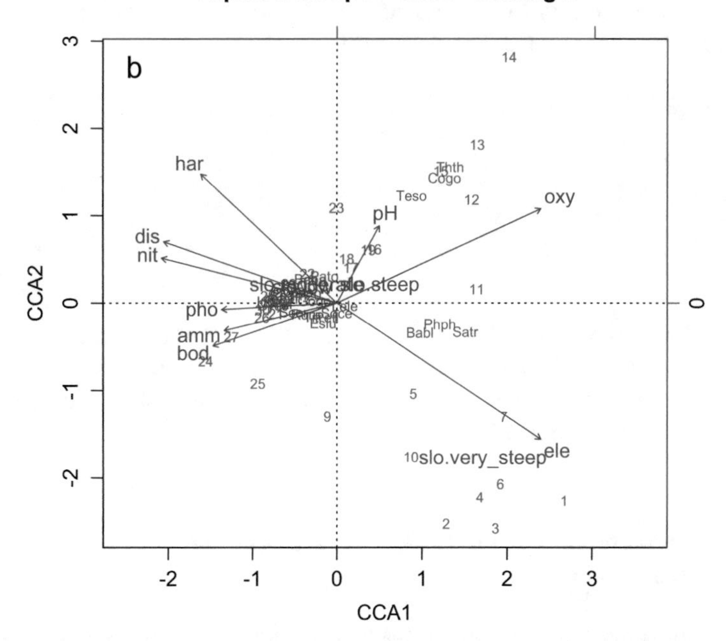

Fig. 6.11 CCA triplot of the Doubs fish species constrained by all environmental variables except `dfs`. (**a**) scaling 1; (**b**) scaling 2

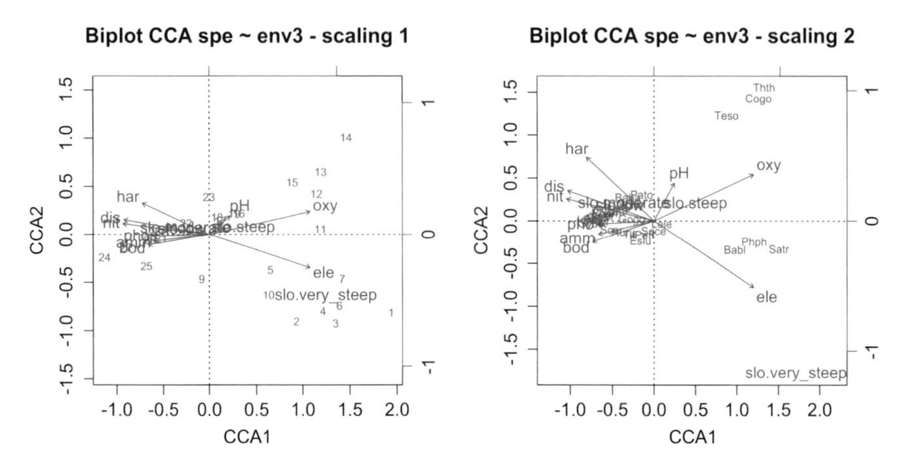

Fig. 6.12 Biplot of CCA. Left: scaling 1 with fitted site scores. Right: scaling 2 with species scores

The triplot displays how the fish community is organized with respect to the environmental constraints.

The scaling 2 triplot (Fig. 6.11b) shows two groups of species: `Thth`, `Cogo` and `Teso` linked to high oxygen concentrations; `Satr`, `Phph` and `Babl` also linked to high oxygen concentration and to high elevation and very steep slope. To help untangle the other species, a biplot without the sites is useful. You can plot it as we did above for the scaling 1 biplot, but this time leaving out the site (`"lc"`) scores and using species (`"sp"`) scores (Fig. 6.12 right).

```
# CCA scaling 2 biplot with species but without sites
plot(spe.cca,
  scaling = 2,
  display = c("sp", "cn"),
  main = "Biplot CCA spe ~ env3 - scaling 2"
)
```

This closer look shows that `Scer`, `Cyca`, `Titi`, `Eslu`, `Gogo` and `Pefl` are linked to high ammonium and phosphate concentrations, as well as high biological oxygen demand; most other species are linked to high nitrate concentrations, moderate to low slopes and high discharge.

6.4.2.3 Permutation Tests in CCA, Forward Selection and Parsimonious CCA

CCA results can be tested for significance by permutation, in the same way as RDA.

```
# Permutation test of the overall analysis
anova(spe.cca, permutations = how(nperm = 999))
# Permutation test of each axis
anova(spe.cca, by = "axis", permutations = how(nperm = 999))
```

The RDA presented in Sect. 6.3.2.2, although globally significant, was not parsimonious. Therefore we computed a forward selection of explanatory variables (Sect. 6.3.2.6). Let us do the same thing in CCA with function **ordistep()**, since **ordiR2step()** and **forward.sel()** can only compute RDA.

```
# CCA-based forward selection using vegan's ordistep()
# This function allows the use of factors like 'slo' in env3
cca.step.forward <-
  ordistep(cca(spe ~ 1, data = env3),
           scope = formula(spe.cca),
           direction = "forward",
           permutations = how(nperm = 199))
```

The result is the same as the most parsimonious one based on RDA. Therefore, we can compute a parsimonious CCA on the basis of the same three explanatory variables: elevation, oxygen concentration and biological oxygen demand.

```
## Parsimonious CCA using ele, oxy and bod
spe.cca.pars <- cca(spe ~ ele + oxy + bod, data = env3)
anova(spe.cca.pars, permutations = how(nperm = 999))
anova(spe.cca.pars, permutations = how(nperm = 999), by = "axis")
# R-square – like statistics
RsquareAdj(spe.cca.pars)
# Compare variance inflation factors
vif.cca(spe.cca)
vif.cca(spe.cca.pars)
```

Hint *Although the explanatory variables are the same, the VIFs differ from those of RDA because CCA is a* ***weighted*** *regression procedure and function* **vif.cca()** *takes into account the weights of the rows (estimated from the response matrix) to compute the VIFs of the explanatory variables. If factors are present among the RDA or CCA explanatory variables, they are decomposed into binary variables before computation of VIF.*

As in RDA, parsimony has paid off. The adjusted explained inertia is barely affected: 0.5128; it was 0.5187 with all explanatory variables. Contrary to RDA, **these values may differ from one run to another, because they are estimated by a bootstrap procedure**. Nevertheless, we have now a clearer model with three significant canonical axes. The largest VIFs of the three remaining variables are around 3, which is far from the value 10 and thus a very reasonable value.

Thanks to the introduction of the computation of R^2_{adj} in CCA, we could also compute a variation partitioning based on CCA. However, we would have to do it step by step or write a function, since no ready-made function is provided in **vegan**. Interested readers are encouraged to write this function as an exercise.

6.4.2.4 Three-Dimensional Interactive Plots

Instead of plotting these parsimonious results as we did before, let us explore a **vegan** function that can be very useful either for a researcher looking for a new perspective on his or her results, or for a teacher: a 3D interactive plot. We will see several options to reveal or combine results in different ways. These 3D plots are run under the **vegan3d** package.

```
# Plot of the sites only (wa scores)
ordirgl(spe.cca.pars, type = "t", scaling = 1)
```

Using the mouse, enlarge the plot by dragging its lower right-hand corner. Then move the plot around by left-clicking on any point in the plot with the left button. Use the scroll wheel to zoom in and out.

```
# Connect weighted average scores to linear combination scores
orglspider(spe.cca.pars, scaling = 1, col = "purple")
```

The purple connections show how well the CCA model fits the data. The shorter the connections, the better the fit.

```
# Plot the sites (wa scores) with a clustering result
# Colour sites according to cluster membership
gr <- cutree(hclust(vegdist(spe.hel, "euc"), "ward.D2"), 4)
ordirgl(spe.cca.pars,
  type = "t",
  scaling = 1,
  ax.col = "black",
  col = gr + 1
)
# Connect sites to cluster centroids
orglspider(spe.cca.pars, gr, scaling = 1)
```

The sites are nicely clustered along their major ecological gradients. Remember that this is an analysis of the fish community data.

```
# Complete CCA 3D triplot
ordirgl(spe.cca.pars, type = "t", scaling = 2)
orgltext(spe.cca.pars,
  display = "species",
  type = "t",
  scaling = 2,
  col = "cyan"
)

# Plot species groups (Jaccard dissimilarity, useable in R mode)
gs <-
  cutree(
      hclust(vegdist(t(spe), method = "jaccard"), "ward.D2"),
      k = 4)
ordirgl(spe.cca.pars,
        display = "species",
        type = "t",
        col = gs + 1)
```

Hint *Three-dimensional plots have many options. Type* **`?ordirgl`** *to explore some of them. It is also possible to draw 3D plots of RDA results, but there is no simple means to draw arrows for the response variables.*

6.5 Linear Discriminant Analysis (LDA)

6.5.1 Introduction

Linear discriminant analysis differs from RDA and CCA in that the response variable is a single variable classifying the sites into groups. This grouping may have been obtained by clustering the sites on the basis of a data set, or it may represent an ecological hypothesis. LDA tries to determine to what extent an *independent* set of quantitative variables can explain this grouping. We insist that the site typology must have been obtained independently from the explanatory variables used in the LDA; otherwise the procedure would be circular and the tests would be invalid.

LDA can provide two types of functions. *Identification functions* are obtained from the original (non-standardized) descriptors and can be used to find the group to which a new object should be attributed. *Discriminant functions* are computed from standardized descriptors. These coefficients quantify the relative contributions of the (standardized) explanatory variables to the discrimination of objects. The example below shows both operations (identification and discrimination).

To perform LDA, one must ensure that the within-group covariance matrices of the explanatory variables are homogeneous, a condition that is frequently violated with ecological data. We will address this step with function **`betadisper()`** of package **`vegan`**. The null hypothesis of the betadisper test is H_0: the multivariate group dispersion matrices are homogeneous. A p-value *larger* than 0.05 indicates a high probability of conformity of the data to H_0. Furthermore, an often-neglected preliminary step is to test if the explanatory variables indeed have different means among the groups defined by the response variable. This test, based on Wilks' lambda, is actually the overall test performed in parametric MANOVA (Legendre and Legendre, 2012). We will compute this test using two different functions in **R**.

6.5.2 Discriminant Analysis Using `lda()`

`lda()` is a function of package **`MASS`**. As a simple example, we can use the four-group classification of sites based on the fish species (`gr` in the 3D plots above), and try to explain this classification using the three environmental variables that have been selected in Sect. 6.3.2.6: `ele, oxy` and `bod`. Function **`lda()`** accepts quantitative and binary dummy variables in the explanatory matrix, but its documentation file warns against the use of explanatory variables of class "factor".

Preliminary steps: compute homogeneity of group dispersions and Wilks' lambda test.

```
# Ward clustering result of Hellinger-transformed species data,
# cut into 4 groups
gr <- cutree(hclust(vegdist(spe.hel, "euc"), "ward.D2"), k = 4)

# Environmental matrix with only 3 variables (ele, oxy and bod)
env.pars2 <- as.matrix(env2[, c(1, 9, 10)])

# Verify multivariate homogeneity of within-group covariance
# matrices using the betadisper() function {vegan}
env.pars2.d1 <- dist(env.pars2)
(env.MHV <- betadisper(env.pars2.d1, gr))
permutest(env.MHV) # Permutational test
```

The within-group covariance matrices are not homogeneous. Let us try a log transformation of variables `ele` *and* `bod`

```
# Log transform ele and bod
env.pars3 <- cbind(log(env$ele), env$oxy, log(env$bod))
colnames(env.pars3) <- c("ele.ln", "oxy", "bod.ln")
rownames(env.pars3) <- rownames(env2)
env.pars3.d1 <- dist(env.pars3)
(env.MHV2 <- betadisper(env.pars3.d1, gr))
permutest(env.MHV2)
```

This time the within-group covariance matrices are homogeneous. We can proceed to the next step: test with Wilks' lambda that the explanatory variables have distinct means. We show two different ways to compute this test (H_0: the multivariate means of the groups are not distinct).

```
# First way: with function Wilks.test() of package rrcov, χ² test
Wilks.test(env.pars3, gr)
# Second way: with function manova() of stats, which uses
#             an F-test approximation
lw <-  manova(env.pars3 ~ as.factor(gr))
summary(lw, test = "Wilks")
```

First example: let us compute identification functions and use them to attribute two new objects to the classification.

```
# Computation of LDA - identification functions (on unstandardized
# variables)
env.pars3.df <- as.data.frame(env.pars3)
(spe.lda <- lda(gr ~ ele.ln + oxy + bod.ln, data = env.pars3.df))
# The result object contains the information necessary to interpret
# the LDA
summary(spe.lda)
# Display the group means for the 3 variables
spe.lda$means
# Extract the unstandardized identification functions (matrix C,
# eq. 11.33 in Legendre and Legendre 2012)
(C <- spe.lda$scaling)

# Classification of two new objects (identification)
# A new object is created with two sites:
#     (1) ln(ele) = 6.8, oxygen = 9 and ln(bod) = 0.8
# and (2) ln(ele) = 5.5, oxygen = 10 and ln(bod) = 1.0
newo <- data.frame(c(6.8, 5.5), c(9, 10), c(0.8, 1))
colnames(newo) <- colnames(env.pars3)
newo
(predict.new <- predict(spe.lda, newdata = newo))
```

The posterior probabilities of the new objects (rows) to belong to groups 1–4 (columns) are given by element "`$posterior`" of the prediction result. Here the result is:

```
$posterior
              1         2            3            4
1 0.1118150083 0.8879697 0.0002152868 2.067257e-09
2 0.0009074883 0.1115628 0.8875164034 1.328932e-05
```

The first object (row 1) has the highest probability (0.88) to belong to group 2, and the second object to group 3. Now you could examine the profiles of the fish species of groups 2 and 3. What you have actually done is to forecast that a site with the environmental values written in data frame **`newo`** should contain this type of fish community.

The second example is based on the same data, but the computation is run on standardized variables to obtain discrimination functions. We complement it with the display or computation of other LDA results. A plot is drawn using our homemade function **`plot.lda.R`** (Fig. 6.13).

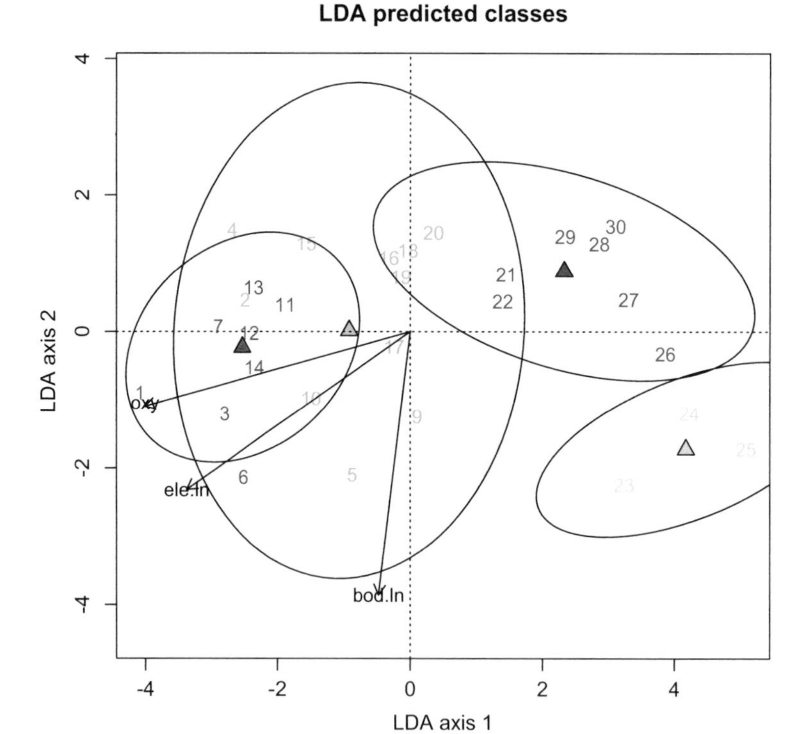

Fig. 6.13 Plot of the first two axes of the LDA of a four-group fish typology explained by three environmental variables

```
# Computation of LDA - discrimination functions (on standardized
# variables)
env.pars3.sc <- as.data.frame(scale(env.pars3.df))
spe.lda2 <- lda(gr ~ ., data = env.pars3.sc)
# Display the group means for the 3 variables
spe.lda2$means
# Extract the classification functions
(C2 <- spe.lda2$scaling)
# Compute the canonical eigenvalues
spe.lda2$svd^2
# Position the objects in the space of the canonical variates
(Fp2 <- predict(spe.lda2)$x)
# Classification of the objects
(spe.class2 <- predict(spe.lda2)$class)
# Posterior probabilities of the objects to belong to the groups
# (rounded for easier interpretation)
(spe.post2 <- round(predict(spe.lda2)$posterior, 2))
# Contingency table of prior versus predicted classifications
(spe.table2 <- table(gr, spe.class2))
# Proportion of correct classification (classification success)
diag(prop.table(spe.table2, 1))
# Plot the LDA results using the homemade function plot.lda()
plot.lda(lda.out = spe.lda2,
  groups = gr,
  plot.sites = 2,
  plot.centroids = 1,
  mul.coef = 2.35
)
```

Play with the numerous arguments of `plot.lda()` *to customize your LDA plot. Some trial and error is needed to adjust the lengths of the arrows with argument* `mul.coef`*.*

LDA can also be associated with cross-validation, to assess the prediction success of the classification. The analysis is repeated numerous times, each time leaving out one observation and verifying where it is classified. This approach is thus prediction-oriented, a desirable property in many real applications. Let us run our example with this option, activated by argument `CV = TRUE` in **`lda()`**.

```
# LDA with jackknife-based classification (i.e., leave-one-out
# cross-validation)
(spe.lda.jac <-
  lda(gr ~ ele.ln + oxy + bod.ln,
      data = env.pars3.sc,
      CV = TRUE))
summary(spe.lda.jac)
# Numbers and proportions of correct classification
spe.jac.class <- spe.lda.jac$class
spe.jac.table <- table(gr, spe.jac.class)
# Classification success
diag(prop.table(spe.jac.table, 1))
```

The classification success in `spe.jac.table` is not as good as the result in `spe.table2`. Remember, however, that `spe.table2` shows an *a posteriori* classification of the objects that have been used in the computations. It is too optimistic. By comparison, cross-validation results are obtained by computing the 'lda' and classification of each object, in turn, with that object taken out of the 'lda' calculation. It is more realistic.

Technical note: in the demonstrations above, we ran two separate analyses for identification and discrimination, to allow the display of the two different sets of functions obtained. Actually, with function **`lda()`** this is not necessary; we did it here for demonstration purpose. Running the function with unstandardized explanatory variables will allow the classification of new objects while producing the exact same discrimination as a run with standardized variables.

6.6 Other Asymmetric Analyses

Not all possible forms of asymmetric multivariate analysis have been presented above. There are several additional methods that may prove useful in some applications. Among them, let us mention the Principal response curves (PRC; Van den Brink and ter Braak 1998, 1999) and the asymmetric form of co-correspondence analysis (ter Braak and Schaffers 2004). We devote short sections to these two methods. The data and applications are those available in the documentation files of the corresponding **R** functions.

6.6.1 Principal Response Curves (PRC)

As community ecology has become more and more experimental, controlled and replicated designs that were previously the domain of single-response (univariate) experiments are now applied to study the response of communities to important ecological stressors. A complex example is a design where a set of experimental units (e.g. mesocosms) is submitted to treatments of various types or intensities, whose outcomes are monitored along time and compared with control units. Such experimental results can be analysed with standard RDA, but the resulting plots and numerical results are complex and the relevant patterns difficult to identify. In particular, the differences between control and treatment units at each time step do not stand out; they should, because they are the most important features of such experiments.

Principal response curves address the problems related to the analysis of multivariate results of designed experiments that involve repeated measurements over time by means of a modified form of RDA. The PRC method focuses on the differences between control and treatments; it provides a clear graphical illustration of treatment effects at the community as well as the species level.

Let D be the total number of treatment levels and T the total number of time points. With a balanced design with R replicates per level, a standard RDA would involve the response data with $n = D \times T \times R$ observations and m species; the explanatory variables would consist in an interaction factor of $D \times T$ levels indicating to which combination of treatment level and time point each observation belongs (Van den Brink and ter Braak 1999) plus, if needed, the main effects D and T. The PRC method, in contrast, specifically focuses on the difference between control and treatment level at each time point. To achieve that, one must remove the overall effect of time, i.e., use the time factor as a covariable. This makes sure that any overall time effect is removed. The explanatory factor, on the other hand, is the one with $D \times T$ levels as above, but with the levels corresponding to the control removed "*so as to ensure that the treatment effects are expressed as deviations from the control*" (Van den Brink and ter Braak 1999). The canonical coefficients resulting from the RDA are plotted against time; curves representing these scores for each treatment along time are called the principal response curves of the community. The species weights can be assessed by means of their regression coefficients against the site scores. They represent "*the multiple by which the principal curves must be multiplied to obtain the fitted response curves of* [each] *species*" (Van den Brink and ter Braak 1999). A high positive weight indicates a high likelihood that the species follows the overall pattern of the PRC. A negative weight shows a tendency of the species to follow an opposite pattern. A weight close to 0 indicates either no pattern or a pattern of a shape differing from the overall one. Note that, in this case, a small weight would not indicate a lack of response of the species to the treatment, but a response pattern that may be strong but different to the overall (community-level) one.

Principal response curves can be computed with the function **`prc()`** of package **`vegan`**. The example available in the documentation file of the function is the one used by Van den Brink and ter Braak (1999) in their paper and consists in the study of the effects of insecticide treatments on aquatic invertebrate communities. We will present it here. It is based on observations on the abundances of 178 invertebrate species (macroinvertebrates and zooplankton) subjected to insecticide treatments in aquatic mesocosms ("ditches"). The species data are log-transformed abundances, $y_{tr} = \ln(10y + 1)$. The experiment involved twelve mesocosms, which were surveyed on eleven occasions, so $n = 12 \times 11 = 132$. Four mesocosms served as controls (dose = 0) and the remaining eight were treated once with the insecticide chlorpyrifos, with dose levels of 0.1, 0.9, 6 and 44 μg L^{-1} in two mesocosms each. Therefore, the explanatory factor has four levels (the four dose levels, excluding control) and the covariable "week" has 11 levels. The code below is extracted from the **prc()** documentation file, with added comments.

```
# Code from the prc() help file, with additional comments
# Chlorpyrifos experiment and experimental design:  Pesticide
# treatment in ditches (replicated) and followed over, from 4 weeks
# before, to 24 weeks after exposure

# Extract the data (available in vegan)
data(pyrifos)

# Create factors for time (week) and treatment (dose). Create an
# additional factor "ditch" representing the mesocosm, for
# testing purposes
week <-
  gl(11, 12,
     labels = c(-4, -1, 0.1, 1, 2, 4, 8, 12, 15, 19, 24))
dose <-
  factor(rep(c(0.1, 0, 0, 0.9, 0, 44, 6, 0.1, 44, 0.9, 0, 6),
         11))
ditch <- gl(12, 1, length = 132)

# PRC
mod <- prc(pyrifos, dose, week)
mod             # Modified RDA
summary(mod)    # Results formatted as PRC

# PRC plot; at the right of it, only species with large total
# (log-transformed) abundances are reported
logabu <- colSums(pyrifos)
plot(mod, select = logabu > 200)

# Statistical test
# Ditches are randomized, we have a time series, and are only
# interested in the first axis
ctrl <-
  how(plots = Plots(strata = ditch, type = "free"),
      within = Within(type = "series"), nperm = 999)
anova(mod, permutations = ctrl, first = TRUE)
```

The analysis results in the following plot (Fig. 6.14).

Figure 6.14 shows that the intensity of community response to the treatment clearly depends on the chlorpyrifos dose. At higher doses, the response is greater and lasts longer. Note, however, that between doses 6 and 44 μg L^{-1} the peak intensities of the responses are comparable; the difference is that dose 44 μg L^{-1} produces a longer effect.

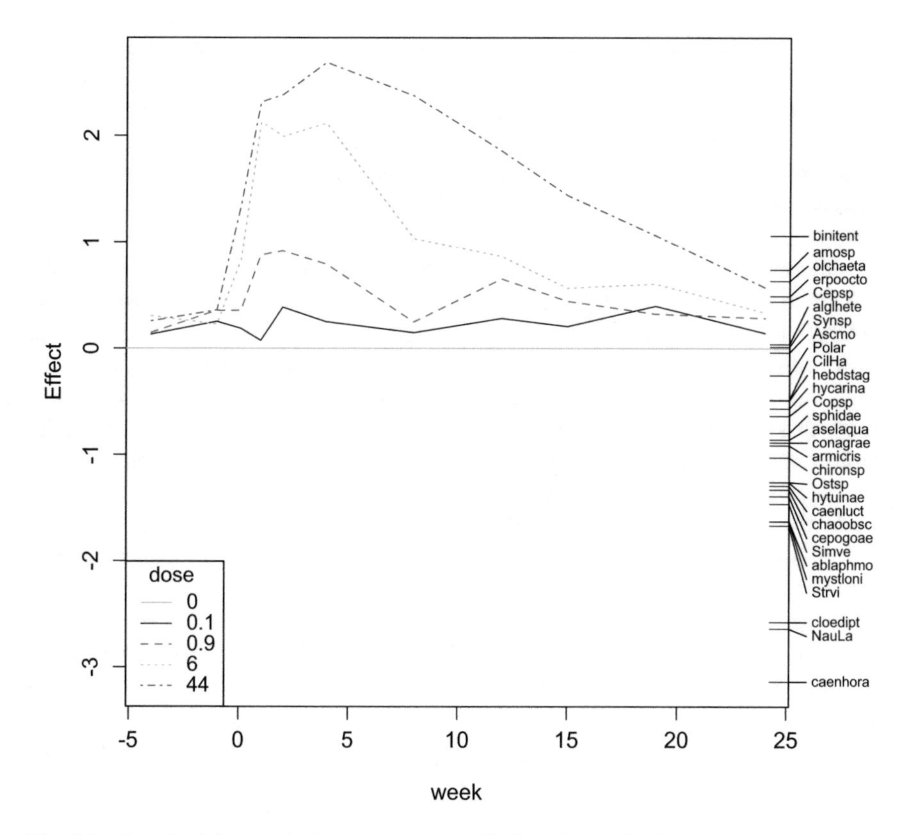

Fig. 6.14 Result of the principal response curves (PRC) analysis. The lines represent differences with respect to the control plots ("dose = 0") expressed as canonical coefficients on the first RDA axis. Along the right-hand margin, species weights show the degree of agreement between the species-level response to the treatment and the overall, community-level response

6.6.2 Co-correspondence Analysis (CoCA)

Co-correspondence analysis is based on correspondence analysis (CA) and is devoted to the simultaneous ordination of two communities sampled at the same sites (ter Braak and Schaffers 2004). Its asymmetric form allows one to predict a community on the basis of the other. The method has been developed to fill a gap: usually ecologists explain community data by means of environmental variables of various types (e.g. climate, soil, chemistry, anthropogenic influence, but not other species data) or, in a reversed calibration and prediction approach, they estimate the values of environmental variables by means of bioindicator species or communities (see Sect. 6.3.2.7 for an example). But in some cases one could be interested in assessing the relationship between two living communities, e.g. invertebrates and plants. This may be done for instance to verify if an invertebrate community is more related to the plant community or to other, e.g. physico-chemical constraints, or else to predict one community that is difficult to sample by means of another, easier one. Co-correspondence analysis can be run in a symmetric way (as Co-inertia analysis (CoIA, Sect. 6.9) does for other types of data, or in an asymmetric way, the one that interests us in this section.

Since both matrices involved contain species data, it is preferable to use a method that respects this particularity also for the explanatory matrix, as co-correspondence analysis does. The two reasons invoked by its authors are (1) CCA (and RDA) can only be used when the number of explanatory variables is smaller than the number of sites (which is often not the case with community data), and (2) the linear combinations of explanatory variables are not well suited to cases where these variables are themselves species data with many zeros and unimodal distributions with respect to the environment. ter Braak and Schaffers (2004) present the mathematics of the method in detail. Very briefly: the first ordination axis of symmetric and asymmetric CoCA is obtained by weighted averages, where (quoting the authors) "*the species scores of one set are obtained as weighted averages of the other set's site scores* [and] *the site scores are weighted averages of the species scores of their own set*". Then, in symmetric CoCA, the next axes are extracted in the same way, with the successive species scores uncorrelated with the previous ones. In asymmetric CoCA, the successive axes are constructed such that "*the site scores derived from the predictor variables* [are] *uncorrelated with the previously derived site scores*". The number of axes to interpret can be assessed either by cross-validation (details in the paper) or by a permutation test.

Symmetric and asymmetric co-correspondence analysis can be computed with an **R** package called **cocorresp**. Here we present one of the examples proposed in the documentation file of the function. The example, also treated in ter Braak and Schaffers' (2004) paper, deals with bryophytes and vascular plants in Carpathian spring meadows. The data set comprises 70 sites and the species of both communities that are present in at least 5 sites, i.e., 30 bryophyte and 123 vascular plant species. The predictive co-correspondence analysis presented here uses the bryophytes as response and the vascular plants as predictor variables.

```
data(bryophyte)
data(vascular)

# Co-correspondence analysis is computed using the function coca()

# The default option is method = "predictive"

carp.pred <- coca(bryophyte ~ ., data = vascular)
carp.pred
# Leave-one-out cross-validation
crossval(bryophyte, vascular)
# Permutation test
(carp.perm <- permutest(carp.pred, permutations = 99))
# Only two significant axes
carp.pred <- coca(bryophyte ~ ., data = vascular, n.axes = 2)
carp.pred
# Extract the site scores and the species loadings used in
# the biplots
carp.scores <- scores(carp.pred)
load.bryo <- carp.pred$loadings$Y
load.plant <- carp.pred$loadings$X
```

> *We have created two plots. As in ter Braak and Schaffers (2004, Fig. 3), in both plots the site scores are derived from the vascular plants (`carp.scores$sites$X`) and the species scores are the "loadings with respect to normalized site scores".*

```
# Printing options:
?plot.predcoca

par(mfrow = c(1, 2))
plot(carp.pred,
  type = "none",
  main = "Bryophytes",
  xlim = c(-2, 3),
  ylim = c(-3, 2)
)
points(carp.scores$sites$X, pch = 16, cex = 0.5)
text(load.bryo,
  labels = rownames(load.bryo),
  cex = 0.7,
  col = "red"
)
plot(carp.pred,
  type = "none",
  main = "Vascular plants",
  xlim = c(-2, 3),
  ylim = c(-3, 2)
)
points(carp.scores$sites$X, pch = 16, cex = 0.5)
text(load.plant,
  labels = rownames(load.plant),
  cex = 0.7,
  col = "blue"
)
# Detach cocorrespo to avoid conflict with ade4
detach("package:cocorresp", unload = TRUE)
```

Figure 6.15 shows the resulting plots, which have been tailored like Fig. 3 in ter Braak and Schaffers (2004). Readers can reproduce them in a larger format by running the R code on their computer, in order to examine them at leisure.

The output of the analysis shows that the two canonical axes together explain 30.4% of the variance of the bryophyte data. The more conservative cross-validatory

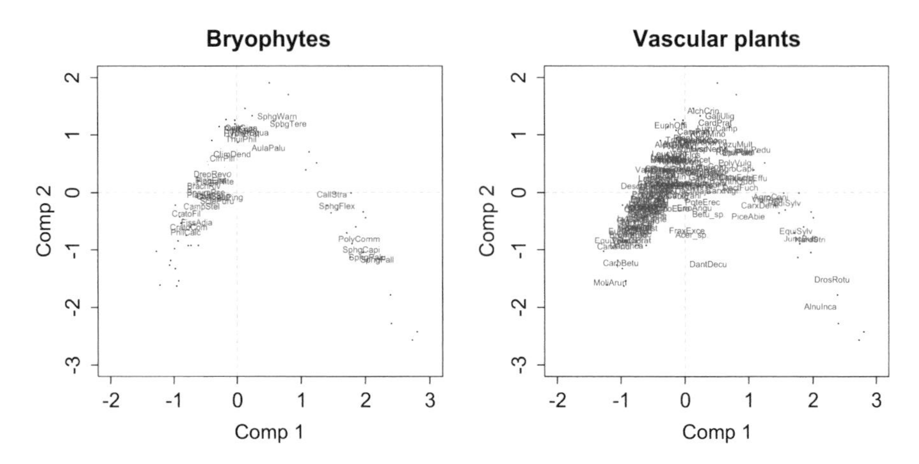

Fig. 6.15 Biplots of a predictive co-correspondence analysis with bryophytes as response variables and vascular plants as explanatory variables

fit is 24.8% for the first two axes. The cross-validatory fit culminates with 5 axes (28.3%); then it decreases because the predictive power decreases with more axes.

6.7 Symmetric Analysis of Two (or More) Data Sets

"Symmetric analysis" means that the two or more matrices involved in the analysis play the same role; there is no "dependent" or "explanatory" matrix. The choice between symmetric and asymmetric ordination methods is akin to the choice between correlation (symmetric) and model I regression analysis (asymmetric analysis). The former is more descriptive or exploratory, and also appropriate when no unidirectional causal hypothesis is embedded in the model, while the latter is more inferential, i.e., oriented at explaining the variation of response variables by means of a (hopefully parsimonious) linear combination of explanatory variables. The two approaches fulfil different research aims and should not be opposed as competitors on the same terrain.

Three symmetric methods are presented here because of their interest in ecology: canonical correlation analysis (CCorA), co-inertia analysis (CoIA) and multiple factor analysis (MFA). Another method, the symmetric form of co-correspondence analysis (whose asymmetric form is presented in Sect. 6.6.2), is devoted to the simultaneous ordination of two communities. As such it is very close to CoIA applied with CA. It can be computed with function **coca()** of package **cocorresp**.

6.8 Canonical Correlation Analysis (CCorA)

6.8.1 Introduction

CCorA is computed from two data tables. The aim of the method is to represent the observations along canonical axes that maximize the correlations between the two tables. The solution is found by maximizing the between-set dispersion, expressed by the covariance matrix between the two sets of variables, with respect to the within-set dispersion (Legendre and Legendre 2012 Sect. 11.4). The two sets of variables must be quantitative and are assumed to be multinormally distributed. The limitation of the method is that the total number of variables in each data table must be smaller than $(n - 1)$.

In CCorA, one can also test the hypothesis of linear independence of the two multivariate data tables. Pillai and Hsu (1979) have shown that Pillai's trace is the most robust statistic to departures from normality.

The availability of RDA and CCA has made the application of CCorA in ecology less frequent, since most ecological problems are stated in terms of control-response hypotheses for which asymmetric ordination should be preferred. CCorA is more appropriate for exploratory purposes and in cases where the two groups of variables are likely to influence each other, which may often occur in real ecological systems. Examples are the study of two groups of competing taxa, a vegetation-herbivore system, and long-term studies of soil-vegetation relationships during a colonization process; CCorA is possible as long as the number of species in each data table is smaller than $(n–1)$.

6.8.2 Canonical Correlation Analysis Using `CCorA()`

In the variation partitioning example of Sect. 6.3.2.8, we used two subsets of environmental variables, chemistry and physiography, to explain the structure of the fish data. Putting aside the variation partitioning of that example, we could study the structure of correlation of the two complete subsets of explanatory variables. How does chemistry relate to physiography?

Since the data should be as close as possible to the condition of multinormality, we will transform some variables in the following example to make them more symmetrically distributed (we used the Shapiro-Wilk normality test, function **`shapiro.test()`** of package **`stats`**; results not shown here). The variables have different physical dimensions. The CCorA equation includes automatic standardization of the variables. See Legendre and Legendre (2012, Sect. 11.4.1) for details. However, asking for standardization, or not, changes the RDA results at the end of the CCorA output file.

The function used for the analysis is **`CCorA()`** of package **`vegan`**.

```
# Preparation of data (transformations to make variable
# distributions approximately symmetrical)
envchem2 <- envchem
envchem2$pho <- log(envchem$pho)
envchem2$nit <- sqrt(envchem$nit)
envchem2$amm <- log1p(envchem$amm)
envchem2$bod <- log(envchem$bod)
envtopo2 <- envtopo
envtopo2$ele <- log(envtopo$ele)
envtopo2$slo <- log(envtopo$slo)
envtopo2$dis <- sqrt(envtopo$dis)

# CCorA (on standardized variables)
chem.topo.ccora <-
  CCorA(envchem2, envtopo2,
        stand.Y = TRUE,
        stand.X = TRUE,
        permutations = how(nperm = 999))
chem.topo.ccora
biplot(chem.topo.ccora, plot.type = "biplot")
```

Hint *The* **`biplot() {vegan}`** *function used here recognizes the structure of a CCorA object. If you type* **`biplot(chem.topo.ccora)`** *without adding the argument* `plot.type = "biplot"`, *you obtain separate plots of sites and variables, i.e., four graphs in a single frame; easier to compare for large data sets.*

The result shows that there is a significant relationship between the two matrices (permutational probability = 0.001). Pillai's trace statistic is the sum of the squared canonical correlations. The canonical correlations are high on the first two axes. The RDA R^2 and adjusted R^2 are not part of the CCorA computations strictly speaking; the two RDAs are computed separately for information. This information is useful to assess whether the canonical axes are likely to express a substantial amount of variation (which is the case here), since canonical correlations may be large even when the common variation is small with respect to the total variation of the two data sets.

In Fig. 6.16, the left-hand biplot shows the standardized chemical variables and the objects projected in their space. The right-hand biplot shows the standardized physiographic variables and objects in their space. Note that the two spaces are "aligned" with respect to one another, i.e. the canonical axes show the same trends expressed by the two sets of variables. The positions of the sites in the two biplots are related, although not similar. The structures displayed are the result of *linear combinations* of variables in each of the biplots. The pair of biplots expresses the fact that oxygenated waters (`oxy`) are related to high elevation (`ele`) and steep

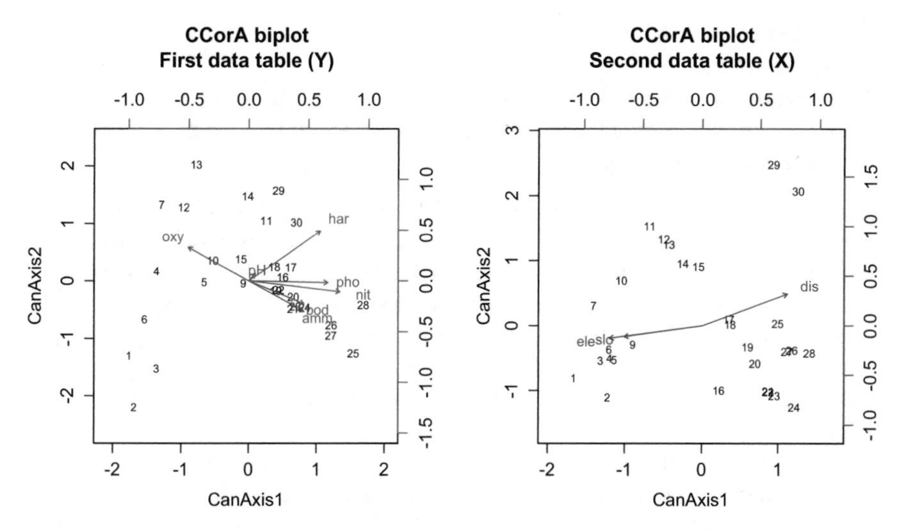

Fig. 6.16 Biplots of a canonical correlation analysis (CCorA) of the chemical (left) and physiographic (right) variables of the Doubs data

slope (`slo`), i.e., upstream conditions, whereas discharge (`dis`) is highly positively correlated with high hardness (`har`), phosphates (`pho`) and nitrates (`nit`); elevation (`ele`) and slope (`slo`) are highly negatively correlated with these same variables.

Canonical correlation analysis is also available in package **`stats`** (function **`cancor()`**) and in a package (unfortunately) called **`CCA`**, a wrapper computing canonical correlations using **`cancor()`** in a function called **`cc()`**, which provides graphical outputs (function **`plt.cc()`**) and extensions to situations where the number of variables exceeds the number of sites (function **`rcc()`**).

6.9 Co-inertia Analysis (CoIA)

6.9.1 Introduction

Dolédec and Chessel (1994) proposed an alternative to CCorA called *co-inertia analysis* (CoIA). Dray et al. (2003) showed that this approach is a very general and flexible way to couple two or more data tables. CoIA is a symmetric approach allowing the use of various methods to model the structure in each data matrix.

The method works as follows (for two data tables):

- Compute the covariance matrix crossing the variables of the two data tables. The sum of the squared covariances is the total co-inertia. Compute the eigenvalues and eigenvectors of that matrix. The eigenvalues represent a partitioning of the total co-inertia.

- Project the objects and variables of the two original data tables on the co-inertia axes. By means of graphs, compare the projections of the two data tables in the common co-inertia space.

Basically, CoIA requires site-by-variables input matrices. The equations are provided by Legendre and Legendre (2012 Sect. 11.5). An attractive feature of CoIA is the possibility to adapt it to the mathematical type of the variables of the two matrices, using the various transformations presented in Chaps. 2 and 3: standardization of the variables expressed in different physical units; Hellinger, chord or other appropriate transformations for species presence-absence or abundance data. If the Euclidean distance among object computed from the raw data is to be preserved, no transformation is necessary.

A function called **`coinertia()`** is available in package **`ade4`** to compute CoIA. However, this function implements the data transformations in a particular way. For internal technical reasons, the two data tables must first be submitted to an **`ade4`** ordination function: **`dudi.pca()`** for PCA, **`dudi.pco()`** for PCoA, and so on[5]. This intermediate step acts as a transformation route. For untransformed data (or pretransformed species data), use **`dudi.pca()`** with argument `scale = FALSE`; to standardize the variables, use the same function with `scale = TRUE`. If the required transformation involves computation of a dissimilarity index, it can be computed by the appropriate function (see Chap. 3); **`dudi.pco()`** is then called to extract its principal coordinates. Function **`coinertia()`** retrieves the centred or transformed data matrices from the **`dudi.xxx()`** output objects and computes coinertia analysis from them.

Note also that the row weights must be equal in the two separate ordinations, a condition that makes the use of CoIA after correspondence analysis (**`dudi.coa`**) difficult. CA is a weighted regression approach, and weights depend on the data, so that the two data tables are likely to have different row weights. To produce a symmetric analysis based on CA, we suggest to use the function **`coca()`** of package **`cocorresp`** with `method = "symmetric"` (see Sect. 6.6.2).

6.9.2 *Co-inertia Analysis Using Function* `coinertia()` *of* `ade4`

In the code below, we apply CoIA to the chemical and physiographic subsets of environmental variables of the Doubs data set. Following the route explained above

---

[5]Note that **ade4** has been developed around a general mathematical framework involving entities that will not be described here, called duality diagrams (Escoufier 1987); hence the **`dudi`** part of the function names. Readers are invited to consult the original publication to learn more about this framework.

(preliminary use of ordination functions to obtain the data, transformed or not), a PCA of the standardized data (correlation matrix) is first performed on each of the two data tables. The proportion of variance accounted for by the axes is then computed to assess the number of axes to be retained in the CoIA. In this example, 3 axes of the chemistry PCA account for 89.8% variation, and 2 axes of the physiography PCA account for 98.9% variation. After having verified that the row weights are equal in the two PCAs, these two results are then submitted to CoIA, which is asked to retain two canonical axes. A permutation test is run to assess the significance of the co-inertia structure of the data tables.

```
# PCA on both matrices using ade4 functions
dudi.chem <- dudi.pca(envchem2,
         scale = TRUE,
         scannf = FALSE)
dudi.topo <- dudi.pca(envtopo2,
         scale = TRUE,
         scannf = FALSE)

# Cumulated relative variation of eigenvalues
cumsum(dudi.chem$eig / sum(dudi.chem$eig))
# Cumulated relative variation of eigenvalues
cumsum(dudi.topo$eig / sum(dudi.topo$eig))
# Are the row weights equal in the 2 analyses?
all.equal(dudi.chem$lw, dudi.topo$lw)

# Co-inertia analysis
coia.chem.topo <-
  coinertia(dudi.chem, dudi.topo,
            scannf = FALSE,
            nf = 2)
summary(coia.chem.topo)
# Relative variation on first eigenvalue
coia.chem.topo$eig[1] / sum(coia.chem.topo$eig)
# Permutation test
randtest(coia.chem.topo, nrepet = 999)
# Plot results
plot(coia.chem.topo)
```

Figure 6.17 gives a visual summary of the results of the CoIA. The numerical output looks like this:

```
Eigenvalues decomposition:
         eig    covar       sdX       sdY      corr
1 6.78059294 2.603957 1.9995185 1.6364180 0.7958187
2 0.05642003 0.237529 0.8714547 0.5355477 0.5089483
```

```
Inertia & coinertia X (dudi.chem):
    inertia      max     ratio
1  3.998074 4.346012 0.9199410
12 4.757508 5.572031 0.8538193    <= "12" means "axes 1 and 2"

Inertia & coinertia Y (dudi.topo):
    inertia      max     ratio
1  2.677864 2.681042 0.9988147
12 2.964675 2.967516 0.9990428

RV:
 0.553804
```

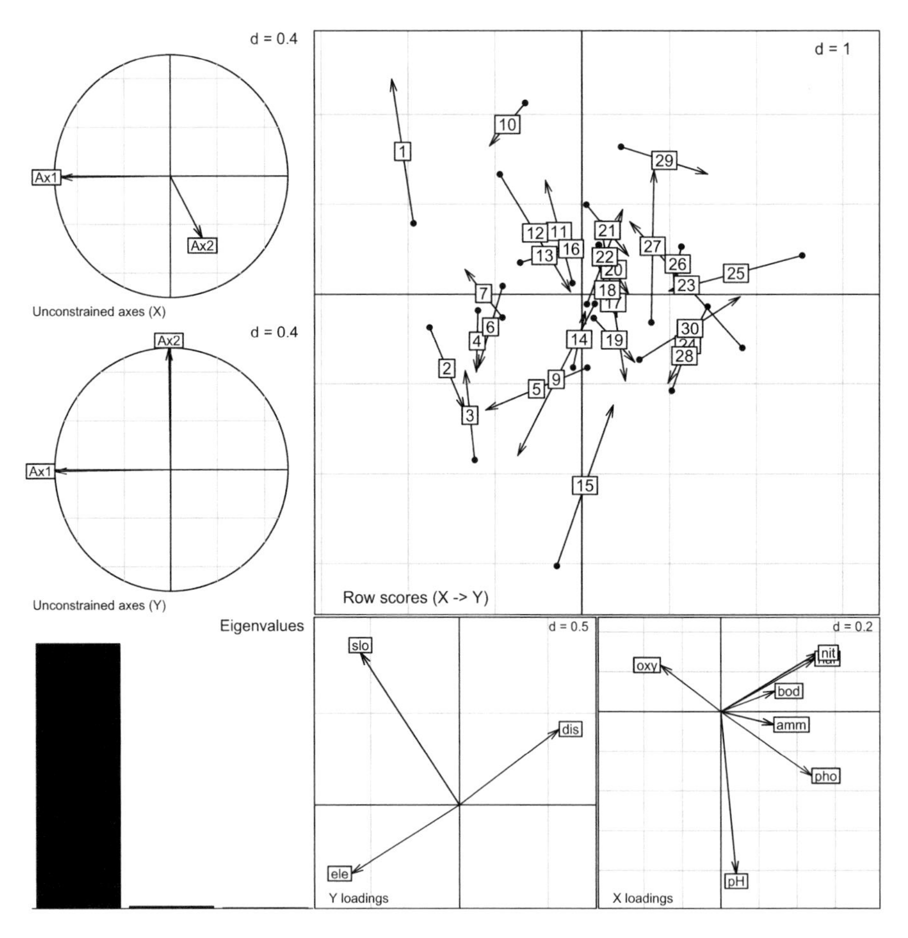

Fig. 6.17 Graphical results of a co-inertia analysis of the chemical and physiographic variables. Details: see text. X refers to the first and Y to the second data table

The numerical results first present the eigenvalue decomposition of the matrix of co-inertia on two axes (rows): eigenvalues (eig), covariance (covar), standard deviation (sdX and sdY) of the two sets of site scores on the co-inertia axes and correlations between the two sets of site scores, computed using the Pearson correlation coefficient.

The second and third blocks of results compare the inertia of the (cumulated) projections of the data tables, called X and Y in the function, as they are projected in the co-inertia analysis ("inertia"), compared to the maximum cumulated inertia of the axes of the separate ordinations ("max"). It also gives the ratio between these values as a measure of concordance between the two projections.

The RV coefficient is the ratio of the total co-inertia to the square root of the product of the total inertias of the separate analyses (Robert and Escoufier, 1976). Ranged between 0 (independent) and 1 (homothetic), it measures the closeness between the two sets of points derived from the separate ordinations of X and Y. For two simple variables $\mathbf{x}_1$ and $\mathbf{x}_2$, RV is the square of their Pearson correlation coefficient.

These results show that the first eigenvalue, which represents 98.9% of the total variation, is overwhelmingly larger than the second one. Most of the common structure of the two data matrices is therefore to be sought along the first axis. The circular plots in Fig. 6.17 show that axes 1 of the two PCAs are almost perfectly aligned on the first CoIA axis. The upper right-hand plot (normed site scores) shows the positions of the sites on the co-inertia axes using the chemistry (origins of the arrows) and physiography (arrowheads) co-inertia weights. The shorter an arrow is, the better the concordance between the two projections of the point. The lower right-hand pair of plots shows the contributions of the two groups of variables to the canonical space. Vectors pointing in the same direction are correlated and longer vectors contribute more to the structure. Oxygen (`oxy`) correlates positively with slope (`slo`), phosphates (`pho`) negatively with slope (`slo`); nitrates (`nit`), hardness (`har`, label masked by nitrates) and biological oxygen demand (`bod`) are all negatively correlated with elevation (`ele`) since these variables have higher values downstream, and positively with discharge (`dis`), which increases downstream.

An extension of CoIA called RLQ analysis (Dolédec et al. 1996; Dray et al. 2002) relates species traits to environmental variables by means of three tables: site-by-species (table L), site-by-environment (table R), and species-by-traits (table Q). Another related method, developed by Legendre et al. (1997) and Dray and Legendre (2008), is also an answer to what these authors have called the fourth-corner problem. The RLQ and fourth-corner analyses are presented in Sect. 6.11.

6.10 Multiple Factor Analysis (MFA)

6.10.1 Introduction

Yet another approach to the symmetric analysis of a data set described by k (usually $k > 2$) subsets or groups of variables is multiple factor analysis (MFA; Escofier and Pagès 1994; Abdi et al. 2013). This analysis is correlative; it does not involve any hypothesis of causal influence of a data set on another. The variables must belong to the same mathematical type (quantitative or qualitative) within each subset. If all variables are quantitative, then MFA is basically a PCA applied to the whole set of variables in which each subset is weighted. Do not confuse it with multiple correspondence analysis (MCA, Sect. 5.4.5) where a single matrix of qualitative variables is submitted to ordination and other matrices may be added as supplementary (passive) information.

MFA computation consists in the following steps:

- a PCA is computed for each (centred and optionally standardized) subset of quantitative variables. PCA is replaced by MCA for subsets of qualitative variables. Each centred table is then weighted so that all receive equal weights in the global analysis, accounting for different variances among the groups. This is done by dividing all variables of each centred table by the first singular value (i.e., the square root of the first eigenvalue) obtained from its PCA (or MCA for qualitative variables) (Abdi et al. 2013);
- the k weighted data sets are regrouped using **cbind()** in R. The resulting table is submitted to a global PCA;
- the different subsets of variables are then projected on the global result; common structures and differences are assessed for objects and variables.

The pairwise similarities between the geometric representations derived from the k groups of variables are measured by the RV coefficient described in Sect. 6.9.2. RV coefficients, which vary between 0 and 1, can be tested by permutations (Josse et al. 2008).

MFA has been mainly used in economics, sensory evaluation (e.g. wine tasting) and chemistry so far, but the potential for ecological applications is promising, as evidenced by a few recent contributions (Beamud et al. 2010; Carlson et al. 2010; Lamentowicz et al. 2010). Indeed, this method is useful to explore the complex relationships among several ecologically meaningful groups of descriptors, whatever their number and type.

MFA can be computed by function **mfa()** of the package **ade4**. In this case, a data frame comprising all blocks of variables must first be assembled and set into class `ktab` by function **ktab.data.frame()**. Here we shall use function **MFA()** of the package **FactoMineR**, which is more straightforward and offers more options.

An extension of MFA to the case where the data are hierarchically organized (e.g. in regions and sub-regions, or questionnaires structured in topics and sub-topics; by extension, species and environmental data obtained from different layers of soil, and so on) has also been developed (Le Dien and Pagès 2003). A first promising ecological application of this Hierarchical Multiple Factor Analysis (HMFA) was devoted to the exploration of the structural relationships among vegetation, soil fauna and humus form in a subalpine forest ecosystem (Bernier and Gillet 2012). Its principle is to compute a MFA at each level of the hierarchy of variables, starting at the lowest (most resolved) level; the (weighted) results (PCA axes) resulting from the MFA at one level are used at the next (higher) level, where a new weighting is done according to the (smaller) number of groups. HMFA can be run with function **HMFA()** of the package **FactoMineR** (Lê et al. 2008)

6.10.2 *Multiple Factor Analysis Using* **FactoMineR**

In the code below, we apply MFA to three subsets of the Doubs data: the species (Hellinger-transformed abundances), the physiographic variables (upstream-downstream gradient), and the chemical variables (water quality). Note that this example is not at all equivalent to a constrained ordination where the species data are explained by environmental variables and where the focus is put on an underlying, one-directional causal model. MFA proposes a symmetric, exploratory point of view, where correlative structures are exposed without any reference to a directionality of possible causal relationships. No formal directional hypothesis is tested, either. This approach is therefore not adapted to the modelling of asymmetric relationships, a task devoted to RDA or CCA. However, MFA could be used in the early stages of a research project as a neutral data exploration technique to help generate causal hypotheses, which could be tested afterwards using independent data sets.

The function **MFA()** includes an important argument `type`, which allows the specification of the mathematical type of each subset: `"c"` for continuous variables (to run a PCA on a covariance matrix), `"s"` for continuous variables requiring standardization (to run a PCA on a correlation matrix) or `"n"` for nominal variables (to run a multiple correspondence analysis (MCA, Sect. 5.4.5). In our case, we have to state that the species subset belongs to type `"c"` whereas the two environmental subsets (chemistry and physiography) belong to type `"s"`.

One can draw a scree plot and a broken stick model of the MFA eigenvalues, but function **screeplot.cca()** of **vegan** does not work on the output of the **MFA()** function, which belongs to **FactoMineR**. This is why we wrote a function called **screestick()**, which uses a vector of eigenvalues. That vector can be retrieved from the output object of an ordination produced by any package.

```
# MFA on 3 groups of variables:
# Regroup the 3 tables (Hellinger-transformed species,
# physiographic variables, chemical variables)
tab3 <- data.frame(spe.hel, envtopo, envchem)
dim(tab3)
# Number of variables in each group
(grn <- c(ncol(spe), ncol(envtopo), ncol(envchem)))

# Compute the MFA without multiple plots
t3.mfa <- MFA(tab3,
 group = grn,
 type = c("c", "s", "s"),
 ncp = 2,
 name.group = c("Fish community", "Physiography", "Water quality"),
 graph = FALSE)
t3.mfa

# Plot the results
plot(t3.mfa,
     choix = "axes",
     habillage = "group",
     shadowtext = TRUE)
plot(
  t3.mfa,
  choix = "ind",
  partial = "all",
  habillage = "group")
plot(t3.mfa,
     choix = "var",
     habillage = "group",
     shadowtext = TRUE)
plot(t3.mfa, choix = "group")

# Eigenvalues, scree plot and broken stick model
ev <- t3.mfa$eig[, 1]
names(ev) <- paste("MFA", 1 : length(ev))
screestick(ev, las = 2)
```

Hint *The* **`MFA()`** *function has a default* `graph = TRUE` *argument, which automatically produces the three graphs presented here. However, this choice does not allow users to plot graphs in external graphic devices with* ***RStudio****. This is why we produced the graphs separately, after having set the* `graph` *argument to* `FALSE`.

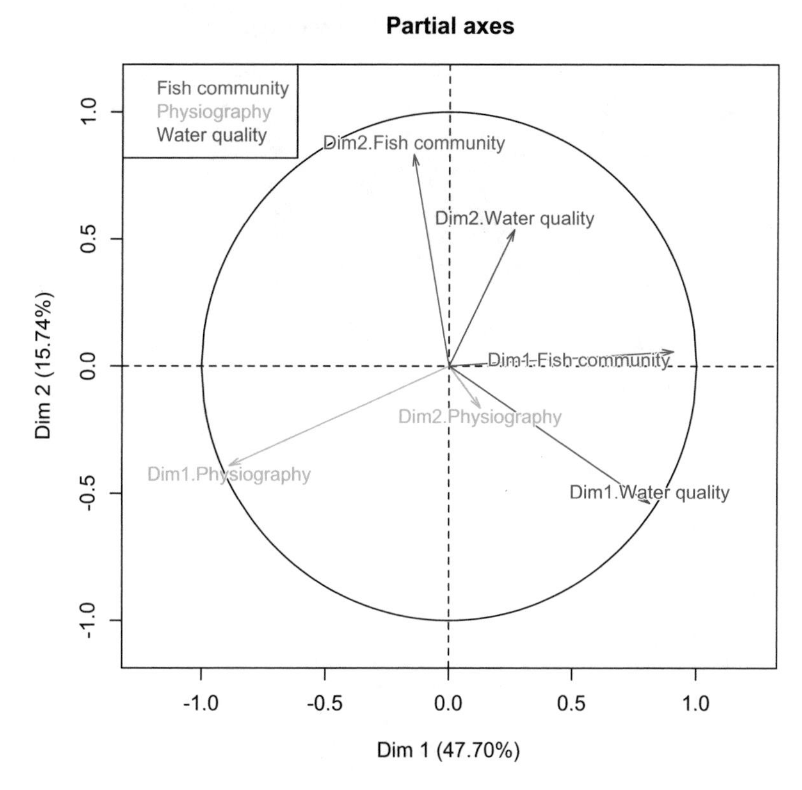

Fig. 6.18 Projection of the PCA axes of each subset on the MFA plane 1 × 2. The circle of radius 1 represents the maximum length of a partial standardized axis

The multiple factor analysis result provides an interesting picture of two main gradients and of the relationships among the three groups of variables. The first two axes represent more than 63% of the total variance. The plot "Partial axes" (Fig. 6.18) represents the projection of the principal components of each separate PCA on the global PCA.

The plot "Individual factor map" (Fig. 6.19) shows the positions of the sites according to four viewpoints: the labelled black points represent the MFA site scores (centroids of the site scores of the three separate PCAs); they are connected by coloured lines to the points representing their scores in the three separate PCAs.

The plot "Correlation circle" (Fig. 6.20) represents the normalized vectors of all quantitative variables.

If we examine Figs. 6.19 and 6.20 together, we can recognize the main upstream-downstream gradient along the first axis and the gradient of water quality along a combination of the first and second axes (from upper left to lower right). For example, the scores of sites 1, 2 and 3 (Fig. 6.19, left-hand part of the graph) correspond (Fig. 6.20) to high elevation and strong slope, as well as high oxygen

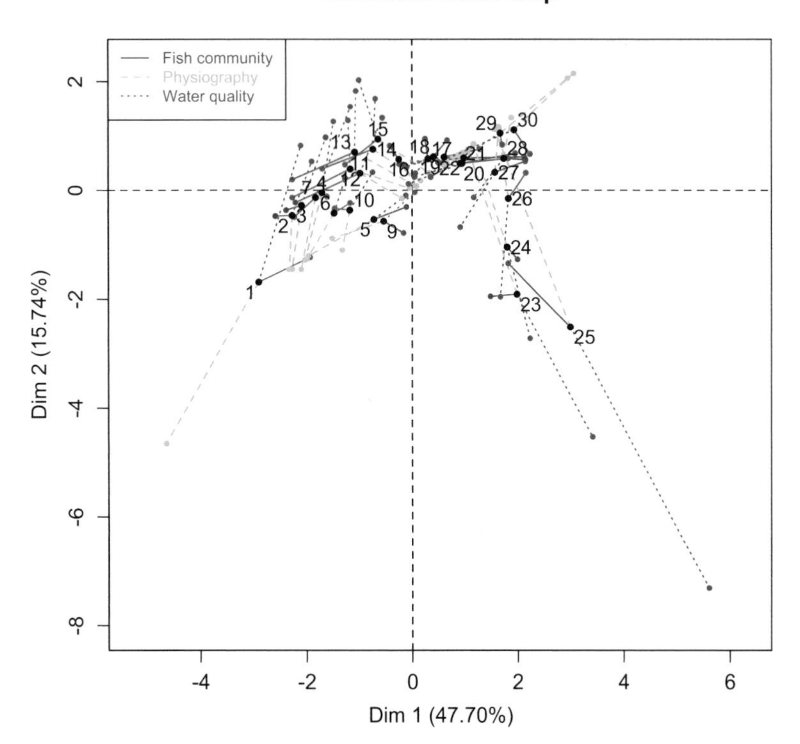

Fig. 6.19 MFA site scores (centroids; black dots) and PCA site scores with their links (colour lines) to the corresponding MFA centroids. Each MFA site score (black dot with site number) is connected to three coloured dots by coloured lines

concentration. Here, close to the source, the ecological conditions are dominated by physiography. The relatively poor fish community is characterized by `Satr`, `Phph` and `Babl`. On the opposite side, sites 23, 24 and 25 show the highest concentrations in phosphates, ammonium and nitrates, and a high biological oxygen demand. These three sites are heavily polluted and their community is characterized by another set of three species: `Alal`, `Ruru` and `Sqce`.

The RV coefficients between the pairs of groups are tested for significance:

```
# RV coefficients with tests (p-values above the diagonal of
# the matrix)
rvp <- t3.mfa$group$RV
rvp[1, 2] <- coeffRV(spe.hel, scale(envtopo))$p.value
rvp[1, 3] <- coeffRV(spe.hel, scale(envchem))$p.value
rvp[2, 3] <- coeffRV(scale(envtopo), scale(envchem))$p.value
round(rvp[-4, -4], 6)
```

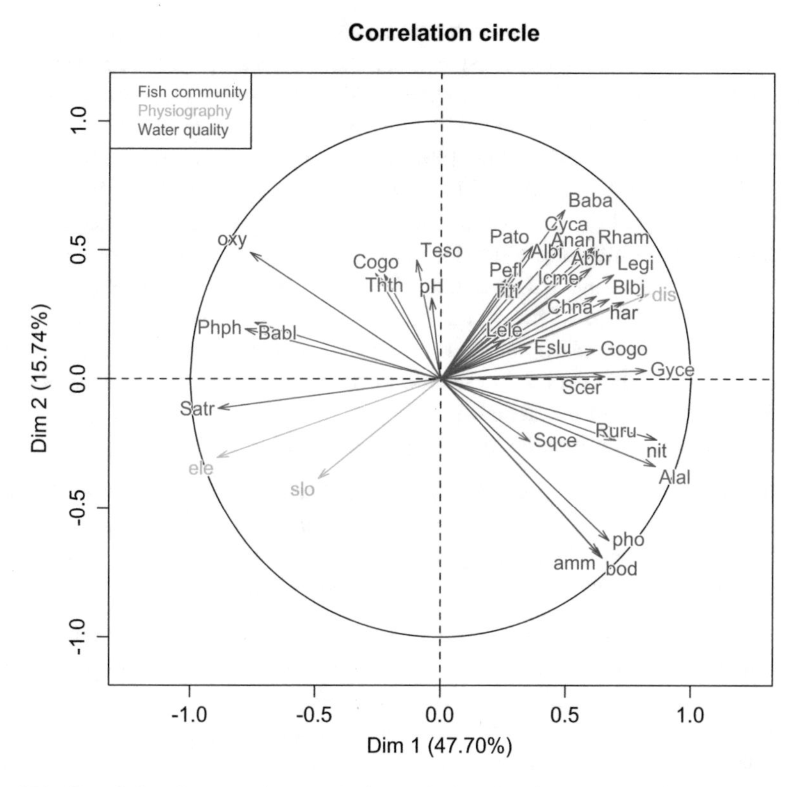

Fig. 6.20 Correlations between the quantitative variables of each subset on MFA axes 1 and 2

Table 6.1 RV coefficients (lower-left triangle) of the pairs of matrices involved in the MFA, and their p-values (upper-right triangle)

	Fish community	Physiography	Water quality
Fish community	1.000000	0.000002	0.000002
Physiography	0.580271	1.000000	0.002809
Water quality	0.505324	0.361901	1.000000

In Table 6.1, the RV coefficients appear in the lower-left triangle, below the main diagonal, while the upper-right triangle contains p-values. These results tell us that fish communities are mostly linked to the physiographic conditions (RV = 0.58), which are themselves partly linked to water chemistry (RV = 0.36).

6.11 Relating Species Traits and Environment

Niche theory predicts that species settle in environments to which their ecological and behavioural characteristics are adapted. Functional ecology is the discipline devoted to the study of the relationship between species traits and the environment.

This type of study poses methodological challenges to researchers because to test hypotheses in this context, one must evaluate the relationships between species traits and environmental characteristics, as mediated by the species presence-absence or abundance data. Two related methods have been proposed to answer this type of questions: the RLQ (Dolédec et al. 1996) and the fourth-corner methods (Legendre et al. 1997, Dray and Legendre 2008, Legendre and Legendre 2012). As explained by Dray et al. (2014), the two methods are complementary: the former is an ordination method allowing the visualization of the joint structure resulting from the three data tables, but with a single global test, whereas the latter consists in a series of statistical tests of individual trait-environment relationships, without consideration of the covariation among traits or environmental variables and no output about the sites and the species. These authors proposed some adjustments to improve the complementarity of the methods.

The data consist of three matrices (Fig. 6.21):

- a matrix of n sites by p species (presence-absence or abundance data), called **A** (Legendre en al. 1997) or **L** (Dray and Legendre 2008; Dray et al. 2014);
- a matrix of p species by s biological or behavioural traits, called **B** or **Q**;
- a matrix of n sites by m habitat characteristics, called **C** or **R**.

Therefore, the data contains information in three matrices, about the relationships between species and environment and about the species traits. The purpose of both methods is to gain knowledge about the relationships between the species traits and the environment. RLQ analyses the joint structure of **R** and **Q** mediated by matrix **L**, which serves as a link between **R** and **Q**. The $s \times m$ matrix crossing traits and environmental variables is the fourth one (hence the name "fourth corner"; see the matrix arrangement in Fig. 6.21), called **D**.

6.11.1 The Fourth-Corner Method

The original fourth-corner method (Legendre et al. 1997) was limited to presence-absence data and the analysis of a single trait and a single environmental variable at a time. An extension to abundance data, several traits and environmental variables, with improvements on the testing procedures, was proposed by Dray and Legendre (2008).

The principle of the fourth-corner method consists in (1) estimating the parameters found in matrix **D** and (2) testing the significance of these parameters, choosing the most appropriate permutational model among 6 possibilities. The parameters of **D** can be estimated by matrix product as follows (Legendre and Legendre 2012), the two approaches producing the same **D** matrix:

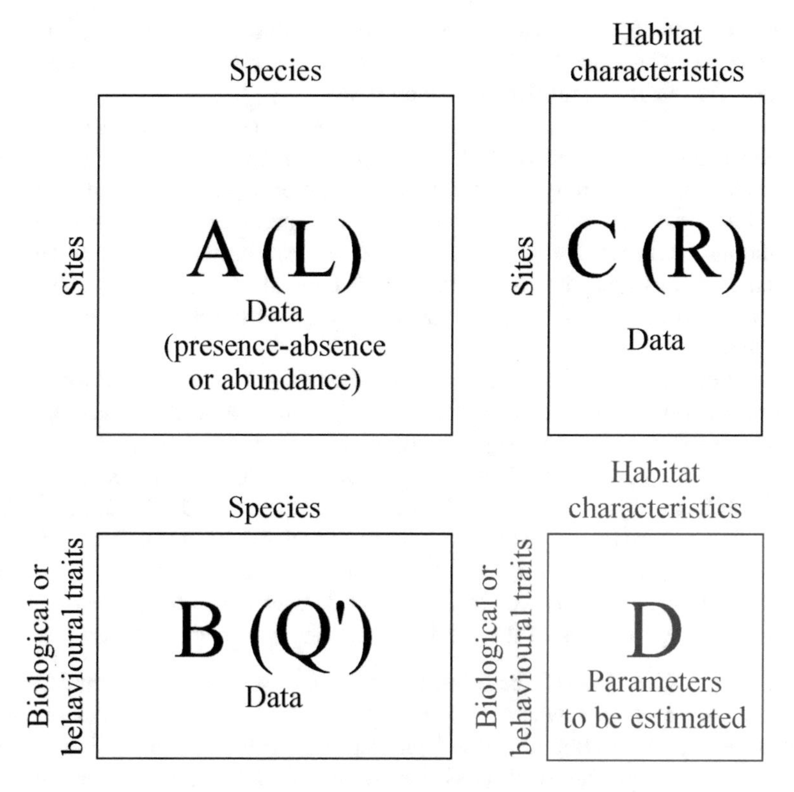

Fig. 6.21 The three data matrices and the fourth-corner matrix, **D,** involved in the fourth-corner problem. After Legendre and Legendre (2012), modified

clockwise: $\mathbf{D} = \mathbf{BA'C}$ or $\mathbf{D} = \mathbf{Q'L'R}$ (6.1)

counter-clockwise: $\mathbf{D} = \mathbf{C'AB'}$ or $\mathbf{D} = \mathbf{R'LQ}$ (6.2)

Numerical examples, as well as explanations about the χ^2, G, F or Pearson r statistics used in the tests, can be found in Legendre and Legendre (2012 Sect. 10.6). The permutation models are the following, considering matrix **L** (or **A**) with observations in the rows and species as columns:

- *Model 1: environmental control over individual species*. Permute the data within each column of matrix **L** (or **A**) independently. This destroys the links between **L** and **Q** (or **A** and **B**) as well as between **L** and **R** (or **A** and **C**). The null hypothesis H_0 states that individuals of a species are randomly distributed with respect to the site conditions.
- *Model 2: environmental control over species assemblages*. Permute entire rows of matrix **L** (**A**). H_0 states that the species compositions of the sites are unrelated to the environmental conditions. This model considers that species assemblages must be taken as whole entities, which are preserved through the permutations, and react as such to environmental constraints.

- *Model 3: lottery*. Permute the species data within each row (site). The null hypothesis of this model is that the distribution of presences of the species at a site is the result of a random allocation process.
- *Model 4: random species attributes*. Permute entire columns of matrix **L** (**A**). H_0 states that species are distributed according to their preferences for site conditions (this is preserved through the permutations), but independently from their traits.
- *Model 5: permute rows and columns*. In turn, permute entire rows then entire columns (or the reverse). H_0 states that the species distributions are not related to their traits or to the site conditions. This is equivalent to permuting the rows of **R** (**C**) and the columns of **Q'** (**B**), as done by Dolédec et al. (1996) in their RLQ method.
- *Model 6*: this is actually a combination of models 2 and 4. A first form of this combination was proposed by Dray and Legendre (2008), who noted, however, that it suffered from a strong inflation of type I error rate when **L** (**A**) is only linked to one other table (**R** or **Q**). ter Braak et al. (2012) proposed to overcome this problem by considering the two tests sequentially and rejecting the overall null hypothesis (i.e., traits and environment unrelated) only if both tests (models 2 and 4) reject H_0 at the α level. The maximum *p*-value becomes the overall *p*-value. These authors showed that this procedure ensures a correct level of type I error and a good power if the number of species is sufficient (at least 30).

Dray et al. (2014) raised the following issue: the fourth-corner method involves multiple tests, a situation where the overall rate of type I error (i.e., the risk of finding at least one false rejection of H_0) is increased. This calls for a correction of the *p*-values, which Legendre et al. (1997) were already advocating. To improve the testing procedure, Dray et al. (2014) proposed a sequential computation of the tests, followed by a correction for multiple testing. The Holm (1979) correction or the false discovery rate method (FDR; Benjamini and Hochberg 1995) can be used.

ter Braak (2017) revisited the fourth-corner analysis by focussing on the fourth-corner correlation statistic used with species abundances, and showing that "*the squared fourth-corner correlation times the total count is precisely the score test statistic for testing the linear-by-linear interaction in a Poisson log-linear model that also contains species and sites as main effects*". Thus, he bridged the gap between the fourth-corner analysis and an alternative approach based on generalized mixed models.

6.11.2 RLQ Analysis

RLQ analysis (Dolédec et al. 1996) is an extension of co-inertia analysis (CoIA, Sect. 6.9) producing a simultaneous ordination of three tables. The method works upon three separate ordinations, one for each data matrix and adapted to its mathematical type, and combines the three to identify the main relationships between the environment and the traits, as mediated by the species. It computes a generalized singular value decomposition of the fourth-corner matrix **D** (Dray et al. 2014). For the first ordination axis, RLQ finds coefficients for the environmental variables and species traits. These coefficients measure the contributions of individual variables and are used to compute site and species scores; they are chosen to maximize the first eigenvalue. The analysis proceeds in the same way for the next, orthogonal ordination axes. For mathematical details see Dray et al. (2014).

6.11.3 Application in R

Due to a lack of significant results, we will refrain from using the Doubs data set in this application. The R code for the Doubs data is provided in the accompanying material, however. The example proposed here is extracted from a tutorial written by Stéphane Dray and provided as a Supplement to the Dray et al. (2014) paper (ESA Ecological Archives E095–002-S1). The example concerns the ecological data analysed in the Dray et al. (2014) paper, which describes the response of plant traits to a snow-melting gradient in the French Alps. The main question is: how does the snow cover duration, with all its consequences, impact the alpine grasslands, as assessed by functional traits of the plant species?

The data come from 75, 5×5 m plots located in the South-Western Alps at about 2700 m elevation. They consist in the three following matrices: community composition (82 species, abundance scale from 0 to 5), traits (8 quantitative variables) and environment (4 quantitative and 2 categorical variables). To simplify this presentation, only some of the variables that are identified as significant are presented below. Readers are referred to Dray et al. (2014) and Choler (2005) for a more complete interpretation of the results.

The data are available in **`ade4`**. The script below shows the computation of an RLQ analysis followed by a fourth-corner analysis. We then conclude with a combined representation of the results of the two analyses.

After having loaded the data, the first step of the RLQ analysis is to compute separate ordinations of the three data sets, which are computed by function **`rlq()`** of **`ade4`** using the same general framework as in co-inertia analysis (Sect. 6.9). The ordination methods are chosen in accordance with the mathematical types of the variables. Here, following Dray et al. (2014), we compute a CA on the species data, a PCA on the (quantitative) trait data; for the environmental data, which are quantitative and categorical, we will apply a special type of PCA that can handle such types of data, called a Hill-Smith analysis (Hill and Smith, 1976).

The RLQ analysis is then computed on the basis of the three ordinations. A single **`plot()`** command allows one to plot all results in a single graphical window, but the individual plots are rather crowded, so we also provide the code to plot the results separately. These plots are assembled in Fig. 6.22. The script below concludes with a global "model 6" test (after ter Braak et al. 2012). The two tests included in "model 6" yielded a combined p-value $= 0.001$, hence the null hypothesis is rejected, which means that both links, $\mathbf{L} - \mathbf{Q}$ and $\mathbf{R} - \mathbf{L}$, are significant.

```
data(aravo)
dim(aravo$spe)
dim(aravo$traits)
dim(aravo$env)
# Preliminary analyses: CA, Hill-Smith and PCA
afcL.aravo <- dudi.coa(aravo$spe, scannf = FALSE)
acpR.aravo <- dudi.hillsmith(aravo$env,
                  row.w = afcL.aravo$lw,
                  scannf = FALSE)
acpQ.aravo <- dudi.pca(aravo$traits,
                  row.w = afcL.aravo$cw,
                  scannf = FALSE)
# RLQ analyses
rlq.aravo <- rlq(
                  dudiR = acpR.aravo,
                  dudiL = afcL.aravo,
                  dudiQ = acpQ.aravo,
                  scannf = FALSE)
plot(rlq.aravo)

# Traits by environment crossed table
rlq.aravo$tab

# Since the plots are crowded, one can plot them one by one
# in large graphical windows.
# Site (L) scores:
s.label(rlq.aravo$lR,
  plabels.boxes.draw = FALSE,
  ppoints.alpha = 0
)
# Species (Q) scores
s.label(rlq.aravo$lQ,
  plabels.boxes.draw = FALSE,
  ppoints.alpha = 0
)
# Environmental variables:
s.arrow(rlq.aravo$l1)
# Species traits:
s.arrow(rlq.aravo$c1)
# Global test:
randtest(rlq.aravo, nrepet = 999, modeltype = 6)
```

Hint *Note that among the preliminary analyses, the CA uses unequal site weights. In the two other* **dudi.xxx()** *calls, the CA site weights* (`afcL.aravo$lw`) *are attributed to the other (Hill-Smith and PCA) analyses.*

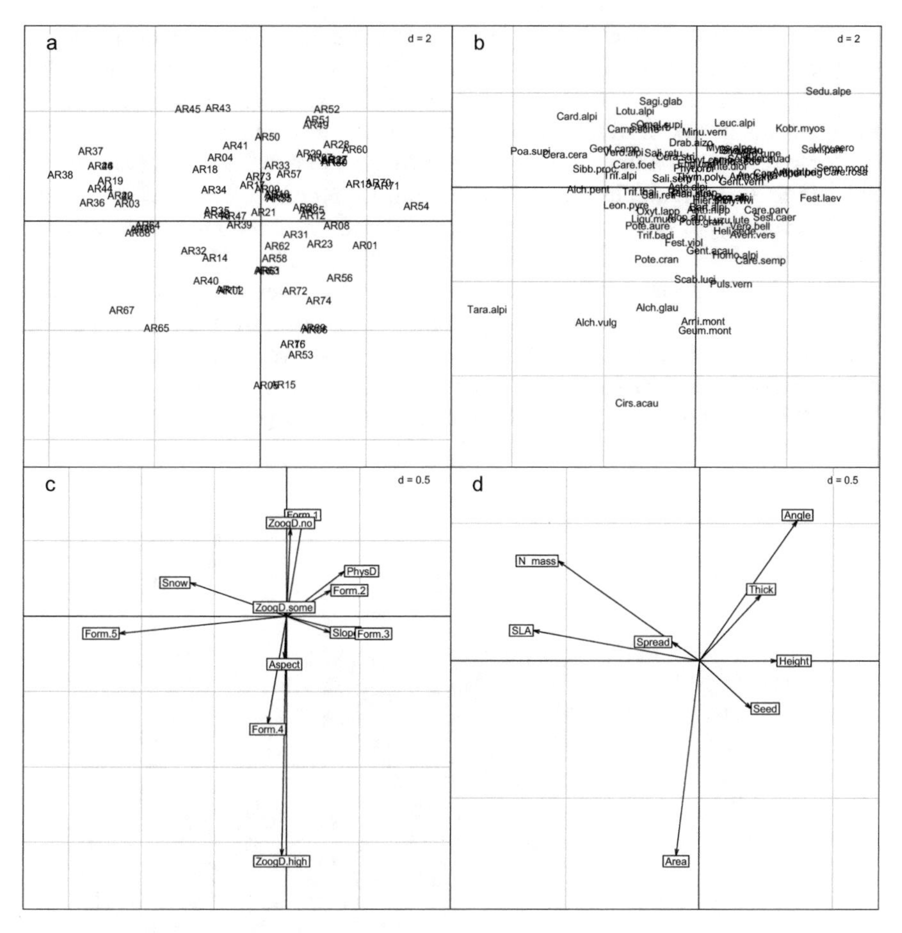

Fig. 6.22 Results of the RLQ analysis of the alpine plant communities: (**a**) site (L) scores, (**b**) species (Q) scores, (**c**) environmental variables and (**d**) species traits

The results are interpreted as follows by Dray et al. (2014): "*The left (negative) part of the first RLQ axis identifies species* (Poa supina, Alchemilla pentaphyllea, *or* Taraxacum alpinum [Fig. 6.22b]) *with higher specific leaf area (SLA) and mass-based leaf nitrogen content (NMass), lower height, and a reduced seed mass* [Fig. 6.22d]. *These species were mostly found in late-melting habitats* [Fig. 6.22a, c]. *The right part of the axis highlights trait attributes (upright and thick leaves) associated with convex landforms, physically disturbed and mostly early-melting sites. Corresponding species are* Sempervivum, montanum, Androsace adfinis, *or* Lloydia serotina. *The second RLQ axis outlined zoogenic disturbed sites located in concave slopes. These habitats were characterized by large-leaved species* (Cirsium acaule, Geum montanum, Alchemilla vulgaris)."

The same data will now be submitted to a fourth-corner analysis, which provides tests at the bivariate level, i.e. one trait and one environmental variable at a time. This is where the correction for multiple tests is necessary. Given the large number of permutations needed to reach an adequately precise estimation of the p-value, the most astute way of computing this analysis consists in a first computation without any correction for multiple testing. The resulting object can be corrected afterwards. So, if several types of corrections must be examined, there is no need to recompute the whole analysis and its large and time-consuming number of permutations.

The fourth-corner analysis is computed by means of the function **fourthcorner()** of **ade4**, using model 6 advocated by Dray et al. (2014). The correction for multiple testing is taken care of by the function **p.adjust.4thcorner()**, which operates on the output object of the analysis. We will first plot the results as a table with coloured cells (Fig. 6.23).

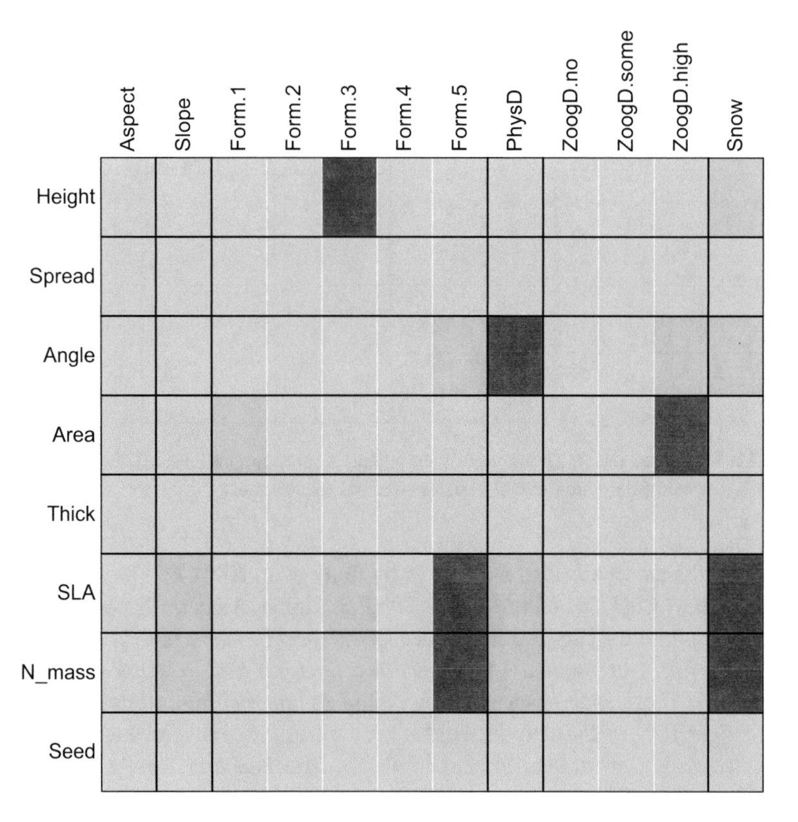

Fig. 6.23 Results of the fourth-corner tests, corrected for multiple testing using the FDR (false discovery rate) procedure. At the $\alpha = 0.05$ level, significant positive associations are represented by red cells and negative ones by blue cells

```
fourth.aravo <- fourthcorner(
                      tabR = aravo$env,
                      tabL = aravo$spe,
                      tabQ = aravo$traits,
                      modeltype = 6,
                      p.adjust.method.G = "none",
                      p.adjust.method.D = "none",
                      nrepet = 49999)
# Correction for multiple testing, here using FDR
fourth.aravo.adj <- p.adjust.4thcorner(
                      fourth.aravo,
                      p.adjust.method.G = "fdr",
                      p.adjust.method.D = "fdr",
                      p.adjust.D = "global")
# Plot
plot(fourth.aravo.adj, alpha = 0.05, stat = "D2")
```

This representation allows a detailed interpretation. For instance, SLA (specific leaf area) and N_mass (mass-based leaf nitrogen content) are positively associated with Snow (mean snow melt date) and Form.5 (concave microtopography), features that can also be observed in Fig. 6.22c, d. This shows that these traits are likely to favour species that tolerate a longer period of snow cover: a higher nitrogen content, partly due to nitrogen storage in snowpacks and partly to the protective effect of snow on soil temperature and water content (Choler 2005), warrants larger reserves, and a larger leaf area allows a larger rate of photosynthesis once the plant is eventually exposed to the sun. Conversely, these two traits are negatively associated with PhysD (physical disturbance due to cryoturbation), which tends to occur more often in areas without snow and therefore more exposed to large temperature oscillations.

The complementarity between RLQ and fourth-corner analyses can further be exploited by representing traits and environmental variables on a biplot (inherited from the RLQ analysis), and adding blue lines for negative and red lines for positive associations (inherited from the fourth-corner tests) (Fig. 6.24):

```
# Biplot combining RLQ and fourth-corner results
plot(fourth.aravo.adj,
  x.rlq = rlq.aravo,
  alpha = 0.05,
  stat = "D2",
  type = "biplot"
)
```

In this biplot, the positive relationships between the traits SLA and N_mass and the environmental characteristics Snow and Form.5, discussed above, show up clearly as a tight group of associations that take place in concave-up sites where snow takes time to melt. Of course, many other relationships can be identified in this graph.

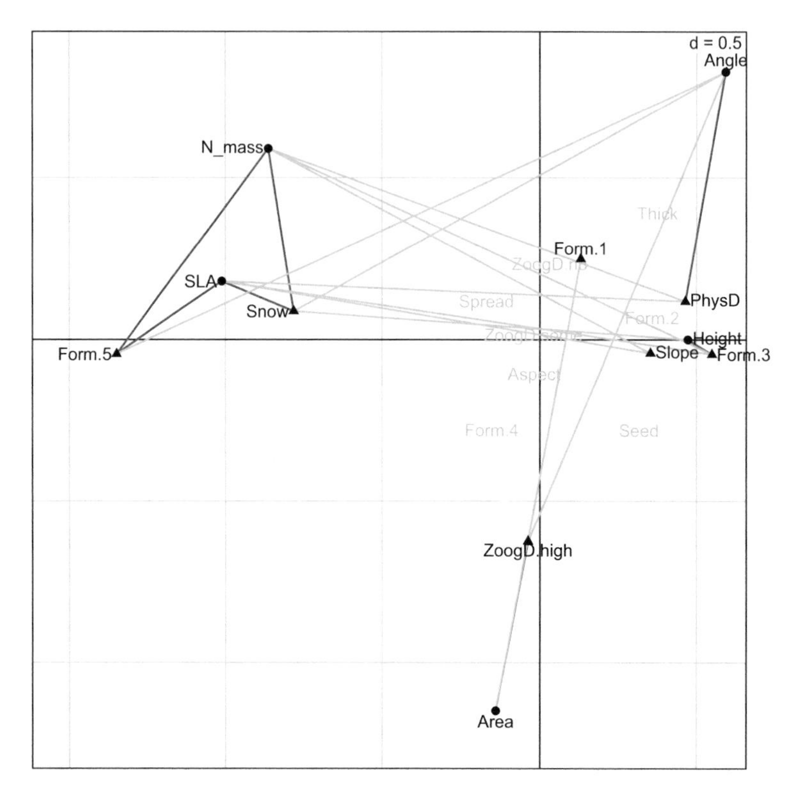

Fig. 6.24 Biplot of the species traits and environmental variables, inherited from the RLQ analysis, with significant negative and positive associations coded blue and red, respectively, inherited from the fourth-corner analysis

6.12 Conclusion

Ordination is a natural and informative way to look at ecological data, and ordination involving more than one data matrix is certainly the most powerful family of methods to reveal, interpret, test and model ecological relationships in the multivariate context. We presented what we think are the most important and useful methods (Table 6.2), illustrating them using examples that highlighted their potentials. This is by no means the end of the story. Researchers continuously propose either new ways of exploiting existing methods, or new methods altogether to answer new types of questions, whose merits and limits must be put under close scrutiny to allow them to be properly integrated into the already powerful statistical toolbox of the ecological community. Readers of this book are encouraged to join this movement.

Table 6.2 Names and some characteristics of the methods described in this chapter

Name, acronym	Use (examples)	**R** functions packages	Data; implementation; limitations
A. Asymmetric analyses			
Redundancy analysis, RDA	Predict **Y** with **X** Variation partitioning	`rda {vegan}` `varpart {vegan}`	All types; species data with prior transformation; $m < (n-1)$. Linear model.
Canonical correspon-dence analysis, CCA	Predict **Y** with **X**	`cca {vegan}`	**Y**: species abundances; **X**: all types; $m < (n-1)$; unimodal response to latent variables.
Linear discriminant analysis, LDA	Explain classification with quantitative variables	`lda{MASS}`	**Y**: classification; **X**: quantitative variables. Linear model.
Principal response curves, PRC	Model community response through time in controlled experiments	`prc {vegan}`	**Y**: community data; factor "treatment"; factor "time".
Co-correspondence analysis (asymmetric form), CoCA	Predict one community on the basis of another	`coca {cocorresp}`	**Y**: data for community 1; **X**: data for community 2; both at the same sites. Unimodal response to latent variables
B. Symmetric analyses			
Co-correspondence analysis (symmetric form), CoCA	Optimized comparison of two communities (descriptive approach)	`coca {cocorresp}`	$\mathbf{Y_1}$: data for community 1; $\mathbf{Y_2}$: data for community 2; both at the same sites. Unimodal response to latent variables.
Canonical correlation analysis, CCorA	Common structures of two data matrices.	`CCorA {vegan}`	Two matrices of quantitative data. Linear model.
Co-inertia analysis, CoIA	Common structures of two or more data matrices.	`coinertia {ade4}`	Very general and flexible; many types of data and ordination methods.
Multiple factor analysis, MFA	Common structures of two or more data matrices.	`mfa {ade4}` `MFA {FactoMineR}`	Simultaneous ordination of 2 or more weighted tables. Mathematical type must be homogeneous within each table.
RLQ analysis, RLQ	Species traits related to environmental variables	`rlq {ade4}`	3 tables: species-by-sites, sites-by-environment; species-by-traits
Fourth-corner analysis	Species traits related to environmental variables	`fourthcorner` `fourthcorner2 {ade4}`	3 tables: species-by-sites, sites-by-environment; species-by-traits

Chapter 7
Spatial Analysis of Ecological Data

7.1 Objectives

Spatial analysis of ecological data is a huge field that could fill several books by itself. To learn about general approaches in spatial analysis in **R**, readers may consult the book by Bivand et al. (2013). The present chapter has a more restricted scope. After a short general introduction, it deals with several methods that were specifically developed for the analysis of scale-dependent structures of ecological data, although they can, of course, be applied to other domains. These methods are based on sets of variables describing spatial structures in various ways, derived from the coordinates of the sites or from the neighbourhood relationships among sites. These variables are used to model the spatial structures of ecological data by means of multiple regression or canonical ordination, and to identify significant spatial structures at all spatial scales that can be perceived by the sampling design. As you will see, the whole analytical process uses many of the techniques covered in the previous chapters.

Practically, you will:

- learn how to compute spatial correlation measures and draw spatial correlograms;
- learn how to construct spatial descriptors derived from site coordinates and from links between sites;
- identify, test and interpret scale-dependent spatial structures;
- combine spatial analysis and variation partitioning;
- assess spatial structures in canonical ordinations by computing variograms of explained and residual ordination scores.

D. Borcard et al., *Numerical Ecology with R*, Use R!,
https://doi.org/10.1007/978-3-319-71404-2_7

7.2 Spatial Structures and Spatial Analysis: A Short Overview

7.2.1 Introduction

As mentioned in Chap. 6, spatial structures play a very important role in the analysis of ecological data. Living communities are spatially structured at many scales, and these structures are the result of several classes of processes. On the other hand, beta diversity is the spatial variation in community composition; so, a study of the factors that can explain the spatial variation of community composition is in every respect an analysis of beta diversity (see Chap. 8). The environmental control model advocates that external forces (climatic, physical, chemical) control living communities. If these factors are spatially structured, their patterns will be reflected on the living communities (examples: patches of desert where the soil is humid enough to support vegetation; gradient of successive communities through an intertidal zone). The biotic control model predicts that intra- and interspecific interactions within communities (examples: social groups of animals; top-down or bottom-up processes), as well as neutral processes such as ecological drift and limited dispersal, may result in spatial patterns that are the cause of spatial autocorrelation in the strict sense. Finally, historical events (e.g. past disturbances like fire or human settlements) may have structured the environment in a way that still influences present-day communities.

In all, ecological data reflect a combination of many structures, spatial or not:

- The overall mean of each response variable (species).
- If the whole sampling area is under the influence of an all-encompassing process that changes the mean in a gradient across the area, then a trend is present. The trend may be due to a process operating at a scale larger than the study area.
- Spatial structures at regional scales: ecological processes of various kinds (biotic or abiotic) and neutral processes influence the data at scales finer than the overall sampling area, producing identifiable spatial patterns.
- Local deterministic structures with no recognizable spatial component because the sampling design is not fine enough to identify such fine-scale patches.
- Random noise (error): this is the residual (stochastic) component of the variation. It can be attributed to local effects operating independently at each sampling site and to sampling variation.

One of the aims of spatial analysis is to discriminate between these sources of variation and model the relevant ones separately.

7.2.2 Induced Spatial Dependence and Spatial Autocorrelation

An important distinction must be made here. As we wrote above, a spatial structure in a response matrix **Y** can result from two main origins: either from the forcing of external (environmental) factors that are themselves spatially structured, or as the result of processes internal to the community itself. In the first case one speaks of **induced spatial dependence**, in the second case of **spatial autocorrelation**.

For value y_j of a response variable y observed at site j, the model for **induced spatial dependence** is the following:

$$y_j = \mu_y + f(\mathbf{X}_j) + \varepsilon_j \tag{7.1}$$

where μ_y is the overall mean of variable y, **X** is a set of explanatory variables, and ε_j is an error term that varies randomly from location to location (residual, stochastic variation). The additional term $[f(\mathbf{X}_i)]$ states that y_j is influenced by external processes represented in the model by explanatory variables. The spatial structure of these variables will be reflected in y. When they form a gradient shape, they represent what Legendre (1993) called "true gradients", that is, gradient-like deterministic structures generated by external forces, whose error terms are not autocorrelated.

The model for **spatial autocorrelation** is:

$$y_j = \mu_y + \sum f(y_i - \mu_y) + \varepsilon_j \tag{7.2}$$

This equation states that y_j is influenced by the values of y at the surrounding sites i. This influence is modelled by a weighted sum of the (centred) values y_i at these sites. The biological context dictates the radius of the zone influencing a given point, as well as the weights to be given to the neighbouring points. These weights are generally dependent on the distance. The spatial interpolation method called *kriging* (Isaaks and Srivastava 1989; Bivand et al. 2013) is based on this model. Kriging is a family of interpolation methods that will not be discussed further in this book. Kriging functions are available in package `geoR`.

Spatial autocorrelation may mimic gradients if the underlying process has a range of influence larger than the sampling area. Legendre (1993) called the resulting structures "false gradients". There is no statistical way to distinguish false from true gradients. One must rely upon biological hypotheses: in some cases one has a strong hypothesis about the processes generating spatial structures, and therefore whether these processes may have produced autocorrelation in the data. In other cases an opinion can be formed by comparing the processes detected at the scale of the study area with those that are likely to occur at the scale of the (larger) target population (Legendre and Legendre 2012).

Spatial correlation measures the fact that close points have either more similar (positive correlation) or more dissimilar values (negative correlation) than randomly selected pairs. This phenomenon, which is generated either by true autocorrelation

(Eq. 7.2) or by spatial structures resulting from spatial dependence (Eq. 7.1), has noxious effects on statistical tests. In spatially correlated data, values at any given site can be predicted, at least partially, from the values at other sites, if the researcher knows the biological process and the locations of the sites. This means that the values are not stochastically independent of one another. The assumption of independence of errors is violated in such cases. In other words, each new observation does not bring with it a full degree of freedom. While the fraction is difficult to determine, the fact is that the number of degrees of freedom used for a parametric test is often overestimated, thereby biasing the test on the "liberal" side: the null hypothesis is rejected too often. Numerical simulations have shown, however, that this statistical problem only occurs when both the response (e.g. species) and the explanatory variables (e.g. environmental) are spatially correlated (Legendre et al. 2002).

7.2.3 Spatial Scale

The term *scale* is used in many senses across different disciplines. It encompasses several properties of sampling designs and spatial analysis.

A sampling design has three characteristics pertaining to spatial observation scale (Legendre and Legendre 2012 Sect. 13.0):

- grain size: size of the sampling units (diameter, surface or volume depending on the study).
- sampling interval, sometimes called lag: average distance between neighbouring sampling units.
- extent (sometimes called range): total length of the transect, surface area or volume (e.g. air, water) included in the study.

These three properties of sampling designs have an influence on the type and size of the spatial structures that can be identified and measured. (1) Sampling units *integrate* the structures occurring in them: one cannot identify structures of sizes equal to or smaller than the grain of the study. (2) The sampling interval determines the size of the finest spatial structures that can be identified (by *differentiation* among sampling units). (3) The extent of the study area sets an upper limit to the size of the measurable patterns. It is therefore essential to match each of these three elements to the hypotheses to be tested and to the characteristics of the system under study (Dungan et al. 2002).

The ecological context of the study dictates the optimal grain size, sampling interval and extent. The optimal grain size (size of the sampling units) should match the size of unit entities of the study (e.g. objects like individual plants or animals, patches of vegetation, lakes, or areas affected by fine-scale processes). The average distance between unit objects or unit processes should be matched by the sampling interval. The extent should encompass the range of the broadest processes targeted by the study. These recommendations are detailed in Dungan et al. (2002).

Note that the expressions "large scale" and "small scale" are somewhat ambiguous because their meanings in ecology and cartography are opposite. In ecology, "small scale" refers to the fine structures and "large scale" to the broadest structures, contrary to cartography where a large-scale map (e.g. 1:25,000) is more detailed than a small-scale map (e.g. 1:1,000,000). Therefore we advocate the use of "broad scale" (phenomena with large grains, large extents) and "fine scale" in ecology (Wiens 1989). Although these terms are not strict antonyms, we feel that they are less ambiguous than "large" and "small scale".

Finally, ecological processes occur at a variety of scales, resulting in complex, multiscale patterns. Therefore, identifying the scale(s) of the patterns and relating them to the appropriate processes are goals of paramount importance in modern ecology. To reach them, the researcher must rely on appropriate sampling designs and powerful analytical methods. The approaches presented in this chapter have been devised for the latter purpose.

7.2.4 Spatial Heterogeneity

A process or a pattern that varies across an area is said to be *spatially heterogeneous*. Many methods of spatial analysis are devoted to the measurement of the magnitude and extent of this heterogeneity and testing for the presence of spatial correlation (in other words, spatial structures of any kind). The latter may be done either to support the hypothesis that no spatial correlation (in the broad sense) is present in the data (if the researcher has statistical tests in mind) or, on the contrary, to show that correlation is present and use that information in conceptual or statistical models (Legendre and Legendre 2012).

Spatial heterogeneity in relation to inter-site distance is most often studied by means of *structure functions*. Examples of these are correlograms, variograms and periodograms. While it is not the purpose of this book to discuss these various functions, it is useful to devote a section to correlograms, since the main underlying measures of spatial correlation will be used later in Sect. 7.4 of this chapter.

7.2.5 Spatial Correlation or Autocorrelation Functions and Spatial Correlograms

The two main statistics used to measure spatial correlation of univariate quantitative variables are Moran's I (Moran 1950) and Geary's c (Geary 1954). The first is constructed in much the same way as the Pearson correlation coefficient:

$$I(d) = \frac{\frac{1}{W}\sum_{h=1}^{n}\sum_{i=1}^{n} w_{hi}\left(y_h - \bar{y}\right)\left(y_i - \bar{y}\right)}{\frac{1}{n}\sum_{i=1}^{n}\left(y_i - \bar{y}\right)^2} \tag{7.3}$$

The expected value of Moran's I for no spatial correlation is

$$E(I) = \frac{-1}{n-1} \tag{7.4}$$

Values below *E(I)* indicate negative spatial correlation, and values above *E(I)* indicate positive correlation. *E(I)* is negative but close to 0 when n (the total number of observations) is large.

Geary's c is more akin to a distance measure:

$$c(d) = \frac{\frac{1}{2W}\sum_{h=1}^{n}\sum_{i=1}^{n} w_{hi}\left(y_h - y_i\right)^2}{\frac{1}{n-1}\sum_{i=1}^{n}\left(y_i - \bar{y}\right)^2} \tag{7.5}$$

The expected value of Geary's c for no spatial correlation is $E(c) = 1$. Values below 1 indicate *positive* spatial correlation, and values above 1 indicate *negative* correlation.

y_h and y_i are the values of variable y at pairs of sites h and i. To compute spatial correlation coefficients, one first constructs a matrix of geographical distances among sites. These distances are then converted to classes d. Both formulas show the computation of the index value for a class of inter-site distance d. The weights w_{hi} have value $w_{hi} = 1$ for pairs of sites belonging to distance class d, and $w_{hi} = 0$ otherwise. W is the number of pairs of points used to compute the coefficient for the distance class considered, i.e., the sum of the w_{hi} weights for that class.

A *correlogram* is a plot of the spatial correlation values against the distance classes. Combined with statistical tests, a correlogram allows a quick assessment of the type and range of the spatial correlation structure of a variable. A typical case is spatial correlation that is positive at short distances, decreases to negative values, and levels out to a point where it becomes nonsignificant. The corresponding distance class sets the distance beyond which a pair of values can be considered as spatially independent. It is important to note that spatial correlograms will display any kind of spatial correlation, i.e. induced spatial dependence (Eq. 7.1) or spatial autocorrelation (Eq. 7.2); so the name "spatial autocorrelogram" which is often given to these plots is too restrictive and therefore somewhat misleading.

Univariate spatial correlograms can be computed using the function **`sp.correlogram()`** of package **`spdep`**. We can apply this function to the variable "Substrate density" of the oribatid mite data set. We will first define neighbourhoods of size $\leq$0.7 m around the points using the function **`dnearneigh()`**. These links can be visualized using our function **`plot.links()`**.

Note that the points do not form a connected graph at this radius. Following that, the function **sp.correlogram()** will find successive lag orders of contiguous neighbours and compute Moran's *I* for each of these lag orders. A lag order is the number of links, or steps in the linkage graph, between two points. It can be construed as a generalized form of distance between points. For instance, if sites A and C are connected through site B, two links (A-B and B-C) are needed to connect A and C, which are then are connected at lag order 2.

Note: Cartesian coordinates can be obtained from latitude-longitude (sometimes abbreviated to Lat/Lon or LatLon) data using the function **geoXY()** of the package **SoDA**.

```
# Load the required packages
library(ape)
library(spdep)
library(ade4)
library(adegraphics)
library(adespatial)
library(vegan)
# Source additional functions
# (files must be in the working directory)
source("plot.links.R")
source("sr.value.R")
source("quickMEM.R")
source("scalog.R")

# Load the oribatid mite data. The file mite.Rdata is assumed
# to be in the working directory.
load("mite.RData")
# Transform the data
mite.h <- decostand (mite, "hellinger")
mite.xy.c <- scale(mite.xy, center = TRUE, scale = FALSE)

## Univariate spatial correlogram (based on Moran's I)
# Search for neighbours of all points within a radius of 0.7 m
# and multiples (i.e., 0 to 0.7 m, 0.7 to 1.4 m and so on).
plot.links(mite.xy, thresh = 0.7)
nb1 <- dnearneigh(as.matrix(mite.xy), 0, 0.7)
summary(nb1)

# Correlogram of substrate density
subs.dens <- mite.env[ ,1]
subs.correlog <-
  sp.correlogram(nb1,
                 subs.dens,
                 order = 14,
                 method = "I",
                 zero.policy = TRUE)
print(subs.correlog, p.adj.method = "holm")
plot(subs.correlog)
```

Hint *We use the* **`print()`** *function to display the correlogram results because it allows for correction of the p-values for multiple testing. In a correlogram, a test is performed for each lag (distance class), so that without correction, the overall risk of type I error is greatly increased. The Holm (1979) correction is applied here.*

This correlogram has a single significant distance class: there is positive spatial correlation at distance class 1 (i.e., 0.0 m to 0.7 m). Negative spatial correlation at distance class 4 (i.e., 2.1 m to 2.8 m) is hinted at, but the coefficient is not significant after Holm (1979) correction for multiple testing (see Sect. 7.2.6). Beyond these marks, no significant spatial correlation exists, which means that for practical purposes measurements taken more than 0.7 m, or (conservatively) 2.8 m apart (the upper limit of class 4) can be considered as spatially independent with respect to substrate density.

Spatial correlation in the multivariate domain can be assessed and tested for by means of a *Mantel correlogram* (Sokal 1986; Oden and Sokal 1986; Borcard and Legendre 2012). Basically, one computes a standardized Mantel statistic r_M (analogous to a Pearson's r coefficient) between a dissimilarity matrix among sites and a matrix where pairs of sites belonging to the same distance class receive value 0 and the other pairs, value 1. The process is repeated for each distance class. Each r_M value can be tested for by permutations. The expectation of the Mantel statistic for no spatial correlation is $r_M = 0$.

A Mantel correlogram can be computed, tested and plotted (Fig. 7.1) by using **`vegan`**'s function **`mantel.correlog()`**. The only data necessary are a response dissimilarity matrix and either the geographical coordinates of the sites or a matrix of geographical distances among sites. Here is an example of a Mantel correlogram for

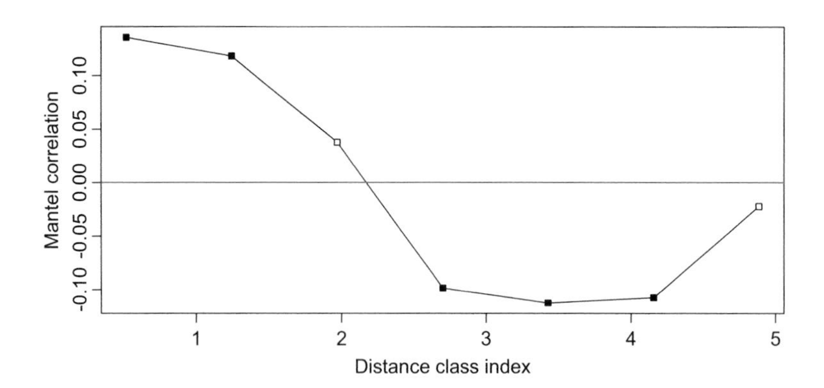

Fig. 7.1 Mantel correlogram of the Hellinger-transformed and detrended oribatid mite species data. Black squares indicate significant multivariate spatial correlation after Holm correction for multiple testing. The abscissa is labelled in metres since this is the unit of the data used to construct the distance classes

the oribatid mite data, which will first be detrended (Sect. 7.3.2) to make them second-order stationary (Sect. 7.2.6).

```
# The species data are first detrended; see Sect. 7.3
mite.h.det <- resid(lm(as.matrix(mite.h) ~ ., data = mite.xy))
mite.h.D1 <- dist(mite.h.det)
(mite.correlog <-
  mantel.correlog(mite.h.D1,
                  XY = mite.xy,
                  nperm = 999))
summary(mite.correlog)

# Plot the Mantel correlogram
plot(mite.correlog)
```

Hint *In this run, the number of classes has been computed automatically using Sturge's rule. Use argument* `n.class` *to provide a user-determined number of classes.*

In this simple run, most default settings have been applied, including Holm's correction for multiple testing (see Section 7.2.6). The number of classes has been computed using Sturge's rule: number of classes $= 1 + (3.3219 \times \log_{10}n)$, where n is the number of elements, here the number of pairwise distances. The resulting number of classes and the corresponding break points can be read in the result object:

```
# Number of classes
mite.correlog$n.class       # or: mite.correlog[2]
# Break points
mite.correlog$break.pts     # or: mite.correlog[3]
```

Hint *The default option* `cutoff = TRUE` *limits the correlogram to the distance classes including all points (the first 7 distance classes in this example); the results for the last 5 distance classes (computed on fewer and fewer points) are not shown.*

The result shows significant positive spatial correlation in the first two distance classes (i.e., between 0.15 m and 1.61 m; see the break points) and negative significant correlation in the fourth to sixth classes (between 2.34 and 4.52 m). Examining the environmental variables allows some speculation about the ecological reasons behind these structures. Close sites tend to show similar communities because the soil conditions are rather similar. On the other hand, any pair of sites whose members are about 2–4 m apart falls into contrasting soil conditions, which in turn explains why their mite communities are different.

7.2.6 Testing for the Presence of Spatial Correlation: Conditions

As shown above, spatial correlation coefficients can be tested for significance. However, conditions of application must be respected. The condition of normality can be relaxed if the test is carried out by permutations. To test the significance of coefficients of spatial correlation, however, the condition of **second-order stationarity** must be met. That condition states that the mean of the variable and its spatial covariance (numerator of Eq. 7.3) are the same over the study area, and that its variance (denominator of Eq. 7.3) is finite. This condition tells us, in other words, that the spatial variation of the data should be adequately described by the same single spatial correlation function in all portions of the study area. Spatial correlation coefficients cannot be tested for significance if an overall trend is present in the data, or if the variable has been measured in a region where several distinct structures should be modelled by different spatial correlation functions. Data displaying simple trends can often be *detrended* by means of a first-degree function of the site geographical coordinates (Sect. 7.3), as we did above before computing the Mantel correlogram of the mite data.

Another, relaxed form of stationarity is called the **intrinsic assumption**, a short form for "hypothesis of intrinsic stationarity of order 2" (Wackernagel 2003). This condition considers only the increments of the values of the variable; it states that the differences $(y_h - y_i)$ for any distance d in the numerator of Eq. 7.5 have zero mean and constant and finite variance over the study area, independently of the location (Legendre and Legendre 2012). This condition allows one to compute and examine correlograms but without tests of significance.

Legendre and Legendre (2012, p. 800) show how to interpret all-directional correlograms (i.e., correlograms built on distance classes defined in the same way in all directions) as well as directional correlograms.

A word is needed here about **multiple testing**. In Sect. 7.2.5 several spatial correlation values were tested simultaneously for significance. In such cases the probability of type I error increases with the number of tests. If k tests are carried out, the binomial law tells us that the overall probability of type I error (technically called the "experimentwise error rate") is equal to $1 - (1 - \alpha)^k$ where α is the nominal value for a single test. For instance, in the Mantel correlogram shown in Fig. 7.1, seven tests are carried out simultaneously. Without correction, the *overall* probability of obtaining *at least* one type I error is equal to $1 - (1–0.05)^7 = 0.302$ instead of the nominal $\alpha = 0.05$. Several methods have been proposed to achieve a correct level of type I error in multiple tests (reviewed in Legendre and Legendre 2012; Wright 1992). The most conservative solution for k independent tests is to divide the significance level by the number of simultaneous tests: $\alpha' = \alpha / k$ and compare the p-values to α'. Conversely one can multiply the p-values by k (i.e., $p' = kp$) and compare the resulting values to the unadjusted α. For non-independent tests, Holm's procedure (Holm 1979) is more powerful. The reason is that Holm's correction consists in applying Bonferroni's correction sequentially as follows. First, order the (uncorrected) p-values in increasing order. Divide the smallest by k. If, and only if the result is smaller than or equal to α, divide the second smallest p-value by $k - 1$. Proceed in the same way, each time relaxing the correcting factor (i.e., $k - 2$, $k - 3$...), until a nonsignificant value is encountered.

Other corrections have been proposed in addition to the two presented above. Several are available in a function called `p.adjust()` in package `stats`. This function can be called whenever one has run several simultaneous tests of significance. The data submitted to that function must be a vector of p-values.

7.2.7 Modelling Spatial Structures

Beyond the methods described above, there are other, more modelling-oriented approaches to spatial analysis. Finding spatial structures in ecological data indicates that some process has been at work to generate them; the most important are environmental forcing (past or present) and biotic processes. Therefore, it is interesting to identify the spatial structures in the data and model them. Spatial structures can then either be related to explanatory variables representing hypothesized causes, or help generate new hypotheses as to which processes may have generated them.

Spatial structures can be present at many different scales. Identifying these scales and modelling the corresponding spatial structures separately is a long-sought goal for ecologists. A first, rather coarse approach in multivariate analysis is the adaptation of trend-surface analysis to canonical ordination. As suggested by ter Braak (1987) and demonstrated by Legendre (1990), response data may be explained by a polynomial function of the (centred) site coordinates. Borcard et al. (1992) have shown how to integrate this method into variation partitioning to identify, among other fractions, the pure spatial component of the ecological variation of species assemblages.

Multivariate trend-surface analysis is limited to the extraction of rather simple broad-scaled spatial structures, because polynomial terms become rapidly cumbersome, and highly correlated if one uses raw polynomials. In practice, its use is restricted to third-degree polynomials. A breakthrough came with the development of eigenvector-based spatial functions, which will be described in Sect. 7.4, after a short example of trend-surface analysis.

7.3 Multivariate Trend-Surface Analysis

7.3.1 Introduction

Most ecological data have been sampled on geographic surfaces. Therefore, the crudest way to model the spatial structure of the response data is to regress the response data on the X-Y coordinates of the sampling sites. Of course, this will only model a *linear trend*; a plane will be fitted through the data in the same way as a straight line would be fitted to data collected along a transect by regressing them on their X coordinates.

A way of allowing curvilinear structures to be modelled is to add polynomial terms of the coordinates to the explanatory data. Second- and third-degree terms are often applied. It is better to centre (but not standardize, lest one distort the aspect-ratio of the sampling design) the X and Y coordinates before computing the polynomial terms, to make at least the second-degree terms less correlated. The first-, second- and third-degree functions are:

$$\hat{z} = f(X, Y) = b_0 + b_1X + b_2Y \tag{7.6}$$

$$\hat{z} = b_0 + b_1X + b_2Y + b_3X^2 + b_4XY + b_5Y^2 \tag{7.7}$$

$$\hat{z} = b_0 + b_1X + b_2Y + b_3X^2 + b_4XY + b_5Y^2 + b_6X^3 + b_7X^2Y + b_8XY^2 + b_9Y^3 \tag{7.8}$$

These polynomial terms can be computed by using function **`poly()`** with argument `raw` = `TRUE`. An alternative method is to compute *orthogonal* polynomial terms with the (default) option `raw` = `FALSE`. In the latter case, for a set of X-Y coordinates, the monomials X, X^2, X^3 and Y, Y^2, Y^3 have a norm of 1 and are orthogonal to their respective lower-order terms. X monomials are not orthogonal to Y monomials, however, except when the points form a regular orthogonal grid; terms containing both X and Y are not orthogonal to one another and their norms differ from 1. Orthogonal polynomials produce the exact same R^2 in regression and canonical analysis as raw polynomials. The orthogonality of orthogonal polynomials presents an advantage when selection of explanatory variables is used to find a parsimonious spatial model because orthogonal terms are uncorrelated.

Trend-surface analysis can be applied to multivariate data by means of RDA or CCA. The result is a set of independent spatial models (one for each canonical axis). One can also use forward selection to reduce the model to its significant components only.

7.3.2 Trend-Surface Analysis in Practice

Our first step in spatial modelling will be to produce some monomials and polynomials of the X and Y coordinates on a grid and visualize the shapes they produce. We will then proceed to apply this technique to the oribatid mite data.

```
## Simple models on a square, regularly sampled surface
# Construct and plot a 10 by 10 grid
xygrid <- expand.grid(1:10, 1:10)
plot(xygrid)
# Centring
xygrid.c <- scale(xygrid, scale = FALSE)
# Create and plot some first, second and third-degree functions
# of X and Y
X <- xygrid.c[ ,1]
Y <- xygrid.c[ ,2]
XY <- X + Y
XY2 <- X^2 + Y^2
XY3 <- X^2 - X * Y - Y^2
XY4 <- X + Y + X^2 + X * Y + Y^2
XY5 <- X^3 + Y^3
XY6 <- X^3 + X^2 * Y + X * Y^2 + Y^3
XY7 <- X + Y + X^2 + X * Y + Y^2 + X^3 + X^2 * Y + X * Y^2 + Y^3
xy3deg <- cbind(X, Y, XY, XY2, XY3, XY4, XY5, XY6, XY7)
s.value(xygrid, xy3deg, symbol = "circle")
```

Try other combinations, for instance with minus signs or with coefficients not equal to 1.

```
# Computation of a raw (non-orthogonal) third-degree polynomial
# function on the previously centred X-Y coordinates
mite.poly <- poly(as.matrix(mite.xy.c), degree = 3, raw = TRUE)
colnames(mite.poly) <-
    c("X", "X2", "X3", "Y", "XY", "X2Y", "Y2", "XY2", "Y3")
```

Function poly produces the polynomial terms in the following sequence: `X, X^2, X^3, Y, XY, X^2Y, Y^2, XY^2, Y^3`*). The original column names give the degree for the two variables. For instance, "1.2" means* `X^1*Y^2`*. Here raw polynomials have been computed. For orthogonal polynomials, which is the default,* `raw = FALSE`*.*

```
# RDA with all 9 polynomial terms
(mite.trend.rda <- rda(mite.h ~ .,
                       data = as.data.frame(mite.poly)))

# Computation of the adjusted R^2
(R2adj.poly <- RsquareAdj(mite.trend.rda)$adj.r.squared)

# RDA using a third-degree orthogonal polynomial of the geographic

# coordinates
mite.poly.ortho <- poly(as.matrix(mite.xy), degree = 3)
colnames(mite.poly.ortho) <-
    c("X", "X2", "X3", "Y", "XY", "X2Y", "Y2", "XY2", "Y3")
(mite.trend.rda.ortho <-
  rda(mite.h ~ .,
      data =as.data.frame(mite.poly.ortho)))
(R2adj.poly2 <- RsquareAdj(mite.trend.rda.ortho)$adj.r.squared)
# Forward selection  using Blanchet et al. (2008a) double stopping
# criterion
(mite.trend.fwd <-
  forward.sel(mite.h, mite.poly.ortho, adjR2thresh = R2adj.poly2))
# New RDA using the 6 terms retained
(mite.trend.rda2 <- rda(mite.h ~ .,
          data = as.data.frame(mite.poly)[ ,mite.trend.fwd[ ,2]]))
# Overall test and test of the canonical axes
anova(mite.trend.rda2)
anova(mite.trend.rda2, by = "axis")

# Plot of the three independent significant spatial structures
# (canonical axes) plus the fourth (p-value around 0.06).
mite.trend.fit <-
  scores(mite.trend.rda2,
         choices = 1:4,
         display = "lc",
         scaling = 1)
s.value(mite.xy, mite.trend.fit, symbol = "circle")
```

Hints *Note that the **fitted site scores in scaling 1** have been used in the plots. We want to display the "pure" spatial model, i.e., the linear combination of spatial variables, in a projection preserving the Euclidean distances among sites.*

If you want to construct a second-degree raw polynomial function directly within the **`rda()`** *call, here is the syntax:*

```
mite.trend.rda <- rda(mite.h ~ X + Y + I(X^2) +
                              I(X * Y) + I(Y^2))
```

Notice how squared variables and product variables are requested to be treated "as they are" by function `I()`*. Otherwise **R** would consider them as ANOVA terms.*

This analysis shows that the oribatid mite community is significantly spatially structured, and that three (or four, depending on the run) significant independent models can be obtained (Fig. 7.2). The first one (first canonical axis, 73.8% of the *explained* variance) displays a strong difference between the upper and the lower half of the area. The next two significant models (12.1% and 8.4% of the explained variance, respectively) display finer-scaled structures. The fourth axis (only 3.0% variance, p $\approx$ 0.05 depending on the run) shows a left-right contrast.

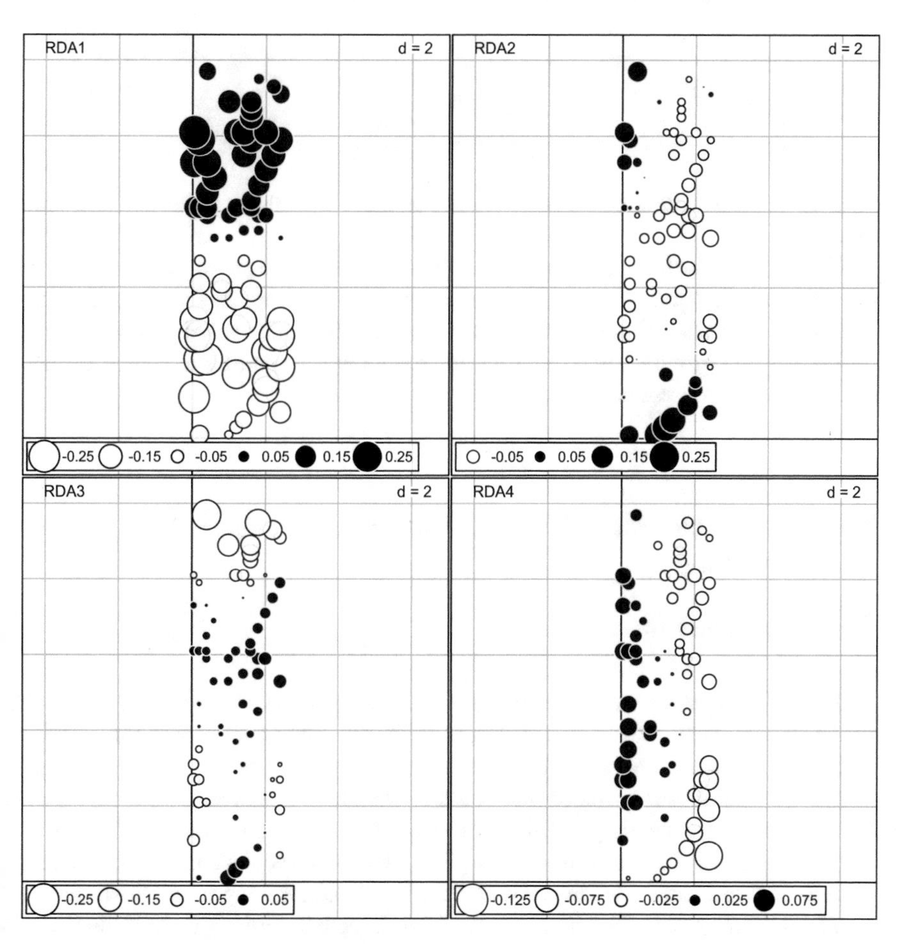

Fig. 7.2 Cubic trend-surface analysis of the Hellinger-transformed oribatid mite data. Three highly significant RDA axes and a fourth, marginally significant one have been retained, representing linearly independent spatial structures.

These models could now be interpreted by regressing them on environmental variables. But we will postpone that step until we can implement it in another spatial modelling framework.

Nowadays, the most useful application of trend-surface analysis is for **detrending**. We have seen in Sect. 7.2.6 that data have to be detrended before spatial correlograms can be tested. We will also see later that most eigenvector-based spatial analyses are best applied to detrended data. Therefore, a handy procedure is to test the response data for linear trends and detrend them if the trend surface is significant. This means to regress all response variables on the *X*-*Y* coordinates and retain the residuals. This can most easily be done using the function `lm()` applied directly to the multivariate matrix of response data.

```
# Is there a linear trend in the mite data?
anova(rda(mite.h, mite.xy)) # Result: significant trend
# Computation of linearly detrended mite data
mite.h.det <- resid(lm(as.matrix(mite.h) ~ ., data = mite.xy))
```

This detrended data set is now ready for more complex spatial analyses and modelling.

Finally, trend surfaces can also be used to model the broad among-group structure of groups of sites that are far from one another on the map. In this case, all sites belonging to a group receive the same *X*-*Y* coordinates (normally those of the centroid of the group). The within-group structures can then be modelled by one of the techniques presented below.

7.4 Eigenvector-Based Spatial Variables and Spatial Modelling

7.4.1 Introduction

Trend-surface analysis is a rather coarse method of spatial modelling. The multiscale nature of ecological processes and data calls for other approaches that can identify and model structures at all scales that can be perceived by a sampling design. Practically, this means methods that could model structures at scales ranging from the broadest, encompassing the whole sampled area, down to the finest, whose sizes are of the order of magnitude of the sampling interval. To achieve this in the context of canonical ordination, we must construct spatial variables representing structures of all relevant scales. This is what the PCNM method (principal coordinates of neighbour matrices; Borcard and Legendre 2002; Borcard et al. 2004) and its offspring the MEM do. These methods will now be studied in detail.

The original PCNM method is now called distance-based MEM, or dbMEM. dbMEM is actually a slightly modified form of PCNM; the difference is explained below. dbMEM is a special case of a wider family of methods now called MEM (Moran's eigenvector maps; Dray et al. 2006). The acronym PCNM should therefore be short-lived in the literature. We advocate the use of the generic acronym MEM for the family, and dbMEM for the method that we originally called PCNM.

7.4.2 Distance-Based Moran's Eigenvector Maps (dbMEM) and Principal Coordinates of Neighbour Matrices (PCNM)

7.4.2.1 Introduction

Borcard and Legendre (2002) and Borcard et al. (2004) proposed to construct eigenvectors of truncated matrices of geographical distances among sites. These eigenvectors have interesting properties that make them highly desirable as spatial explanatory variables. Let us briefly expose the principle of the original PCNM method and see how it has been refined to become today's dbMEM technique.

The "classical" (first-generation) PCNM method worked as follows:

- Construct a matrix of Euclidean (geographic) distances among sites. A larger distance means that the difficulty of communication between two sites increases.
- Truncate this matrix to retain only the distances among close neighbours. The threshold *thresh* depends on the site coordinates. In most cases, it is chosen to be as short as possible, but all points must remain connected by links smaller than or equal to the truncation distance. Otherwise, different groups of eigenfunctions are created, that model the spatial variation within separate subgroups of points but not among these groups. How to choose the truncation threshold distance will be described below. All pairs of points more distant than the threshold receive an arbitrary "large" distance value corresponding to four times the threshold, i.e., $4 \times thresh$, or any larger multiplier.
- Compute a PCoA of the truncated distance matrix.
- In most studies, retain the eigenvectors that model positive spatial correlation (Moran's I larger than $E(I)$, Eq. 7.4). This step requires the computation of Moran's I for all eigenvectors, since this quantity cannot be directly derived from the eigenvalues.

- Use these eigenvectors as spatial explanatory variables in multiple regression or RDA.

In the classical PCNM method described above, the diagonal values of the truncated distance matrix are 0, indicating that a site is connected to itself. However, Dray et al. (2006) showed that setting the diagonal values to 4 × *thresh* instead of 0 resulted in interesting properties. In particular, the eigenvalues of the resulting spatial eigenvectors, now called **dbMEM** (acronym for distance-based Moran's eigenvector maps) are proportional to Moran's *I* coefficient computed on these eigenfunctions using the pairs of sites that remain connected after truncation (Legendre and Legendre 2012). This makes it easy to identify the eigenvectors that model positive spatial correlation, which are the ones used in most ecological studies: they are those with eigenvalues larger than Moran's *I* expectation (Eq. 7.4) to within a multiplicative constant. Note, however, that the eigenfunctions obtained by PCNM and dbMEM are the same.

The dbMEM method presents great advantages over trend-surface analysis. It produces *orthogonal* (linearly independent) spatial descriptors and covers a *much wider range* of spatial scales. It allows the modelling of *any type* of spatial structures, as Borcard and Legendre (2002) have demonstrated through extensive simulations.

The dbMEM method can work for any sampling design, but the spatial variables are easier to interpret in the case of regular designs, as will be seen below. When the design is irregular, it may happen that a large truncation value must be chosen to allow all site-points to remain connected on the map. A large truncation value means a loss of the finest spatial structures. Therefore, ideally, even an irregular sampling design should ensure that the minimum distance allowing all points to be connected is as short as possible. The most commonly applied solution is to compute the minimum spanning tree of a single-linkage clustering of the site coordinates (see Sect. 4.3.1) and retain the largest edge value. In cases where this distance is too large, which may happen if the sampling sites are clustered on the territory, Borcard and Legendre (2002) suggested (1) to add a limited number of supplementary points to the spatial data to cut down the threshold distance, (2) compute the dbMEM variables, and (3) remove the supplementary points from the dbMEM matrix. This ensures that the finest scales are better modelled. The trade-off is that the resulting dbMEM variables are no longer totally orthogonal to one another, but if the number of supplementary points is small with respect to the number of true points, the departure from orthogonality remains small. Another possibility arises when the largest gap or gaps correspond to the separation between to or more distinct entities in the study area (e.g. separate patches of forest; different islands in an archipelago). This problem will be discussed in Sect. 7.4.3.5, which addresses nested sampling designs.

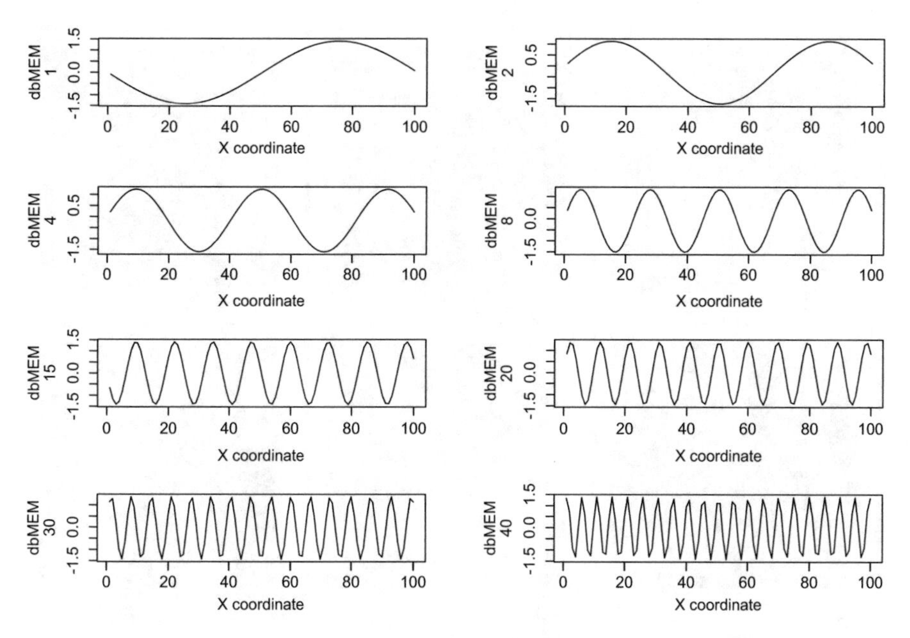

Fig. 7.3 Some of the 49 dbMEM variables with positive eigenvalues built from a transect with 100 equispaced points

7.4.2.2 dbMEM Variables on Regular Sampling Designs

When the spatial coordinates correspond to points that are equispaced along a transect or across a surface, the resulting dbMEM variables represent a series of sinusoids of decreasing periods. For a transect with n regularly spaced points and sampling interval s, the wavelength λ_i of the eigenfunction with rank i is: $\lambda_i = 2(n + s)/(i + 1)$ (Guénard et al. 2010, Eq. 3)[1]. Let us construct and illustrate a one-dimensional (Fig. 7.3) and a two-dimensional (Fig. 7.4) example, both equispaced. Computation of the dbMEM will be done using function **dbmem()** of package **adespatial**.

[1]A simple function to find the wavelength of rank i along a transect with n points for an intersite distance $s = 1$ is:

```
wavelength <- function(i, n) {2 * (n + 1) / (i + 1)}
```

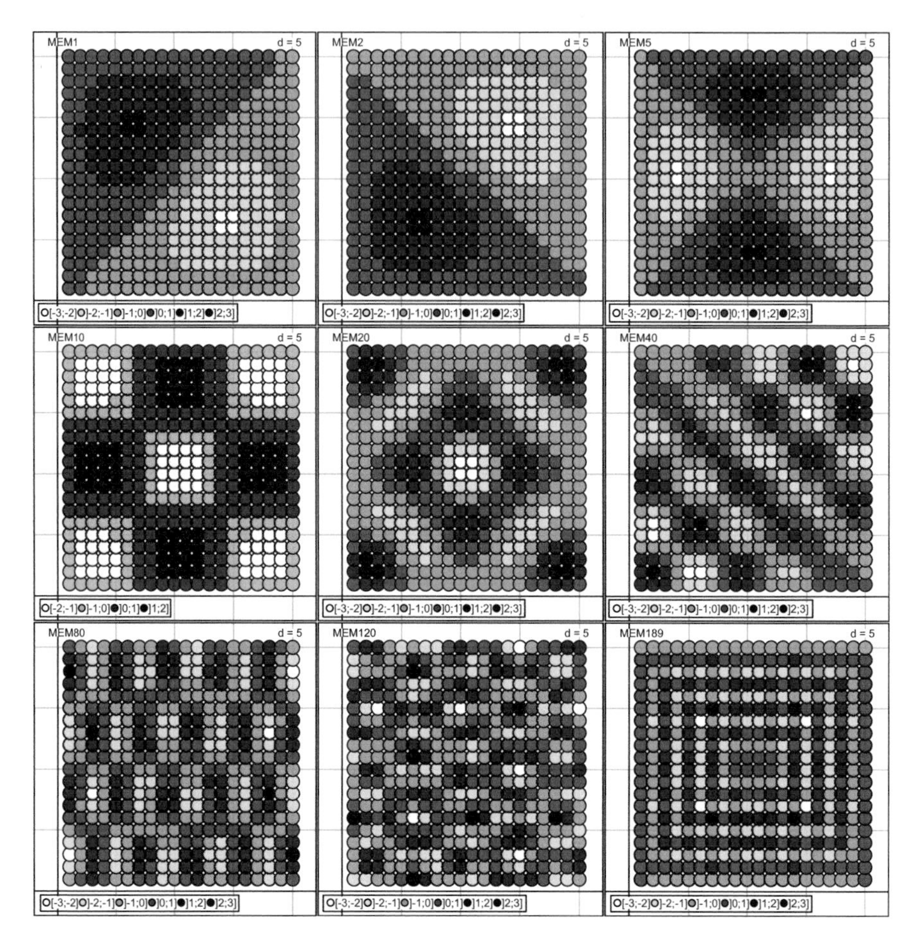

Fig. 7.4 Some of the 189 dbMEM variables with positive eigenvalues built from a grid comprising 20 by 20 equispaced points

```
# 1. One-dimensional sampling: transect with 100 equispaced points.

# Generate transect points
tr100 <- 1 : 100

# Creation of the dbMEM eigenfunctions with Moran's I corresponding
# to positive spatial correlation: argument
# MEM.autocor = "positive" which is the default
tr100.dbmem.tmp <- dbmem(tr100, silent = FALSE)
tr100.dbmem <- as.data.frame(tr100.dbmem.tmp)
# Display the eigenvalues
attributes(tr100.dbmem.tmp)$values
# Number of (positive) eigenvalues
length(attributes(tr100.dbmem.tmp)$values)
```

The distance between adjacent points is 1. Function **dbmem()** *automatically computes the threshold value, 1 in this case.*

```
# Plot some dbMEM variables modelling positive spatial correlation
#    along a transect (Fig. 7.3)
par(mfrow = c(4, 2))
somedbmem <- c(1, 2, 4, 8, 15, 20, 30, 40)
for(i in 1:length(somedbmem)){
    plot(tr100.dbmem[ ,somedbmem[i]],
         type = "l",
         xlab = "X coordinate",
         ylab = c("dbMEM", somedbmem[i]))
}

# 2. Two-dimensional sampling: grid of equispaced points with
#    smallest distance between points equal to 1 (Fig. 7.4).

# Generate grid point coordinates
xygrid2 <- expand.grid(1:20, 1:20)
# Creation of the dbMEM eigenfunctions with positive Moran's I
xygrid2.dbmem.tmp <- dbmem(xygrid2)
xygrid2.dbmem <- as.data.frame(xygrid2.dbmem.tmp)
# Count the eigenvalues
length(attributes(xygrid2.dbmem.tmp)$values)
# Plot some dbMEM variables using s.value {adegraphics}
somedbmem2 <- c(1, 2, 5, 10, 20, 40, 80, 120, 189)
s.value(xygrid2, xygrid2.dbmem[ ,somedbmem2],
  method = "color",
  symbol = "circle",
  ppoints.cex = 0.5
)
```

Hints *Function* **dbmem()** *has an argument called* `MEM.autocor` *to compute all or a subset of the spatial eigenfunctions. Default* `MEM.autocor="positive"` *computes the dbMEM with Moran's I larger than their expected value (eq. 7.4), i.e. those modelling structures with positive spatial correlation. Usually, there are also one 0 eigenvalue and several negative ones; the corresponding eigenvectors can be produced by calling one of the alternative choices in* `MEM.autocor`: `"non-null"`, `"all"`, `"negative"`.

Here we applied function **dbmem()** *to the matrix of geographical coordinates of the sites. The function also accepts matrices of geographical distances.*

In the code that produced Fig. 7.4, see how the function **s.value()** *of* **adegraphics** *can produce nine plots with a single call.*

In the case of regular sampling designs, the number of dbMEM with positive Moran's I is close to $n/2$. The figures show that these variables are cosines and that they range from broadest to finest scales. As mentioned above, this does not imply that only periodical structures can be modelled by dbMEM analysis, however. Even short-range spatial correlation can be modelled by fine-scaled dbMEM variables. This topic will be addressed later.

7.4.2.3 dbMEM Analysis of the Mite Data

dbMEM analysis is not restricted to regular sampling designs. The drawback with irregular designs is that the dbMEM variables lose the regularity of their shapes, making the assessment of scale more difficult at times.

Now it is time to apply dbMEM analysis to the Hellinger-transformed oribatid mite dataset. In the code below, as in the examples above, dbMEM variables are constructed using function **dbmem()** of package **adespatial**[2].

```
## Step 1. Construct the matrix of dbMEM variables
mite.dbmem.tmp <- dbmem(mite.xy, silent = FALSE)
mite.dbmem <- as.data.frame(mite.dbmem.tmp)
# Truncation distance used above:
(thr <- give.thresh(dist(mite.xy)))

# Display and count the eigenvalues
attributes(mite.dbmem.tmp)$values
length(attributes(mite.dbmem.tmp)$values)
```

Hint *The argument* `silent=FALSE` *of function* **dbmem()** *allows the function to display the truncation level onscreen; this level is a little bit larger than the largest distance among sites that keeps all sites connected in a minimum spanning tree. Here we computed this distance separately by means of function* **give.thresh()** *of package* **adespatial**.

As one can see, there are 22 dbMEM eigenvectors with positive spatial correlation. Prior to our RDA, we will apply forward selection with the Blanchet et al. (2008a) double stopping criterion.

[2]To compute "classical" PCNM eigenfunctions, users can apply function **pcnm()** of the **vegan** package. However, that function does not provide the values of the Moran's I indices to identify the eigenvectors modelling positive spatial correlation. These would have to be computed separately.

```
## Step 2. Run the global dbMEM analysis on the detrended
##    Hellinger-transformed mite data
(mite.dbmem.rda <- rda(mite.h.det ~., mite.dbmem))
anova(mite.dbmem.rda)

## Step 3. Since the R-square is significant, compute the adjusted
##    R2 and run a forward selection of the dbmem variables
(mite.R2a <- RsquareAdj(mite.dbmem.rda)$adj.r.squared)
(mite.dbmem.fwd <- forward.sel(mite.h.det, as.matrix(mite.dbmem),

    adjR2thresh = mite.R2a))
(nb.sig.dbmem <- nrow(mite.dbmem.fwd))    # Number of signif. dbMEM

# Identity of the significant dbMEM in increasing order
(dbmem.sign <- sort(mite.dbmem.fwd[ ,2]))
# Write the significant dbMEM to a new object
dbmem.red <- mite.dbmem[ ,c(dbmem.sign)]

## Step 4. New dbMEM analysis with 8 significant dbMEM variables
##    Adjusted R-square after forward selection: R2adj = 0.2418
(mite.dbmem.rda2 <- rda(mite.h.det ~ ., data = dbmem.red))
(mite.fwd.R2a <- RsquareAdj(mite.dbmem.rda2)$adj.r.squared)
anova(mite.dbmem.rda2)
(axes.test <- anova(mite.dbmem.rda2, by = "axis"))
# Number of significant axes
(nb.ax <- length(which(axes.test[ , ncol(axes.test)] <= 0.05)))

## Step 5. Plot the significant canonical axes
mite.rda2.axes <-
  scores(mite.dbmem.rda2,
         choices = c(1:nb.ax),
         display = "lc",
         scaling = 1)
par(mfrow = c(1,nb.ax))
for(i in 1:nb.ax){
    sr.value(mite.xy, mite.rda2.axes[ ,i],
             sub = paste("RDA",i),
             csub = 2)
             }
```

Hint *In the* **scores()** *call above, be careful to set* `display = "lc"`. *The default is* `"wa"`, *but here we want to display the canonical model scores.*

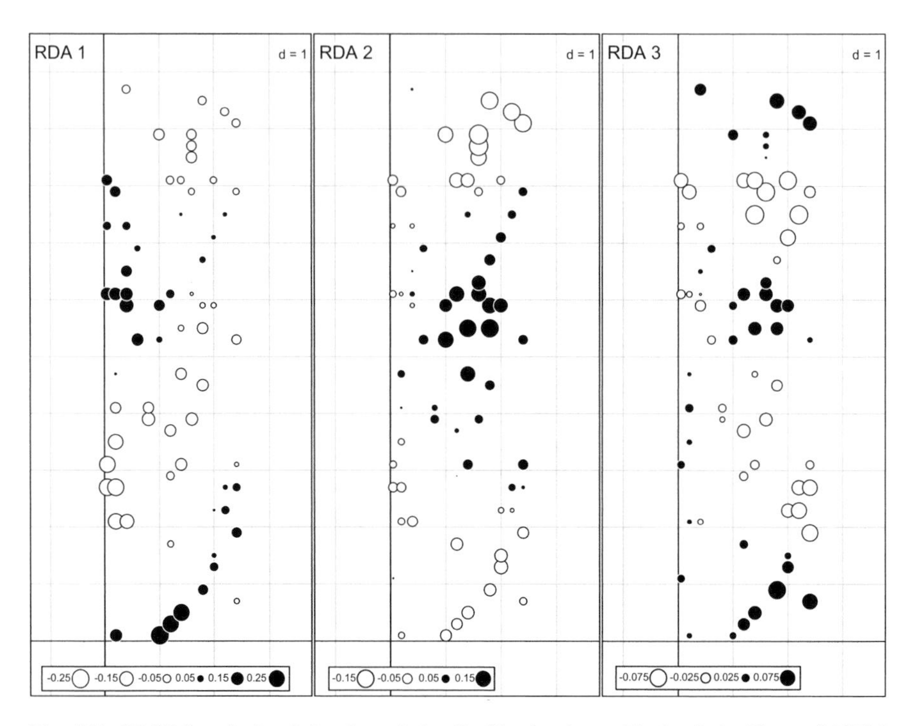

Fig. 7.5 dbMEM analysis of the detrended oribatid mite data with the 8 significant dbMEM variables. Maps of the fitted site scores of the first three canonical axes, which are significant at the $\alpha = 0.05$ threshold

The dbMEM analysis of the detrended mite data explained 24.18% of the variance (see **mite.fwd.R2a**, the adjusted R^2 obtained with the 8 dbMEM variables retained by forward selection). Three canonical axes, explaining 24.18×0.8895 [3] $= 21.5\%$ of the total variance, are significant; their fitted site scores have been plotted on a map of the sites. These three plots (Fig. 7.5) represent the spatially structured variation of the detrended oribatid mite data. Now, is this variation related to any of the environmental variables? A simple way to assess this is to regress the fitted site scores of these three canonical axes on the environmental data by means of function **lm()**. In the process, we make sure that the parametric tests of the regression coefficients are valid, by verifying that the residuals of the regressions are normally distributed. This is done by computing Shapiro-Wilk tests (function **shapiro.test()** of package **stats**).

[3]The value 0.8895 is found in the section "Accumulated constrained eigenvalues" of the RDA output. It is the proportion of variance explained by the first three canonical axes with respect to the total *explained* variance. Therefore, if the total *adjusted* explained variance (in the 8 canonical axes) is 24.18%, then the variance explained by axes 1, 2 and 3 together is $24.18\% \times 0.8895 = 21.5\%$.

```
# Interpreting the spatial variation: regression of the significant
# canonical axes on the environmental variables, with Shapiro-Wilk
# normality tests of residuals
mite.rda2.axis1.env <- lm(mite.rda2.axes[ ,1] ~ ., data = mite.env)
shapiro.test(resid(mite.rda2.axis1.env))
summary(mite.rda2.axis1.env)

mite.rda2.axis2.env <- lm(mite.rda2.axes[ ,2] ~ ., data = mite.env)
shapiro.test(resid(mite.rda2.axis2.env))
summary(mite.rda2.axis2.env)

mite.rda2.axis3.env <- lm(mite.rda2.axes[ ,3] ~ ., data = mite.env)
shapiro.test(resid(mite.rda2.axis3.env))
summary(mite.rda2.axis3.env)
```

As one can see, the three spatial axes are not related to the same environmental variables (except for shrubs). They are not fully explained by them either. A precise assessment of the portions explained will require variation partitioning.

This dbMEM analysis produced spatial models combining all the spatial variables forward-selected from the set of 22 dbMEM with positive spatial correlation. Here, the three significant canonical axes are a combination of dbMEM variables ranging from broad (dbMEM1) to fine scale (dbMEM20). While this may be interesting if one is interested in the global spatial structure of the response data, it does not allow one to discriminate between broad, medium and fine-scaled structures since all significant dbMEM are combined.

To address this question, one possible approach consists in computing *separate* RDAs constrained by subsets of the significant dbMEM variables. The dbMEM variables being linearly independent of one another, any submodel that contains a subset of dbMEM is also independent of any other submodel containing another subset. These subsets can be defined in such a way as to model different scales. The choices are arbitrary: there is no strict rule to identify what is broad, medium or fine scale. How should one decide? Here are several ways.

- Predefine these limits, using the sizes of the patterns corresponding to the dbMEM variables.
- Try to identify groups of eigenfunctions by examining a scalogram (Fig. 7.6) showing in ordinate the R^2 of the dbMEM eigenfunctions ordered along the abscissa by decreasing eigenvalues (Legendre and Legendre 2012 p. 864). This scalogram can be drawn by applying our homemade function **`scalog.R()`** to an RDA result object produced by function **`rda()`**. The RDA must have been computed with the formula interface.

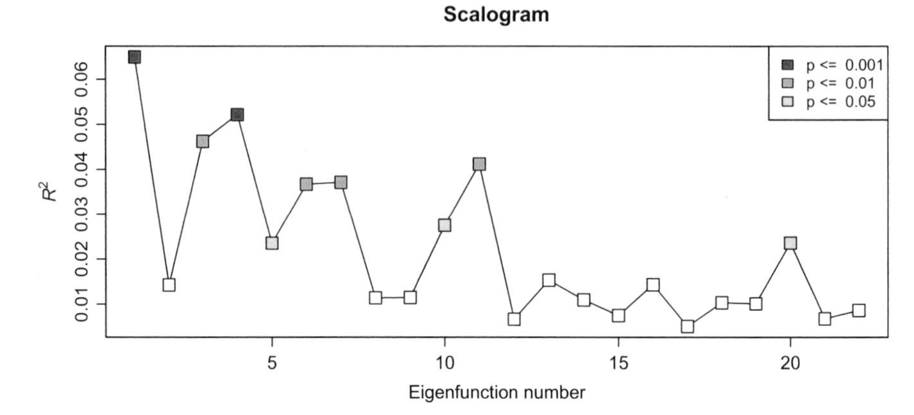

Fig. 7.6 Scalogram showing the explained variance (unadjusted R^2) of the detrended, Hellinger-transformed mite data explained by the dbMEM eigenfunctions, with color-coded permutation test results. The p-values are more liberal than the ones obtained with the forward selection based on the Blanchet et al. (2008a) double stopping criterion.

```
# Scalogram of the variance explained by all dbMEM eigenfunctions,
# computed with our homemade function scalog()
scalog(mite.dbmem.rda)
```

- Run forward selection and define submodels corresponding to groups of more or less consecutive dbMEM variables retained by the procedure. This approach is more conservative than the tests of significance in the scalogram because selection of explanatory variables stops when the adjusted R^2 of the global model is reached..
- Draw maps of the significant dbMEM variables (Fig. 7.7) and group them visually according to the scales of the patterns they represent:

```
# Maps of the 8 significant dbMEM variables with the homemade
# function sr.value()
par(mfrow = c(2, 4))
for(i in 1 : ncol(dbmem.red))
{
    sr.value(mite.xy,
             dbmem.red[ ,i],
             sub = paste("dbMEM", dbmem.sign[i]),
             csub = 2)
}
```

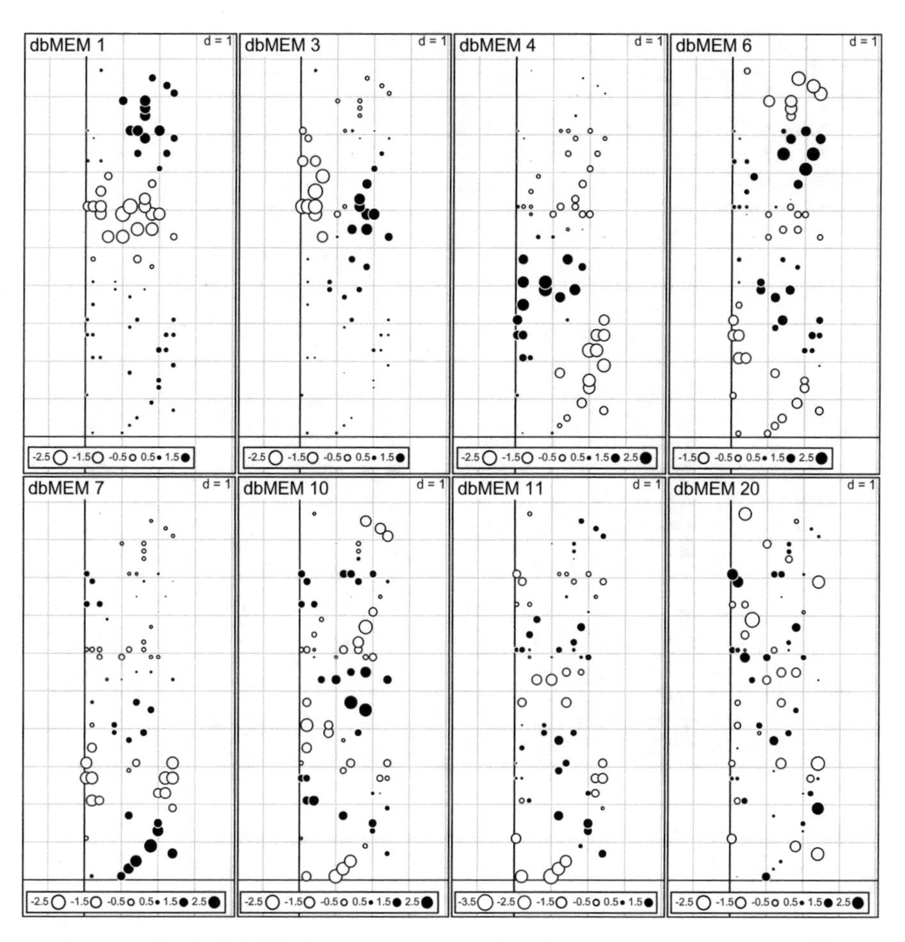

Fig. 7.7 The 8 significant dbMEM variables with positive spatial correlation used in the dbMEM analysis of the detrended oribatid mite data.

On these bases, one could for instance define dbMEM 1, 3, and 4 as "broad scale", dbMEM 6, 7, 10 and 11 as "medium scale", and dbMEM 20 as "fine scale" descriptors. Separate RDAs with these subsets will model broad, medium and fine-scaled patterns, respectively.

```
## dbMEM analysis of the mite data - broad scale
(mite.dbmem.broad <-
  rda(mite.h.det ~ ., data = mite.dbmem[ ,c(1,3,4)]))
anova(mite.dbmem.broad)
(axes.broad <- anova(mite.dbmem.broad, by = "axis"))
# Number of significant axes
(nb.ax.broad <-
  length(which(axes.broad[ , ncol(axes.broad)] <=  0.05)))
# Plot of the two significant canonical axes
mite.dbmembroad.axes <-
  scores(mite.dbmem.broad,
         choices = c(1,2),
         display = "lc",
         scaling = 1)
par(mfrow = c(1, 2))
sr.value(mite.xy, mite.dbmembroad.axes[ ,1])
sr.value(mite.xy, mite.dbmembroad.axes[ ,2])

# Interpreting the broad-scaled spatial variation: regression of
# the two significant spatial canonical axes on the environmental
# variables
mite.dbmembroad.ax1.env <-
  lm(mite.dbmembroad.axes[ ,1] ~ ., data = mite.env)
summary(mite.dbmembroad.ax1.env)
mite.dbmembroad.ax2.env <-
  lm(mite.dbmembroad.axes[ ,2] ~ ., data = mite.env)
summary(mite.dbmembroad.ax2.env)
```

The broad-scale relationships are clearly related to microtopography and the absence of shrubs.

```
## dbMEM analysis of the mite data - medium scale
(mite.dbmem.med <-
  rda(mite.h.det ~ ., data = mite.dbmem[ ,c(6,7,10,11)]))
anova(mite.dbmem.med)
(axes.med <- anova(mite.dbmem.med, by = "axis"))
# Number of significant axes
(nb.ax.med <- length(which(axes.med[ ,ncol(axes.med)] <= 0.05)))
# Plot of the significant canonical axes
mite.dbmemmed.axes <-
  scores(mite.dbmem.med, choices = c(1,2),
         display = "lc",
         scaling = 1)
par(mfrow = c(1, 2))
sr.value(mite.xy, mite.dbmemmed.axes[ ,1])
sr.value(mite.xy, mite.dbmemmed.axes[ ,2])
```

```
# Interpreting the medium-scaled spatial variation: regression of
# the significant spatial canonical axes on the environmental
# variables
mite.dbmemmed.ax1.env <-
  lm(mite.dbmemmed.axes[ ,1] ~ ., data = mite.env)
summary(mite.dbmemmed.ax1.env)
mite.dbmemmed.ax2.env <-
  lm(mite.dbmemmed.axes[ ,2] ~ ., data = mite.env)
summary(mite.dbmemmed.ax2.env)
```

The medium-scale features correspond to two types of soil coverage, to variation in shrub density and to microtopography.

```
## dbMEM  analysis of the mite data - fine scale
(mite.dbmem.fine <-
  rda(mite.h.det ~  ., data = as.data.frame(mite.dbmem[ ,20])))
anova(mite.dbmem.fine)
```

The analysis stops here, since the RDA is not significant.

Something has occurred here, which is often found in fine-scale dbMEM analysis. The fine-scaled dbMEM model is not significant. When significant, the fine-scaled dbMEM variables are mostly signatures of local spatial correlation generated by community dynamics. This topic will be addressed later.

7.4.2.4 Hassle-Free dbMEM Analysis: Function `quickMEM()`

A single-step dbMEM analysis can be performed easily with our function **`quickMEM()`**. This function only requires two arguments: a response data table (pre-transformed if necessary) and a table containing the site geographic coordinates, which can be one- or two-dimensional. When the default arguments are applied, the function performs a complete dbMEM analysis: it checks whether the response data should be detrended and does it if a significant trend is identified; it constructs the dbMEM variables and tests the global RDA; it runs forward selection, using the dbMEM with positive spatial correlation; it runs RDA with the retained dbMEM variables and tests the canonical axes; it delivers the RDA results (including the set of dbMEM variables) and plots maps of the significant canonical axes.

```
mite.dbmem.quick <- quickMEM(mite.h, mite.xy)
summary(mite.dbmem.quick)
# Eigenvalues
mite.dbmem.quick[[2]]      # OR mite.dbmem.quick$eigenvalues
# Results of forward selection
mite.dbmem.quick[[3]]      # OR mite.dbmem.quick$fwd.sel
```

Function **quickMEM()** provides several arguments to respond to various needs. For instance, detrending is done by default if a significant trend is found, but this option can be disabled (detrend = FALSE). The truncation threshold is computed automatically (largest value of minimum spanning tree) unless the user provides another value (e.g. thresh = 1.234). Computation of the dbMEM variables is overridden if the user provides a ready-made matrix of spatial variables (myspat = ...).

quickMEM() provides a composite output object (of class "list") containing many results. The summary shows all the component names. The components can be retrieved as in the code block above (e.g. eigenvalues: mite.dbmem.quick [[2]]). To draw a biplot of the RDA results, the code is the following:

```
# Extract and plot RDA results from a quickMEM output (scaling 2)
plot(mite.dbmem.quick$RDA, scaling = 2)
sp.scores2 <-
  scores(mite.dbmem.quick$RDA,
         choices = 1:2,
         scaling = 2,
         display = "sp")
arrows(0, 0,
  sp.scores2[ ,1] * 0.9,
  sp.scores2[ ,2] * 0.9,
  length = 0,
  lty = 1,
  col = "red"
)
```

Hint *In this call, plotting is done by the* **plot.cca()** *function of* **vegan** *since the* $RDA *element is a* **vegan** *output object.*

The scaling 2 shows the relationship of some species with some dbMEM variables. These correlations can be explored to reveal at which scale the species distributions are spatially structured.

7.4.2.5 Combining dbMEM Analysis and Variation Partitioning

A clever and global approach to assess the environmental variation related to all scales of spatial variation is to perform a **variation partitioning** analysis with an environmental data set and up to three subsets of spatial variables. Function **`varpart()`**, studied in Sect. 6.3.2.8, can only handle numerical variables (not factors), however, so that we will have to recode environmental variables 3–5 into dummy binary variables.

Variation partitioning aims at quantifying the various unique and combined fractions of variation explained by several sources. In this context, a linear trend can be considered to represent a source of variation like any other. The trend is likely to manifest itself on the response as well as the environmental explanatory variables. Therefore, in this application we advocate **not** to detrend the response data prior to variation partitioning, but rather to test for a linear trend and incorporate it explicitly in the partitioning procedure if it is significant. Note that we consider the linear trend, represented by a pair of *X*-*Y* coordinates, as a whole. If it is significant we incorporate it into the variation partitioning without forward-selecting the coordinates. This ensures that any linear trend, large or small, is included.

We have seen in Sect. 6.3.2.8 that, prior to variation partitioning, forward selection must be done independently in each block of explanatory variables. Therefore, it would seem that in the present case forward selection should be computed separately on the environmental variables and the dbMEM variables, all this with the undetrended response variables. However, a technical point will lead to recommend to slightly derogate from this procedure. We will indeed forward-select the environmental variables using the undetrended response variables. The selection of the dbMEM variables, however, will be run on the response variables that have been *detrended* by the *X*-*Y* coordinates (if the trend is significant). We are proceeding in this way because dbMEM, among other properties, are also able to model linear trends, in addition to other structures (Borcard et al. 2002). Consequently, if a linear trend is present in the data, it is bound to be modelled by a subset of the dbMEM as well as by the *X*-*Y* coordinates. Contrary to the environment vs dbMEM case, this does not tell us anything more about the response data: it is only a consequence of using two different types of spatial variables in the variation partitioning[4].

In this example, we will arbitrarily split the significant dbMEM variables into a broad and a fine scale fraction. The partitioning results are presented in Fig. 7.8.

---

[4]One may consider forward-selecting the dbMEM on the undetrended response data, using the *X*-*Y* coordinates as covariables, by means of function **`ordiR2step()`** of **`vegan`**. However, this technique implies that the dbMEM variables are residualized on the *X*-*Y* coordinates. As a consequence, the residualized dbMEM variables would no longer be orthogonal, and they would not be those that will finally be used in the variation partitioning. Therefore, we do not recommend this procedure.

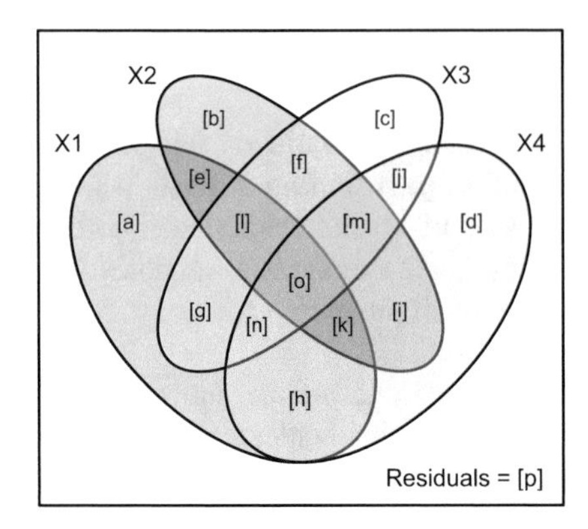

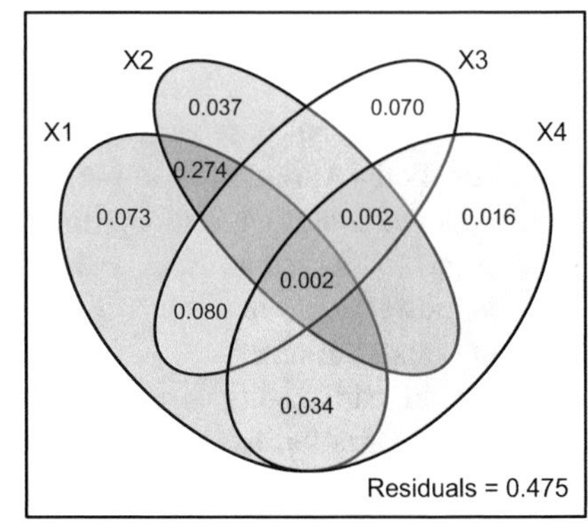

Fig. 7.8 Variation partitioning of the undetrended oribatid mite data into an environmental component (X1), a linear trend (X2), a broad scale (X3) and fine scale (X4) dbMEM spatial components. The empty fractions in the plots have small negative R^2_{adj} values. All values are given in the screen output table.

```
# 1. Test trend
mite.XY.rda <- rda(mite.h, mite.xy)
anova(mite.XY.rda)

# 2. Test and forward selection of the environmental variables
# Recode environmental variables 3 to 5 into dummy binary variables
substrate <- model.matrix( ~ mite.env[ ,3])[ ,-1]
shrubs <- model.matrix( ~ mite.env[ ,4])[ ,-1]
topography <- model.matrix( ~ mite.env[ ,5])[ ,-1]
mite.env2 <- cbind(mite.env[ ,1:2], substrate, shrubs, topography)
colnames(mite.env2) <-
  c("SubsDens", "WatrCont", "Interface", "Litter", "Sphagn1",
    "Sphagn2", "Sphagn3", "Sphagn4", "Shrubs_Many", "Shrubs_None",
    "topography")
# Forward selection of the environmental variables
mite.env.rda <- rda(mite.h ~., mite.env2)
(mite.env.R2a <- RsquareAdj(mite.env.rda)$adj.r.squared)
mite.env.fwd <-
  forward.sel(mite.h, mite.env2,
              adjR2thresh = mite.env.R2a,
              nperm = 9999)
env.sign <- sort(mite.env.fwd$order)
env.red <- mite.env2[ ,c(env.sign)]
colnames(env.red)

# 3. Test and forward selection of the dbMEM variables
# Run the global dbMEM analysis on the *detrended* mite data
mite.det.dbmem.rda <- rda(mite.h.det ~., mite.dbmem)
anova(mite.det.dbmem.rda)
# Since the analysis is significant, compute the adjusted R2
# and run a forward selection of the dbMEM variables
(mite.det.dbmem.R2a <-
  RsquareAdj(mite.det.dbmem.rda)$adj.r.squared)
```

```
(mite.det.dbmem.fwd <-
  forward.sel(mite.h.det,
              as.matrix(mite.dbmem),
      adjR2thresh = mite.det.dbmem.R2a))
# Number of significant dbMEM
(nb.sig.dbmem <- nrow(mite.det.dbmem.fwd))
# Identify the significant dbMEM sorted in increasing order
(dbmem.sign <- sort(mite.det.dbmem.fwd$order))
# Write the significant dbMEM to a new object (reduced set)
dbmem.red <- mite.dbmem[ ,c(dbmem.sign)]
# 4. Arbitrarily split the significant dbMEM into broad and
#    fine scale
# Broad scale: dbMEM 1, 3, 4, 6, 7
dbmem.broad <- dbmem.red[ , 1 : 5]
# Fine scale: dbMEM 10, 11, 20
dbmem.fine <- dbmem.red[ , 6 : 8]

## 5. Mite - environment - trend - dbMEM variation partitioning
(mite.varpart <-
  varpart(mite.h, env.red, mite.xy, dbmem.broad, dbmem.fine))
# Show the symbols of the fractions and plot their values
par(mfrow = c(1,2))
showvarparts(4, bg = c("red", "blue", "yellow", "green"))
plot(mite.varpart,
     digits = 2,
     bg = c("red", "blue", "yellow", "green")
)
```

The default option of function **`plot.varpart()`** *leaves the fractions with negative* R^2_{adj} *empty in the graph. To display the negatives values, set the argument* `cutoff = -Inf`.

```
# Tests of the unique fractions [a], [b], [c] and [d]
# Fraction [a], pure environmental
anova(
  rda(mite.h, env.red, cbind(mite.xy, dbmem.broad, dbmem.fine)))
# Fraction [b], pure trend
anova(
  rda(mite.h, mite.xy, cbind(env.red, dbmem.broad, dbmem.fine)))
# Fraction [c], pure broad scale spatial
anova(rda
  (mite.h, dbmem.broad, cbind(env.red, mite.xy, dbmem.fine)))
# Fraction [d], pure fine scale spatial
anova(
  rda(mite.h, dbmem.fine, cbind(env.red, mite.xy, dbmem.broad)))
```

All unique fractions are significant.

When interpreting such a complex variation partitioning diagram, keep in mind that the R^2 adjustment is done for the fractions that are directly fitted by a linear RDA model, without resorting to partial RDA or multiple regression (here the first 15 rows of the table of results); the individual fractions [a] to [p] are then computed by subtraction. Small negative R^2_{adj} values frequently appear in this process. Small negative R^2_{adj} values generally correspond to explanatory components that explain less of the response variables' variation than would be expected by chance for random normal deviates; so, for all practical purposes, they can be interpreted as zeros and neglected during interpretation, although they must be taken into account when computing sums of subsets, or sums of all fractions (the latter is equal to 1)[5].

The whole set of environmental and spatial variables explains 52.5% of the variation of the undetrended mite data (see the R^2_{adj} for "All" fractions). The environmental variables alone (matrix X1 in the partitioning results) explain 40.8% of the variation, of which a mere 7.3% is not spatially structured (fraction [a]). This fraction represents species-environment relationships associated with local environmental conditions.

The fractions involving environmental and spatial variables (in this example, essentially fractions [e], [g] and [h]) represent spatially structured environmental variation. Fraction [e] (27.4% variation explained) is common to the environmental data and the linear (*X*-*Y*) gradient. This is a typical case of **induced spatial dependence**, where the spatial structure of environmental factors produces a similar spatial structure in the response data. Fraction [g] (8.0% variation explained) and fraction [h] (3.4% variation explained) are common to the environmental and, respectively, the broad-scale and fine-scale dbMEM variables. These fractions might also be interpreted as signatures of induced spatial dependence, but with spatial patterns more complex than linear gradients. However, when some variation is explained jointly by environmental and spatial variables, one should be careful when inferring causal species-environment relationships: the correlations may be due to a direct influence of the environmental variables on the species (direct induced spatial dependence), or to some unmeasured underlying process that is spatially structured and is influencing both the mite community and the environmental variables (e.g., spatial variation induced by a historical causal factor).

The variation partitioning results also show that the four sources of variation have unequal but significant unique contributions: the environment alone ([a], 7.3%) and the broad-scaled ([c], 7.0%) variation are the largest, while the trend alone ([b], 3.7%) and the fine-scaled variation ([d], 1.6%) play smaller roles.

The variation explained by spatial variables independently of the environment is represented by fractions [b], [c], [d], [f], [i], [j] and [m]. Together, these fractions explain 11.6% variation. Most likely, some of this variation, especially at broad and

[5]A negative common fraction of variation [b] may also occur if two variables or groups of variables **X** and **W**, together, explain **Y** better than the sum of their individual effects (Legendre and Legendre 2012, p. 573). This happens, for instance, when the correlation between **X** and **W** is negative, but both are correlated positively with **Y**.

medium scales, could be explained by unmeasured environmental variables, although one cannot exclude the influence of past events that could still show their marks in the mite community (Borcard and Legendre 1994). Fine-scale structures are more likely explainable by spatial correlation produced by neutral biotic processes. Neutral processes include ecological drift (variation in species demography due to random reproduction and random survival of individuals due to competition, predator-prey interactions, etc.) and random dispersal (migration in animals, propagule dispersion in plants). Controlling for spatial correlation by means of dbMEM variables when testing species-environment relationships will be briefly addressed in Sect. 7.4.3.4.

Finally, note that the broad and fine scale dbMEM variables have a non-null intersection despite the fact that the dbMEM variables are orthogonal: fraction [j + m + n + o] totals −0.7%. This occurs because other variables (environment and trend), which are not orthogonal to the dbMEM, are involved in the partitioning, and also because the variation partitioning procedure involves subtractions of R^2 that have been adjusted on the basis of different numbers of explanatory variables.

7.4.3 MEM in a Wider Context: Weights Other than Geographic Distances

7.4.3.1 Introduction

The dbMEM method provides an elegant way of constructing sets of linearly independent spatial variables. Since its publication, it has gained a wide audience and has been applied in many research papers. But it is not the end of the story.

Dray et al. (2006) have greatly improved the mathematical formalism of the original PCNM analysis by showing that it is a particular case of a wider family of methods that they called Moran's eigenvector maps (MEM). They demonstrated the link between the eigenvalues of the MEM eigenvectors and Moran's spatial correlation index, I (Eq. 7.3).

They reasoned that the relationship among sites, which is the basis for any spatial eigenvector decomposition, actually has two components: (1) a list of **links** among objects, represented by a *connectivity matrix*, and (2) a matrix of **weights** to be applied to these links. In the simplest case, the weights are binary (i.e., either two objects are linked, or they are not). In more complex models, non-negative weights can be placed on the links; these weights represent the easiness of exchange (of organisms, energy, information, etc.) between the points connected by the links. For instance, link weights can be made to be inversely proportional to the Euclidean or squared Euclidean distances among sites.

Furthermore, Dray et al. (2006) showed that (1) by using similarities instead of distances among sites, (2) setting the relationship of the sites with themselves to null similarity, and (3) avoiding a square-root standardization of the eigenvectors within the PCoA procedure, one obtains a family of flexible methods (MEM) that can be modulated to optimize the construction of spatial variables. The MEM method produces $n - 1$ spatial variables with positive and negative eigenvalues, allowing the construction of a wide range of variables modelling positive and negative spatial correlation. As in the special dbMEM case, the eigenvectors maximize Moran's I index, the eigenvalues being equal to Moran's I multiplied by a constant. Therefore, the spatial structures of the data are extracted in such a way that the axes first optimally display the positively autocorrelated structures in decreasing order of importance, and then the negatively autocorrelated structures in increasing order.

The general MEM method consists in defining two matrices describing the relationships among the sites:

- a binary connectivity matrix **B** defining which pairs of sites are connected (1) and which are not (0);
- a weighting matrix **A** providing the intensity of the connections.

The final *spatial weighting matrix* **W** results from the Hadamard (i.e., term-by-term) product of these two matrices, **B** and **A**.

The connectivity matrix **B** can be constructed on the basis of distances (by selecting a distance threshold and connecting all points that are within that distance) or by other connection schemes such as Delaunay triangulation, Gabriel graph or others (described by Legendre and Legendre 2012 Sect. 13.3). The connection matrix can of course be customized to fit special needs — for instance by only allowing connections among sites along the littoral zone of a lake (not across water) or along the shoreline of an island.

Matrix **A** is not mandatory, but it is often used to weight the connections according to distance, e.g. by inverse distance or inverse squared distance, since it is ecologically realistic to assume that a process influences a community with an intensity decreasing with distance. The choice of both matrices is very important because it greatly affects the structure of the spatial variables obtained. These variables, in turn, condition the results of the spatial analysis, especially in the case of irregular sampling: "*In the case of* ***regular*** *sampling (e.g., a regular grid), structures defined by eigenvectors are roughly similar for different definitions of* ***W****. For* ***irregular*** *distributions of sites, however, the number of positive/negative eigenvalues and the spatial structures described by their associated eigenvectors are greatly influenced by the spatial relationships defined in* ***W****.*" (Dray et al. 2006). These authors provide the following general recommendations:

"*The choice of the spatial weighting matrix* ***W*** *is the most critical step in spatial analysis. This matrix is a model of the spatial interactions recognized among the sites, all other interactions being excluded. In some cases, a theory-driven specification can be adopted, and the spatial weighting matrix can be constructed based upon biological considerations [...]. In most situations, however, the choice of a particular matrix may become rather difficult and a data-driven specification could then be applied. Under this latter approach, the objective is to select a configuration of* ***W*** *that results in the optimal performance of the spatial model.*"

For data-driven model specification, the authors proposed a procedure starting with a user-defined set of possible spatial weighting matrices. For *each* candidate, one computes the MEM eigenfunctions, reorders them according to their explanatory power, enters them one by one into the model and retains the model with the lowest AIC_c (corrected Akaike information criterion). When this is done for all candidates, one retains the **W** matrix yielding the lowest AIC_c.

The AIC_c-based selection is but one possibility. One could also forward-select the MEM within each candidate model using Blanchet et al. (2008a)'s double stopping criterion and retain the model with the highest R^2_{adj}. This alternative, which had not yet been devised when the Dray et al. (2006) paper was published, addresses the concerns raised by these authors in their conclusion about the drawbacks of forward selection procedures.

7.4.3.2 Generalized MEM Analysis of the Mite Data

Dray et al. (2006) used the oribatid mite data to illustrate MEM analysis. As an example, we will duplicate their analysis, exploring some choices along the steps of the method. Several packages will be used. The following example is based on Stéphane Dray's tutorial on MEM analysis, with our thanks to the author.

The functions available in **`adespatial`** and used below should make model selection relatively easy. Of course, the final result depends upon a proper choice of a class of model. The function **`test.W()`** is particularly useful as it combines construction of MEM variables and model selection; examine the documentation file of that function.

We will experiment with three classes of models:

- The first class is based on Delaunay triangulation with binary weights.
- The second class starts from the same connectivity matrix, to which weights are added. The weighting function is based on Euclidean distances among the sites:
- $f_2 = 1 - (d/d_{max})^\alpha$ where d is a geographic distance value and d_{max} is the maximum value in the distance matrix. This ensures that easiness of communication ranges from 1 (easy) to 0 (isolation).
- The third class evaluates a series of models based on a range of distances around the points. All pairs of points within the distance considered are linked, the others not. What should the range of distances be? This can be assessed by means of a multivariate variogram of the response data. Variograms are plots of semivariances against distance classes. Semivariance is a distance-dependent measure of variation, which is used in the same way as in the correlograms presented earlier (e.g. Bivand et al. 2013). A variant of this approach will weight the links by the function of inverse distance that was used in the second model class above. This last variant will duplicate the results presented in the Dray et al. (2006) paper.

To be consistent with the dbMEM analysis, we shall restrict our model selection to MEM with Moran's I larger than its expectation (i.e., positive spatial correlation). To compute an analysis with all MEM variables, change argument `MEM.autocor` to `"all"` in all applications of function **`test.W()`**.

First and second classes of MEM models: unweighted (binary) and distance-weighted Delaunay triangulation.

```
# Selection of an optimal spatial weighting matrix

# 1. Search based on Delaunay triangulation.
#    We use mite.h.det as response data and mite.del as Delaunay
#    triangulation data.
#    No weighting matrix (binary weights only): 1 means connected
#    (easy communication), 0 = not connected (no exchange
#    possible). Function test.W() selects among the MEM variables
#    constructed on the basis of the Delaunay triangulation.
# Delaunay triangulation and model selection
(mite.del <- tri2nb(mite.xy))
mite.del.res <-
  test.W(mite.h.det,
         mite.del,
         MEM.autocor = "positive")
```

The screen output says that the best model has an AICc value of -93.87 and is based on 6 MEM variables.

```
# Summary of the results for the best model
summary(mite.del.res$best)
# Unadjusted R^2 of the model with the smallest AICc value
(R2.del <-
  mite.del.res$best$AIC$R2[which.min(mite.del.res$best$AIC$AICc)])
# Adjusted R^2 of the model with the smallest AICc value
RsquareAdj(
  R2.del,
  n = nrow(mite.h.det),
  m = which.min(mite.del.res$best$AIC$AICc)
)

# 2. Delaunay triangulation weighted by a function of distance.
#    Distances are ranged to maximum 1, and raised to power y.
#    After transformation of the distances by function f2, values
#    near 1 are attributed to pairs of sites with easy exchange,
#    values near 0 mean difficult communication.
f2 <- function(D, dmax, y)
{
  1 - (D/dmax)^y
}

# Largest Euclidean distance on links belonging to the Delaunay
# triangulation
max.d1 <- max(unlist(nbdists(mite.del, as.matrix(mite.xy))))
# Power y is set from 2 to 10
mite.del.f2 <-
  test.W(mite.h.det, mite.del,
         MEM.autocor = "positive",
         f = f2,
         y = 2:10,
         dmax = max.d1,
         xy = as.matrix(mite.xy))
```

The screen output says that the best model has an AICc value of -95.32 and is based on 4 MEM variables.

```
# Unadjusted R^2 of best model
(R2.delW <-
  mite.del.f2$best$AIC$R2[which.min(mite.del.f2$best$AIC$AICc)])
# Adjusted R^2 of best model
RsquareAdj(
  R2.delW,
  n = nrow(mite.h.det),
  m = which.min(mite.del.f2$best$AIC$AICc)
)
```

Third class of MEM model: connectivity matrix based on distances.

```
# 3a. Connectivity matrix based on a distance (radius around
#     points)
# Assessment of the relevant distances based on a multivariate
# variogram of the detrended mite data, with 20 distance classes.
(mite.vario <- variogmultiv(mite.h.det, mite.xy, nclass = 20))
plot(
  mite.vario$d,
  mite.vario$var,
  ty = 'b',
  pch = 20,
  xlab = "Distance",
  ylab = "C(distance)"
)
```

The multivariate variogram is presented in Fig. 7.9. It consists in the sum of univariate variograms computed over all species. The variance increases from 0 to 4 m. Since the shortest distance to keep all sites connected is 1.011187 m (see dbMEM analysis), we will explore a range of 10 evenly distributed distances ranging

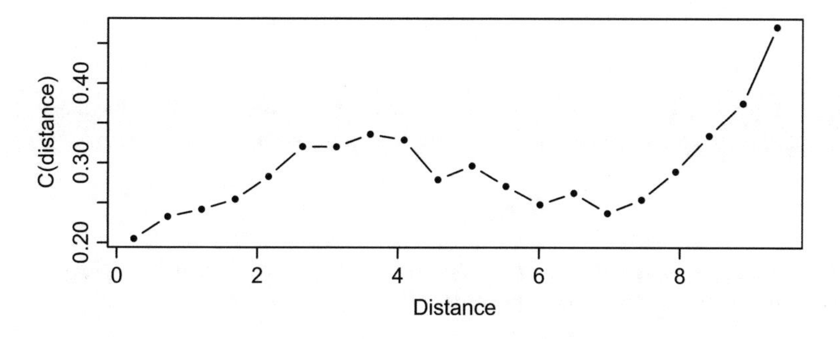

Fig. 7.9 Multivariate variogram of the detrended oribatid mite data. 20 distance classes

from this threshold up to 4.0 m (i.e., approximately 4 times the threshold, used in the dbMEM analysis).

```
# Construction of 10 neighbourhood matrices (class nb)
# Vector of 10 threshold distances
(thresh10 <- seq(give.thresh(dist(mite.xy)), 4, le = 10))
# Create 10 neighbourhood matrices.
# Each matrix contains all connexions with lengths smaller or equal
# to the threshold value
list10nb <-
  lapply(thresh10,
         dnearneigh,
         x = as.matrix(mite.xy),
         d1 = 0)
# Display an excerpt of the first neighbourhood matrix
print(
  listw2mat(nb2listw(list10nb[[1]], style = "B"))[1:10, 1:10],
  digits = 1
)
# Now we can apply the function test.W() to the 10 neighbourhood
# matrices. There are no weights on the links.
mite.thresh.res <-
  lapply(list10nb,
    function(x) test.W(x,
                       Y = mite.h.det,
                       MEM.autocor = "positive")
)
# Lowest AICc, best model, threshold distance of best model
mite.thresh.minAIC <-
  sapply(mite.thresh.res,
         function(x) min(x$best$AIC$AICc, na.rm = TRUE))
# Smallest AICc (best model among the 10)
min(mite.thresh.minAIC)
# Number of the model among the 10
which.min(mite.thresh.minAIC)
# Truncation threshold (distance)
thresh10[which.min(mite.thresh.minAIC)]
```

Hint **`dnearneigh()`** *requires 2 geographic dimensions. Add a constant column (e.g. a column of 1) if you only have 1 dimension, e.g. a transect or a time series.*

Identify the best model, that is, the model with the lowest AICc. What is the range of distances in that model? How many MEMs were selected?

This result is more interesting than that of the weighted Delaunay MEM. The AICc of the best model, obtained with a threshold of 2 m, is −99.85 with a model consisting of 4 MEM variables. Let us see if we can improve this result by weighting the connections by an inverse distance function.

```
# 3b. Variant: same as above, but connections weighted by the
#     complement of the power of the distances, 1-(d/dmax)^y.
#     Again, after transformation of the distances by function f2,
#     values near 1 are attributed to pairs of sites with easy
#     exchange, values near 0 mean difficult communication.
mite.thresh.f2 <-
  lapply(list10nb,
         function(x) test.W(x, Y = mite.h.det,
         MEM.autocor = "positive",
         f = f2,
         y = 2:10,
         dmax = max(unlist(nbdists(x, as.matrix(mite.xy)))),
         xy = as.matrix(mite.xy)))
# Lowest AIC, best model
mite.f2.minAIC <-
  sapply(mite.thresh.f2,
         function(x) min(x$best$AIC$AICc, na.rm = TRUE))
# Smallest AICc (best model among the 10)
min(mite.f2.minAIC)
# Number of the model among the 10
(nb.bestmod <- which.min(mite.f2.minAIC))
# Actual dmax of best model
(dmax.best <- mite.thresh.f2[nb.bestmod][[1]]$all[1, 2])
```

Hints *The actual* d_{max} *value found by the function is often smaller than the* d_{max} *provided to the function by means of the vector of user-selected threshold distances, because the output of the function shows the largest actual distance within the limit provided by each threshold value. In the example, the 6th value in vector* `thresh10`, *which contains the list of user-selected threshold distances, is 2.671639. There is no such distance in the mite geographic distance matrix; the function found that the largest distance smaller than or equal to that threshold is 2.668333.*

In some applications, the computations are followed by one or more warnings:

```
1: In nb2listw(nb, style = "B", glist = lapply(nbdist, f),
zero.policy = TRUE) :
zero sum general weights
```

This means that one or more points have no neighbours under the definition of neighbourhood provided to the analysis. This may or may not be important depending on the context of the analysis.

With an AICc of −100.96, this is the best result of all our attempts in terms of AICc. We can therefore extract this champion model, which contains 6 selected MEM variables, from the output object:

```
# Extraction of the champion MEM model
mite.MEM.champ <-
  unlist(mite.thresh.f2[which.min(mite.f2.minAIC)],
         recursive = FALSE)
summary(mite.MEM.champ)
# Number of MEM variables in best model
(nvars.best <- which.min(mite.MEM.champ$best$AIC$AICc))
# MEM variables by order of added R2
mite.MEM.champ$best$AIC$ord
# MEM variables selected in the best model
MEMid <- mite.MEM.champ$best$AIC$ord[1:nvars.best]
sort(MEMid)
MEM.all <- mite.MEM.champ$best$MEM
MEM.select <- mite.MEM.champ$best$MEM[ , sort(c(MEMid))]
colnames(MEM.select) <- sort(MEMid)
# Unadjusted R2 of best model
R2.MEMbest <- mite.MEM.champ$best$AIC$R2[nvars.best]
# Adjusted R2 of best model
RsquareAdj(R2.MEMbest, nrow(mite.h.det), length(MEMid))
# Plot the links using the function plot.links()
plot.links(mite.xy, thresh = dmax.best)
```

The very best MEM model among those tested contains 6 MEM variables (1, 2, 3, 6, 8, 9) and $R^2_{adj} = 0.258$. Readers who want to avoid overfitting could use the result of the AIC-based MEM analysis and run a forward selection using the Blanchet et al. double stopping rule. The MEM variables from which the selection must be made has been saved above under the name `MEM.all`. Applying the double criterion to this example yields a model with MEM 1, 2, 3 and 6 ($R^2_{adj} = 0.222$).

RDA of the detrended mite data with the 6 MEM variables can be computed in a similar fashion as in the dbMEM analysis:

```
# RDA of the mite data constrained by the significant MEM,
# using vegan
(mite.MEM.rda <- rda(mite.h.det~., as.data.frame(MEM.select)))
(mite.MEM.R2a <- RsquareAdj(mite.MEM.rda)$adj.r.squared)
anova(mite.MEM.rda)
(axes.MEM.test <- anova(mite.MEM.rda, by = "axis"))
# Number of significant axes
(nb.ax <-
  length(which(axes.MEM.test[ ,ncol(axes.MEM.test)] <= 0.05)))

# Plot maps of the significant canonical axes
mite.MEM.axes <-
  scores(mite.MEM.rda,
         choices = 1:nb.ax,
         display = "lc",
         scaling = 1
)
par(mfrow = c(2, 2))
for(i in 1 : ncol(mite.MEM.axes))
{
sr.value(mite.xy, mite.MEM.axes[ ,i])
}
```

The R^2_{adj} of the MEM and dbMEM models are similar (approximately 0.26 and 0.24, respectively), but the dbMEM model requires 8 variables to reach this value and is thus less parsimonious than the MEM model (which needs only 6). The graphical result (not reproduced here) closely resembles that of the dbMEM analysis, showing that the structures revealed by the two analyses are the same.

For the sake of comparison with the dbMEM variables, one can plot the 6 MEM variables on a map of the sampling area:

```
# Maps of the significant MEM variables
par(mfrow = c(3, 3))
for(i in 1 : ncol(MEM.select))
{
 sr.value(mite.xy,
          MEM.select[ ,i],
          sub = sort(MEMid)[i],
          csub = 2)
}
```

The MEM differ more from the dbMEM than one would have expected. Indeed, the two groups of spatial variables are rather weakly correlated:

```
# Correlation of the retained MEM and dbMEM variables
cor(MEM.select, dbmem.red)
```

These MEM results show that the whole process of selecting and fine-tuning a spatial model, cumbersome as it may seem, can end up with an efficient and parsimonious set of spatial variables.

7.4.3.3 Other Types of Connectivity Matrices

In some special cases where one has a specific model of spatial connections in mind, it is useless to go through the automatic procedure shown above, which finds the best model among multiple possibilities. The present section shows how to construct connectivity matrices of several types by hand.

Depending on the context (hypotheses, data), researchers may need connecting schemes that are more or less dense or even locally customized. In such cases, the use of an automatic procedure including the computation of a minimum spanning tree as in the dbMEM procedure presented in Sect. 7.4.2.1, or of a Delaunay connectivity matrix, may not be appropriate. Ecological reasons include topographical structure of the sampling area (including possible barriers), dispersion ability of the organisms, permeability of some types of substrates, and so on.

In addition to the Delaunay triangulation used in the example above, the package **spdep** offers many possibilities for the definition of connectivity matrices. The ones constructed below and shown in Fig. 7.10 are described in Legendre and Legendre (2012) Sect. 13.3. They are presented in decreasing order of connectivity and are nested, i.e., the edges (connections) of a minimum spanning tree are all included in the relative neighbourhood graph, and so on.

```
# Other connectivity matrices

# Examples of connectivity matrices in decreasing order of
# connectivity. All these neighbourhood matrices are stored in
# objects of class nb Delaunay triangulation (as in the
previous
# example)
mite.del <- tri2nb(mite.xy)
# Gabriel graph
mite.gab <- graph2nb(gabrielneigh(as.matrix(mite.xy)), sym = TRUE)
# Relative neighbourhood
mite.rel <- graph2nb(relativeneigh(as.matrix(mite.xy)), sym = TRUE)
# Minimum spanning tree
mite.mst <- mst.nb(dist(mite.xy))
```

All these neighbourhood matrices are stored in objects of class nb.

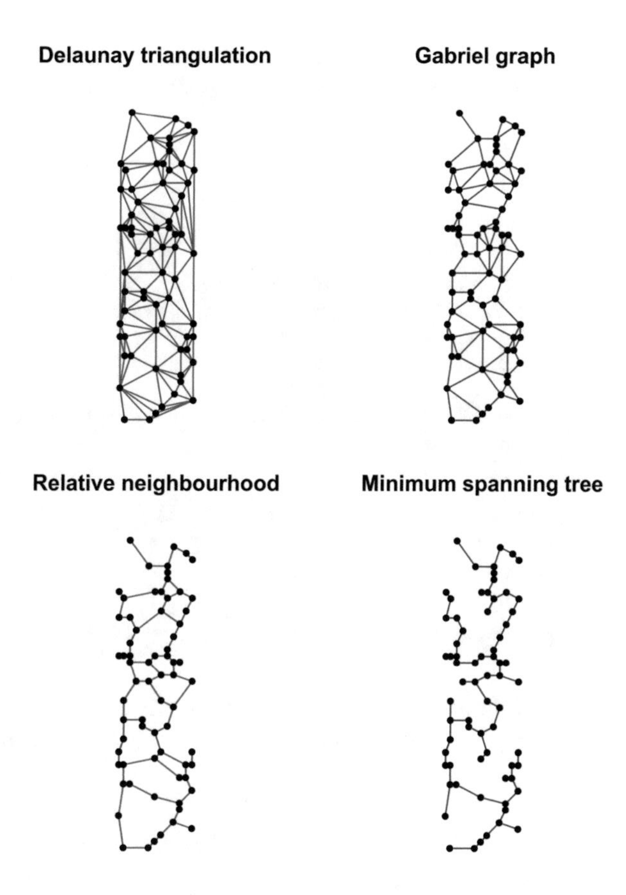

Fig. 7.10 Four types of connectivity matrices applied to the oribatid mite sampling plot. They are presented in decreasing order of connectivity. The links in each model are a subset of the links in the previous one

```
# Plots of the connectivity matrices
par(mfrow = c(2,2))
plot(mite.del, mite.xy, col = "red", pch = 20, cex = 1)
title(main = "Delaunay triangulation ")
plot(mite.gab, mite.xy, col = "purple", pch = 20, cex = 1)
title(main = "Gabriel graph")
plot(mite.rel, mite.xy, col = "dark green", pch = 20, cex = 1)
title(main = "Relative neighbourhood")
plot(mite.mst, mite.xy, col = "brown", pch = 20, cex = 1)
title(main = "Minimum spanning tree")
```

Some of these matrices may contain unwanted links (for instance along the borders of the areas). These can be edited out either interactively or by command lines:

```
# Link editing
# 1. Interactive:
dev.new(title = "Delaunay triangulation")
plot(mite.del, mite.xy, col = "red", pch = 20, cex = 2)
title(main = "Delaunay triangulation")
mite.del2 <- edit.nb(mite.del, mite.xy)
```

To delete a link, click on its two nodes. Follow on-screen instructions. Wait until you have finished editing before entering the next command line.

```
# 2. Alternatively, links can also be removed by command lines,
# after having converted the nb object into an editable matrix:
mite.del.mat <- nb2mat(mite.del, style = "B")
# Remove connection between objects 23 and 35:
mite.del.mat[23,35] <- 0
mite.del.mat[35,23] <- 0
# Back-conversion into nb object:
mite.del3 <- neig2nb(neig(mat01 = mite.del.mat))
plot(mite.del3, mite.xy)

# Example: list of neighbours of site 23 for the Delaunay
#          triangulation:
mite.del[[23]]       # Before editing
mite.del2[[23]]      # After interactive editing - depends on what
                     # you have edited above.
mite.del3[[23]]      # After command line editing
```

The following code shows how to construct connectivity matrices based on a distance: pairs of sites within a given radius are connected, the others are not.

```
# Connectivity matrix based on a distance (radius around points)
# Using the same truncation distance dmin as in the dbMEM
# example (1.011187).
dmin = 1.011187
mite.thresh4 <- dnearneigh(as.matrix(mite.xy), 0, dmin * 4)
# Display some values
nb2mat(mite.thresh4)[1:10, 1:10]

# Using a shorter distance (1 * dmin, 2 * dmin)
mite.thresh1 <- dnearneigh(as.matrix(mite.xy), 0, dmin * 1)
mite.thresh2 <- dnearneigh(as.matrix(mite.xy), 0, dmin * 2)
# Using a longer distance
mite.thresh8 <- dnearneigh(as.matrix(mite.xy), 0, dmin * 8)

# Plot of some connectivity matrices
par(mfrow = c(1,2))
plot(mite.thresh1, mite.xy, col = "red", pch = 20, cex = 0.8)
title(main = "1 * dmin")
plot(mite.thresh4, mite.xy, col = "red", pch = 20, cex = 0.8)
title(main = "4 * dmin")
```

*The 1*dmin version shows one disconnected point (7). To avoid such problems, use a slightly larger dmin value. In this case 1.0011188 is enough. The 4*dmin version is very crowded. Many links are possible within a little more than 4 meters around each point.*

These connectivity matrices belong to class "nb". To use them further we need to convert them into another class called "listw". The function doing this conversion is called **nb2listw()**.

In the simplest case, one of the binary matrices above can be directly converted as follows (including a matrix-class representation of the connectivity matrix for convenience, using function **listw2mat()**):

```
# Conversion of a "nb" object into a "listw" object
# Example: mite.thresh4 created above. "B" is for "binary"
mite.thresh4.lw <- nb2listw(mite.thresh4, style = "B")
print(listw2mat(mite.thresh4.lw)[1:10, 1:10], digits = 1)
```

This binary (unweighted) matrix could be used directly to create MEM variables using the function **scores.listw()**; see below.

Now, if you want to apply weights onto a binary matrix (matrix **A**) on the basis of Euclidean distances, you need two additional steps: (1) replace all values "1" in the connectivity matrix by the corresponding Euclidean distances [function **nbdists()**], and (2) define weights as a function of inverse distances in this example, so that weights reflect the facility of exchange between sites (weights may be different in other examples):

```
# Creation of a spatial weighting matrix W = Hadamard product of
# B and A.
# Replace "1" by Euclidean distances in the connectivity matrix
mite.thresh4.d1 <- nbdists(mite.thresh4, as.matrix(mite.xy))
# Weights as function of inverse distance
mite.inv.dist <-
  lapply(mite.thresh4.d1,
         function(x) 1-x/max(dist(mite.xy))
        )
# Creation of spatial weighting matrix W. Argument "B" stands for
# "binary" but concerns the links themselves, not their weights
mite.invdist.lw <-
  nb2listw(mite.thresh4,
           glist = mite.inv.dist,
           style = "B")
print(listw2mat(mite.invdist.lw)[1:10, 1:10], digits = 2)
```

All nonzero links have been replaced by weights proportional to the inverses of the Euclidean distances between the points.

Now it is time to compute the MEM spatial variables. This can be done by function **scores.listw()** of the package **adespatial**. We will do it on the inverse distance matrix created above. The MEM will then be tested for spatial correlation (Moran's *I*).

```
# Computation of MEM variables (from an object of class listw)
mite.invdist.MEM <- scores.listw(mite.invdist.lw)
summary(mite.invdist.MEM)
attributes(mite.invdist.MEM)$values
barplot(attributes(mite.invdist.MEM)$values)

# Store all MEM vectors in new object
mite.invdist.MEM.vec <- as.matrix(mite.invdist.MEM)

# Test of Moran's I of each eigenvector
mite.MEM.Moran <-
  moran.randtest(mite.invdist.MEM.vec, mite.invdist.lw, 999)

# MEM with significant spatial correlation
which(mite.MEM.Moran$pvalue <=  0.05)
length(which(mite.MEM.Moran$pvalue <=  0.05))
```

MEM 1, 2, 3, 4, 5 and 6 have significant spatial correlation.

```
# MEM with positive spatial correlation
MEM.Moran.pos <-
  which(mite.MEM.Moran$obs > -1/(nrow(mite.invdist.MEM.vec)-1))
mite.invdist.MEM.pos <- mite.invdist.MEM.vec[ ,MEM.Moran.pos]
# MEM with positive *and significant* spatial correlation
mite.invdist.MEM.pos.sig <-
  mite.invdist.MEM.pos[ , which(mite.MEM.Moran$pvalue <= 0.05)]
```

To show that the MEM variables are directly related to Moran's *I*, let us draw a scatterplot of the MEM eigenvalues and their corresponding Moran's *I*:

```
# Plot of MEM eigenvalues vs Moran's I
plot(attributes(mite.invdist.MEM)$values,
     mite.MEM.Moran$obs,
     ylab = "Moran's I",
     xlab = "Eigenvalues"
)
text(0, 0.55,
     paste("Correlation = ",
           cor(mite.MEM.Moran$obs,
               attributes(mite.invdist.MEM)$values
               )
           )
)
```

As in the case of the automatic model selection presented before, these MEM variables can now be used as explanatory variables in RDA or multiple regression, in the same way as dbMEM variables were.

These are only a few examples. We suggest users to explore the manual of the package **adespatial**, which presents in great detail the use of many other options to construct, present, and use various types of connectivity matrices.

7.4.3.4 Controlling for Spatial Correlation Using MEM

Peres-Neto and Legendre (2010) explored the potential use of polynomials and MEM eigenfunctions to control for spatial correlation in statistical tests. Their main conclusion is that MEM, but not polynomials, can adequately achieve this goal. They propose the following procedure: (1) Test for the presence of a spatial structure using all positive MEM variables. (2) If the global test is significant, proceed to forward-select MEM variables, but (a novelty) do this individually for each species, and retain the union of the MEM selected, i.e., retain all MEM that have been selected at east once. (3) Proceed to test the species-environment relationships, controlling for spatial correlation by placing the retained MEM variables in a matrix of covariables. The authors demonstrate that this procedure yields correct type I error for tests of significance in linear models, in the presence of spatial correlation.

7.4.3.5 MEM on Sampling Designs with Nested Spatial Scales

The hierarchical structure of many natural entities (e.g. metapopulations or metacommunities; landscapes at various scales) sometimes calls for nested sampling designs. An example is found in Declerck et al. (2011), where the authors studied cladoceran metacommunities in wetland pools found in several valleys of the High Andes. The authors analysed the metacommunity spatial structure among- and within-valleys by means of a two-level spatial model. The among-valley component was modelled by a set of dummy variables. For the within-valley component, where several pools had been sampled in each valley, a set of MEM variables was computed for each valley. All dummy and MEM variables were assembled into a single staggered matrix. The MEM variables were arranged in blocks corresponding to each valley. Within each block, all pools belonging to other valleys received the value 0, in a way similar to the one presented in Appendix C of Legendre *et al.* (2010) in the context of space-time analysis. Declerck et al. (2011) provided a function called `create.MEM.model()` to construct the staggered spatial matrix from a set of Cartesian coordinates and information about the number of groups and number of sites per group. An updated version of that function, called `create.dbMEM.model()`, is available in package **adespatial**.

7.4.4 MEM with Positive or Negative Spatial Correlation: Which Ones should Be Used?

In the course of the examples above, dbMEM and MEM eigenfunctions have been produced, some with positive and some with negative spatial correlation. The question therefore arises: should one use all the (significant) eigenfunctions as explanatory variables in the following regression or canonical analyses, or only those that model positive spatial correlation?

There is no single answer to this question. Ecologically speaking, one is generally more interested in features that are positively correlated at various ranges, simply because they are the signature of contagious processes that are frequent in nature. On the other hand, our experience shows that with real data the significant and negatively correlated variables are either related to very local, almost "accidental" data structures, or they belong to the pure spatial fraction of variation in partitioning, i.e., they are caused by biotic interactions. If these are of interest, then all eigenfunctions should be considered in the analyses.

The dbMEM procedure generates a maximum of $n - 1$ eigenfunctions (n = number of sites), with roughly the first $n/2$ modelling positive spatial correlation on regular sampling designs, so a forward selection procedure including all variables cannot be conducted with the Blanchet et al. (2008a) double stopping criterion, which involves the computation of the R^2_{adj} of the global analysis. Indeed, the $n - 1$ spatial variables saturate the regression model if they are all considered together. This is why Blanchet *et al.* (2008a) proposed to run separate selections on the MEM with positive and negative eigenvalues and then apply the Sidák (1967) correction to the probability values: $P_S = 1 - (1 - P)^k$ where P is the P-value to be corrected and k is the number of tests (here $k = 2$). Of course, this is useful only in the specific cases where negative spatial correlation is of interest.

7.4.5 Asymmetric Eigenvector Maps (AEM): When Directionality Matters

7.4.5.1 Introduction

The dbMEM and MEM analyses presented above are designed for situations where the physical processes generating the response structures (e.g. in communities) do not present any directionality. In other words, the influence of any given point on its surroundings does not depend on the direction.

There are other situations, however, where directionality matters. The most obvious ones are the cases of streams or rivers. Consider community effects driven by current: the physical process is geographically asymmetrical, the influence of a site onto another following an upstream-downstream direction. Colonization of the stream network by fish from the river mouth represents a different process, which follows the opposite direction. dbMEM or MEM variables are computed on distance or connectivity matrices where no directionality is specified. Therefore, information about directionality is lost and the modelling, although adequate to reveal major spatial structures, does not exploit all the potential of directional data. Trends must not be extracted from the data prior to AEM analysis because directional processes are expected to produce trends in the response data; so, a trend is a part of the response data that one wants to model in AEM analysis.

This is the reason why Blanchet et al. (2008b) developed the Asymmetric Eigenvector Maps modelling method. AEM is an eigenfunction-based technique that uses information about the direction of the physical process, plus the same information as MEM (spatial coordinates of sites, connection diagram, optional weights) if needed. It works best on tree-like structures like river networks, on two-dimensional sampling designs like series of cross-river traps, or on sampling sites located in a large river or marine current. Depending on the process under study, the origin(s), or root(s), in a river network may be located upstream (e.g. flow of dissolved chemical substances, plankton dispersal) or downstream (fish invasion routes).

For spatial transects or time series, AEM and MEM regression and canonical models are very similar, and in most cases they explain the response data with similar (although not strictly equal) R^2. The AEM eigenfunctions are cosine-like, just like MEM eigenfunctions, although the AEM have longer wavelengths than MEM along transects. If the n observations are regularly spaced along the transect and the sampling interval is s, the wavelength λ_i of the AEM with rank i is $\lambda_i = 2\ ns/i$. AEM analysis should be preferred when modelling gradients and other spatial structures generated by directional physical processes.

AEM analysis was devised for cases where *physical* forces drive the communities in such a way that the causal relationships are directional. This is not the same as a simple ecological gradient, where an ecological factor is spatially structured but the communities can still interact in any direction. In the latter case, dbMEM and MEM modelling are appropriate.

7.4.5.2 Principle and Application of the Method

The basic piece of information needed is a table where each site is described by the connections (hereafter called “edges”, following graph-theory vocabulary) it has with other sites located in the direction of the root(s) or origin(s) of the directional structure. The result is a rectangular sites-by-edges Table **E** where the sequence of edges connecting each site to the "root" of the network receive code "1" and the others get code "0".

Legendre and Legendre (2012, Sect. 14.3) give an example for fish dispersal from the river mouth in a group of lakes interconnected by a river arborescence. In other cases, for instance a two-dimensional grid consisting of rows of sampling devices placed across a large river or a marine current at regular or irregular intervals, each sampling point may influence (and hence may be connected to) the one directly downstream of it, plus the two adjacent to the latter. If the process is assumed to originate upstream, an imaginary point "0" is created upstream of the sampling area, representing the root of the process, with connections to each of the points in the first row of sites. All links present in the network are numbered. In Table **E**, the rows (i) are the sites and the columns (j) are the edges. The construction rule for AEM is that $E(i,j) = 1$ for links j connecting site i to the root (or site 0) of the graph; otherwise, $E(i,j) = 0$.

The edges (columns) of Table **E** may be weighted if deemed necessary, e.g. if the transmission of the directional effects are supposed to be more difficult through some paths than others.

The next step consists in transforming Table **E** into eigenfunctions. This can be done in different ways, but the simplest is to compute a PCA of Table **E** and use the matrix of principal components in scaling type 1 as explanatory variables. The AEM method produces $n - 1$ eigenvectors with positive eigenvalues and none with negative eigenvalues. The corresponding eigenfunctions, however, are divided in two groups depicting positive or negative spatial correlation, so that the selection of significant variables must be run separately for these two groups, in the same way as for MEM variables.

A more detailed explanation about AEM construction is provided by Blanchet et al. (2008b). The authors address the various issues related to edge definition and weighting, which can greatly influence the results of AEM analysis.

As a first example, let us construct a fictitious set of AEM variables based on the river arborescence shown by Legendre and Legendre (2012 Sect. 14.3 p. 889). This example shows how to construct AEM variables in the simplest case, when one can easily produce a matrix of edges by hand. The 8 nodes are 6 lakes connected by a river arborescence, plus two river junction points (Fig. 7.11). Consider that samples have been taken at these 8 nodes. The construction is done by function `aem()` of the package `adespatial`.

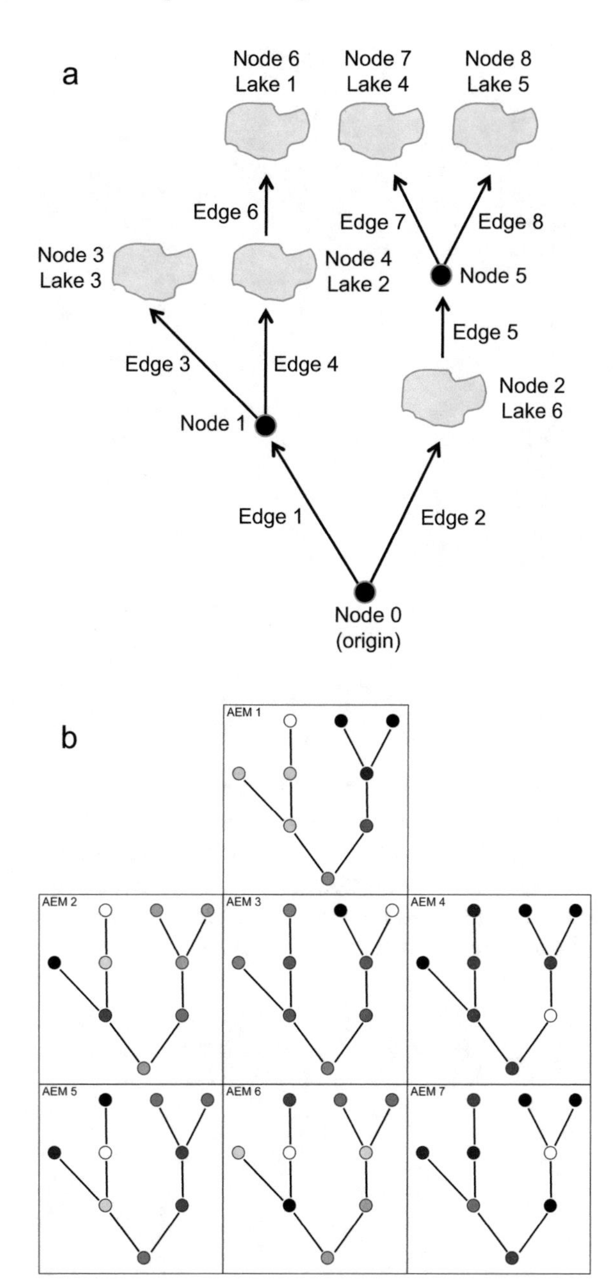

Fig. 7.11 Construction of AEM variables. (**a**) River network; (**b**) the seven corresponding AEM variables produced by function **aem()** (Redrawn after Legendre and Legendre 2012)

```
# Coding of a river arborescence.
# See Legendre and Legendre (2012, p. 889).
node1 <- c(1, 0, 0, 0, 0, 0, 0, 0)
n2lk6 <- c(0, 1, 0, 0, 0, 0, 0, 0)
n3lk3 <- c(1, 0, 1, 0, 0, 0, 0, 0)
n4lk2 <- c(1, 0, 0, 1, 0, 0, 0, 0)
node5 <- c(0, 1, 0, 0, 1, 0, 0, 0)
n6lk1 <- c(1, 0, 0, 1, 0, 1, 0, 0)
ln7k4 <- c(0, 1, 0, 0, 1, 0, 1, 0)
n8lk5 <- c(0, 1, 0, 0, 1, 0, 0, 1)
arbor <- rbind(node1, n2lk6, n3lk3, n4lk2, node5,
               n6lk1, ln7k4, n8lk5)

# AEM construction
(arbor.aem <- aem(binary.mat = arbor))
arbor.aem.vec <- arbor.aem$vectors

# AEM eigenfunctions can also be obtained directly by singular
# value decomposition (function svd()), which is what the function
# aem() does:
arbor.c <- scale(arbor, center = TRUE, scale = FALSE)
arbor.svd <- svd(arbor.c)
# Singular values of the construction above
arbor.svd$d[1:7]
# AEM eigenfunctions of the construction above
arbor.svd$u[ ,1:7]
```

Let us now construct AEM variables in a case where the number of data points and edges is too large to allow the use of the simple procedure presented above. The sampling design will consist of 10 cross-river transects with 4 traps per transect and the edges will be weighted proportional to inverse squared distance (Fig. 7.12). The procedure involves function **cell2nb()** of the package **spdep** to construct a list of neighbours from a grid of predefined dimensions.

```
# Coding of sampling design: 10 cross-river transects, 4 traps
# per transect. Edges weighted proportional to inverse squared
# distance.
# X-Y coordinates
xy <- cbind(1:40, expand.grid(1:4, 1:10))
# Object of class nb (spdep) containing links of chess type "queen"
nb <- cell2nb(4, 10, "queen")
# Site-by-edges matrix (produces a fictitious object "0")
edge.mat <- aem.build.binary(nb, xy)
# Matrix of Euclidean distances
D1.mat <- as.matrix(dist(xy))
# Extract the edges, remove the ones directly linked to site 0
edges.b <- edge.mat$edges[-1:-4,]
# Construct a vector giving the length of each edge
length.edge <- vector(length = nrow(edges.b))
for(i in 1:nrow(edges.b))
{
      length.edge[i] <- D1.mat[edges.b[i,1], edges.b[i,2]]
}
```

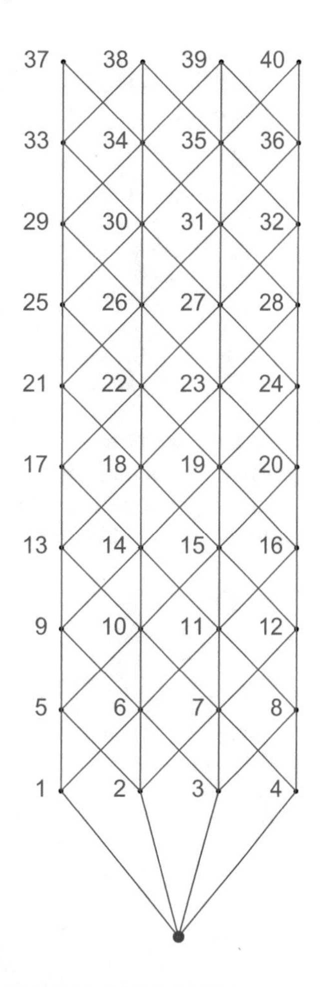

Fig. 7.12 Fictitious directional sampling design for AEM analysis: 10 rows of 4 capture devices along a stream. A site "0" has been added upstream (bottom of the figure) to set the direction of the flow.

```
# Weighting of edges based on inverse squared distance
weight.vec <- 1-(length.edge/max(length.edge))^2
# Construction of AEM eigenfunctions from edge.mat, of class
# build.binary
example.AEM <-
  aem(aem.build.binary = edge.mat,
      weight = weight.vec,
      rm.link0 = TRUE)
example.AEM$values
ex.AEM.vec <- example.AEM$vectors
```

Let us now construct a set of 5 fictitious species observed at these 40 sites:

```
# Construction of 5 fictitious species
# Two randomly distributed species
sp12 <- matrix(trunc(rnorm(80,5,2),0),40)
# One species restricted to the upper half of the stream
sp3 <- c(trunc(rnorm(20,8,2.5),0), rep(0,20))
# One species restricted to the left-hand half of the transect
sp4 <- t(matrix(c(trunc(rnorm(20,8,3),0), rep(0,20)),10))
sp4 <- as.vector(sp4)
# One species restricted to the 4 upper left-hand sites
sp5 <- c(4,7,0,0,3,8, rep(0,34))
# Build the species matrix
sp <- cbind(sp12, sp3, sp4, sp5)
colnames(sp) <- c("sp1", "sp2", "sp3", "sp4", "sp5")
```

We are ready to proceed with the AEM analysis, using the first half (20, with positive spatial correlation) of the AEM variables generated earlier. Note that 4 out of the 5 species generated in this example have a random component, so the actual result of the AEM analysis that follows will vary from run to run.

```
# Global AEM analysis with 20 first AEM variables (for
# computation of R2a)
AEM.20 <- rda(sp ~ ., as.data.frame(ex.AEM.vec[ ,1:20]))
(R2a.AEM <- RsquareAdj(AEM.20)$adj.r.squared)
# Forward selection of the AEM variables
AEM.fwd <- forward.sel(sp, ex.AEM.vec, adjR2thresh = R2a.AEM)
(AEM.sign <- sort(AEM.fwd[ ,2]))
# Write significant AEM in a new object
AEM.sign.vec <- ex.AEM.vec[ ,c(AEM.sign)]
# RDA with signif. AEM
(sp.AEMsign.rda <- rda(sp ~ ., data = as.data.frame(AEM.sign.vec)))
anova(sp.AEMsign.rda)
(AEM.rda.axes.test <- anova(sp.AEMsign.rda, by = "axis"))
# Number of significant axes
(nb.ax.AEM <- length(which(AEM.rda.axes.test[ ,4] <=  0.05)))
# Adjusted R-square
RsquareAdj(sp.AEMsign.rda)
# Plot of the significant canonical axes
AEM.rda.axes <-
  scores(sp.AEMsign.rda,
         choices = c(1,2),
         display = "lc",
         scaling = 1)
par(mfrow = c(1,nb.ax.AEM))
for(i in 1:nb.ax.AEM) sr.value(xy[ ,c(2,3)], AEM.rda.axes[ ,i])
```

In most of the runs, depending on the random generation of species abundances, this small example shows that AEM analysis reveals the patterns formed by the species present only in the upper half of the stream, as well as the left-right contrast created by the species present only in the left-hand part of the stream. The pattern of the more restricted species 5 is less obvious.

Applications of AEM modelling are presented in Blanchet et al. (2008a, b, 2011) as well as in Gray and Arnott (2011), Sharma et al. (2011), and other papers.

7.5 Another Way to Look at Spatial Structures: Multiscale Ordination (MSO)

7.5.1 Principle

Wagner (2003, 2004) took an entirely different path towards integration of spatial information and MEM eigenfunctions into canonical ordination. Under the well-known fact that autocorrelated residuals can alter the results of statistical tests, she used geostatistical methods to devise diagnostic tools allowing (1) the partitioning of ordination results into distance classes, (2) the distinction between induced spatial dependence and spatial autocorrelation, and (3) the use of variograms to check important assumptions such as independence of residuals and stationarity. The principle of MSO is the following[6]:

- Analyse the species by RDA. The explanatory variables can be of any kind (environmental, spatial, and so on). This provides the matrix of fitted values and its eigenvectors, as well as the matrix of residuals and its eigenvectors.
- By way of a variogram matrix computed for the fitted values, obtain the spatial variance profiles of the canonical ordination axes (see below).
- By way of a variogram matrix computed for the residuals, obtain the spatial variance profiles of the residual ordination axes.
- Plot the variograms of the explained and residual variances. Permutation tests may be used to identify significant spatial correlation in the distance classes.

A variogram matrix is a three-dimensional array containing a multivariate variance-covariance matrix for each distance class (Wagner 2003 Fig. 2.2; Legendre and Legendre 2012 Fig. 13.11). The diagonal of each matrix quantifies the contribution of the corresponding distance class to the variance of the data. MSO computes a variogram matrix on the fitted values of a constrained ordination, thereby allowing its spatial decomposition. Multiplying this variogram matrix with the matrix of constrained eigenvectors provides the spatial decomposition of each eigenvalue (variance profiles). The same holds for the residuals.

[6]Wagner (2004) describes the method for CCA, but the principle is the same for RDA.

7.5.2 Application to the Mite Data – Exploratory Approach

Let us use the oribatid mite data as an example. Wagner (2004) also used these data, but in a CCA context, so that the results of the RDA below will differ from hers. MSO can be computed using function **mso()** of the package **vegan**. This function uses a result object produced by functions **rda()** or **cca()**, plus the table of geographical coordinates and a value for the interval size (argument 'grain') of the distance classes of the variograms. The first example applies MSO in the exploratory way proposed by Wagner. An MSO plot of direct ordination can show whether the spatial structure of the response data can be explained by the explanatory (environmental) variables alone. In such a case, no detrending is necessary (H. Wagner, pers. comm.), but the confidence interval of the variogram is only indicative since a variogram should be computed on stationary data.

Hereunder MSO is run using the RDA result of the (Hellinger-transformed) oribatid mite data explained by the environmental variables. The "grain" of the variogram (size of a distance classes) is chosen to be the truncation threshold used in the dbMEM analysis, 1.011187.

```
## MSO of the undetrended mite data vs environment RDA
mite.undet.env.rda <- rda(mite.h~., mite.env2)
(mite.env.rda.mso <-
  mso(mite.undet.env.rda,
      mite.xy,
      grain = dmin,
      perm = 999))
msoplot(mite.env.rda.mso, alpha = 0.05/7)
```

The resulting plot (Fig. 7.13) is rich in information. In the upper part of the diagram, the dashed line with the plus signs represents the sum of the explained and residual empirical variograms. The continuous lines represent the confidence envelopes of the variogram of the data matrix. The monotonic increase of the dashed line is the signature of the strong linear gradient present in the data. Note, however, that the variogram of the residuals (squares, bottom of the graph) shows no distance class with significant spatial correlation (after a global Bonferroni correction for 7 simultaneous tests, where the rejection threshold is divided by the number of classes), and that variogram is essentially flat. This means that the broad scale linear gradient is well explained by the environmental variables.

However, an intriguing feature appears. When the species-environment correlations do not vary with scale, the dashed line remains within the boundaries of the confidence envelopes (full lines). This is not the case here (see classes 1, 2 and 5, which correspond to distances 0, 1 and 4 along the abscissa), suggesting that it is not appropriate to run a non-spatial, global species-environment analysis with the implicit assumption that the relationships are scale-invariant. On the contrary, we can expect the regression parameters to vary with scale, so that a global estimation is meaningless unless one controls for the regional-scale spatial structure causing the problem.

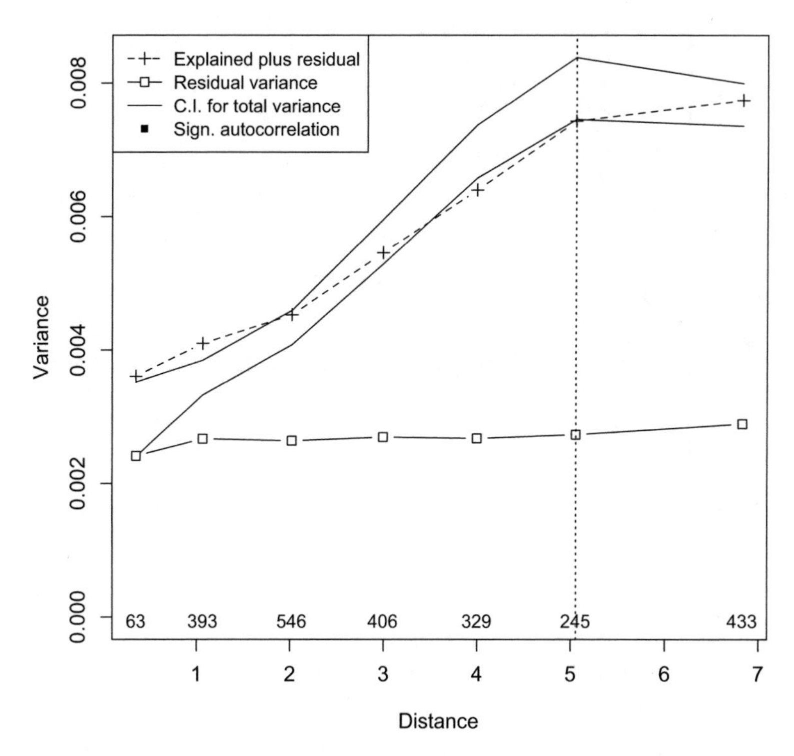

Fig. 7.13 Plot of the MSO of an RDA of the Hellinger-transformed oribatid mite data explained by the environmental variables. Explanations: see text.

As an attempt in this direction, let us run a MSO on a partial RDA of the mite species explained by the environment, controlling for the spatial structure, here represented by the 6 MEM variables of our best model (object `MEM.select` obtained in Sect. 7.4.3.2) (Fig. 7.14).

```
# MSO of the undetrended mite data vs environment RDA, controlling
# for MEM
mite.undet.env.MEM <-
  rda(mite.h,
      mite.env2,
      as.data.frame(MEM.select))
(mite.env.MEM.mso <-
  mso(mite.undet.env.MEM,
      mite.xy,
      grain = dmin,
      perm = 999))
msoplot(mite.env.MEM.mso,
        alpha = 0.05/7,
        ylim = c(0, 0.0045)  # Expanded height to clear legend
)
```

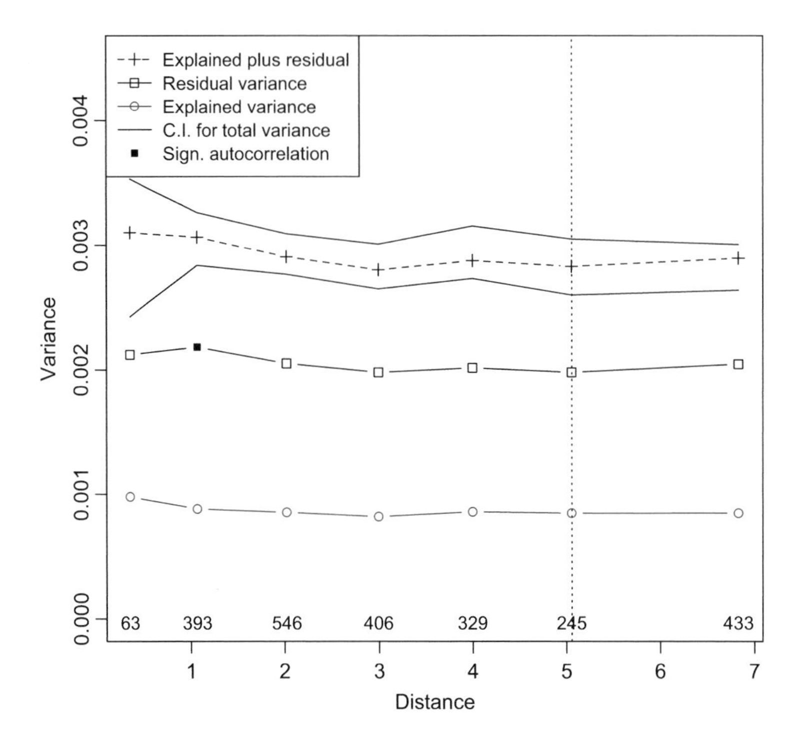

Fig. 7.14 Plot of the MSO of an RDA of the Hellinger-transformed oribatid mite data explained by the environmental variables, controlling for spatial structure (6 MEM variables).

This analysis shows that the problem of scale-dependence in the model has been quite properly addressed. There is one single significant spatial correlation left in the residuals, and the variogram of the explained plus residual species-environment relationship ("+" symbols, after taking the MEM spatial structure into account) stays within the confidence interval across all scales. Furthermore, the MEM variables have also controlled for the major gradient in the data, resulting in a globally flat empirical variogram. The console message stating that the "Error variance of regression model [is] underestimated by −1.3%" refers to the difference between the total residual variance and the sill of the residual variance. When the value is negative or NaN (not a number), no significant autocorrelation causes an underestimation of the global error value of the regressions. A positive value (e.g., 10%) occurring if the residuals were significantly autocorrelated, would act as a warning that the condition of independent residuals is violated, thereby invalidating the statistical tests (see Sect. 7.2.2).

7.5.3 Application to the Detrended Mite and Environmental Data

Let us apply an MSO analysis to the detrended mite and environmental data, as an effort to meet the conditions of application of the calculation of the variogram confidence intervals. Preliminary computations (not shown) revealed that the goal of completely eliminating spatial autocorrelation in the residuals could be reached by modelling the trend by means of the *Y* coordinate only. We shall therefore detrend the mite and environmental data on the *Y* coordinate before running the RDA.

```
# Detrend mite data on Y coordinate
mite.h.det2 <- resid(lm(as.matrix(mite.h) ~ mite.xy[ ,2]))
# Detrend environmental data on Y coordinate
env2.det <- resid(lm(as.matrix(mite.env2) ~ mite.xy[ ,2]))
# RDA and MSO
mitedet.envdet.rda <- rda(mite.h.det2 ~., env2.det)
(miteenvdet.rda.mso <-
  mso(mitedet.envdet.rda,
      mite.xy,
      grain = dmin,
      perm = 999))
msoplot(miteenvdet.rda.mso, alpha = 0.05/7, ylim = c(0, 0.006))
```

The result (Fig. 7.15) tells us the same story as our first MSO analysis, less the broad-scale gradient that has been removed prior to the analysis by detrending. The residual variance shows no spatial correlation, and the second, fourth and fifth classes of the variogram of explained plus residual data fall outside the confidence intervals. So the overall variogram shows no trend, but some regional spatial variance is present. Can the MEM control successfully for this spatial variance?

```
## MSO of the detrended mite data vs detrended environment RDA,
## controlling for MEM
mite.det.env.MEM <-
  rda(mite.h.det2,
      env2.det,
      as.data.frame(MEM.select))
(mite.env.MEM.mso <-
  mso(mite.det.env.MEM,
      mite.xy,
      grain = dmin,
      perm = 999))
msoplot(mite.env.MEM.mso, alpha = 0.05/7, ylim = c(0, 0.005))
```

The answer is "yes" (Fig. 7.16). As in the undetrended example, one can see no spatial variance in the residuals or in the data. Compare with Fig. 7.14: the variograms are very similar. The MEM variables have successfully controlled for the spatial variance unexplained by the environmental data.

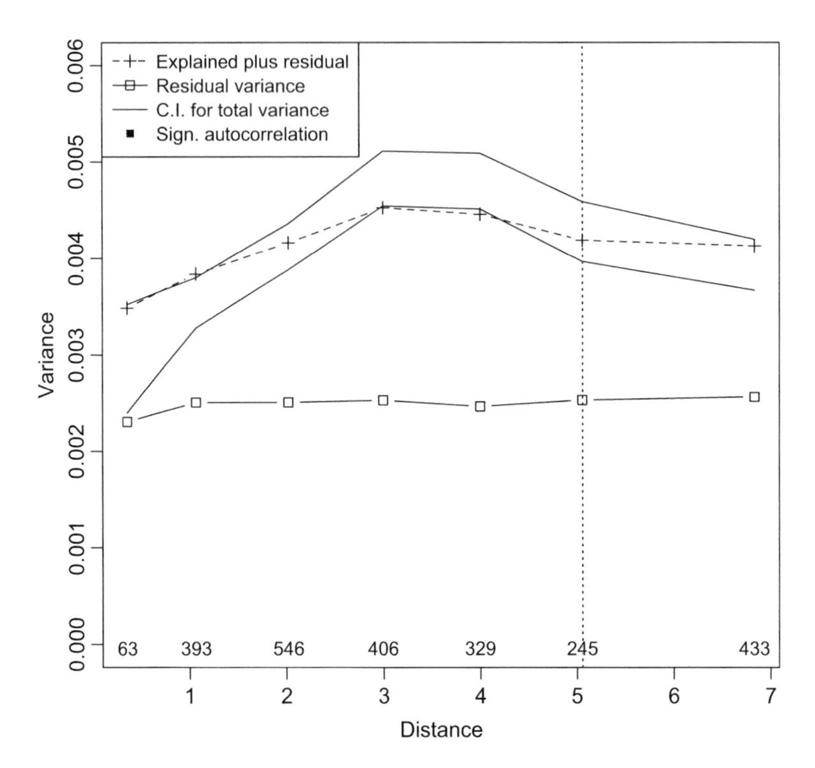

Fig. 7.15 Plot of the MSO of an RDA of the Hellinger-transformed and detrended oribatid mite data explained by the detrended environmental variables. Explanations: see text

This example shows the potential of combining multivariate geostatistical methods with canonical ordination and MEM covariables when the aim of the study is to test for and model species-environment relationships while discriminating between the two major sources of concern related to spatial structures: spatial dependence (Eq. 7.1) and spatial autocorrelation (Eq. 7.2). Some aspects of this approach remain to be explored, however. Wagner (2004) notes "*an important discrepancy between the results presented here and those by Borcard* et al. *(*1992*)* [authors' note: in this 1992 paper, a cubic polynomial of the spatial coordinates was used for spatial modelling]. *Borcard found that 12.2% of the total inertia was spatially structured but could not be explained by the environmental variables. In the spatial partitioning of CCA results by multi-scale ordination (MSO), however, spatial autocorrelation appeared to be limited to distances smaller than 0.75 m, and there was no evidence of any cyclic pattern that could account for such a large portion of inertia. The large portion of nonenvironmental spatial structure identified by Borcard* et al. *(*1992*) may partly be due to a confounding of the effects of space and environment (Meot* et al. 1998*)*". Arguing from an opposite point of view, we believe that the pure spatial structures revealed by canonical ordination (and especially in the MEM framework which would give an even larger pure spatial fraction)

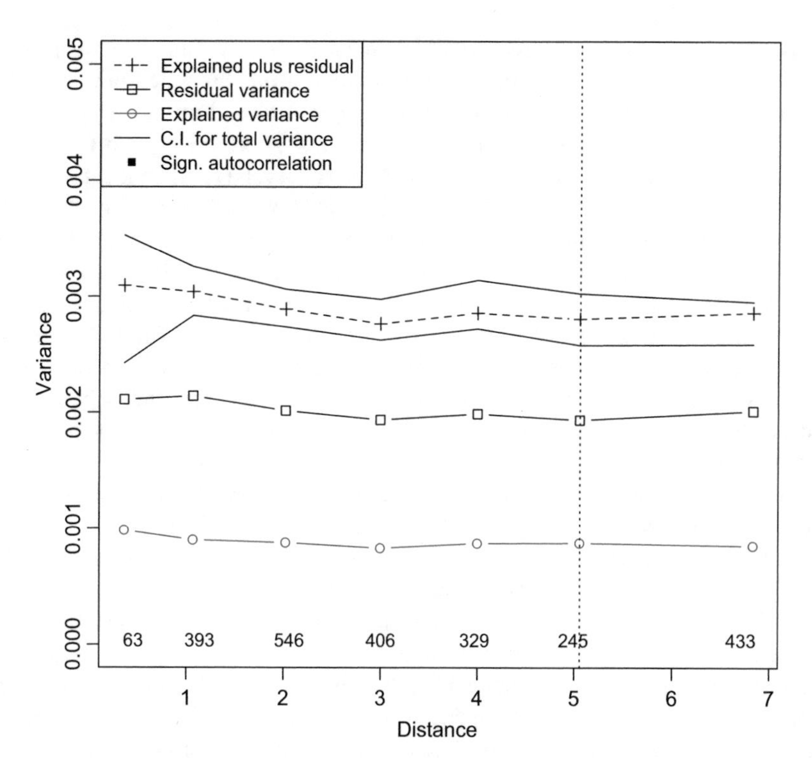

Fig. 7.16 Plot of the MSO of an RDA of the Hellinger-transformed and detrended oribatid mite data explained by the detrended environmental variables, controlling for spatial structure (6 MEM variables). Further explanations: see text.

are real and not due to confounding effects. In the latter case, they would have shown up in the common fraction of variation, not the pure spatial fraction. The question is rather: why did the MSO *not* reveal these structures? This may be due to the fact that no formal way of quantifying variance components in MSO has been devised as yet (H. Wagner, pers. comm.). However, this does by no means invalidate the whole approach which, combined with the powerful tools developed in this chapter, increases our control over the complex process of extracting meaningful spatial information from ecological data sets.

7.6 Space-Time Interaction Test in Multivariate ANOVA, Without Replicates

7.6.1 Introduction

The title of this section may seem outrageous: everybody knows that one needs replicates in an experimental design to be able to test an interaction term in ANOVA, since it is the within-cell replicates that provide the degrees of freedom for the test.

Consequently, the method presented here must involve a "trick" to spare some degrees of freedom from the main factors. How can this be done? In this section, we are examining the case where s sites have been surveyed t times.

In a two-way crossed and balanced ANOVA design involving two factors with s and t levels (hence there are $s \times t$ combinations of levels) and r replicates per cell, the number of degrees of freedom of the explainable part of the variance is equal to $(s \times t) - 1$. The tests of the main factors take up $(s - 1)$ and $(t - 1)$ d.f. respectively. When there are r replicates within each cell ($r > 1$), then the interaction term uses $(s - 1) \times (t - 1)$ d.f. and $st(r - 1)$ d.f. remain for the error term. In other words, it is the within-cell replicates that provide the d.f. of the residuals. Without replication, $r = 1$ and $st(r - 1) = 0$. Therefore, no test of the interaction is possible since the $(s - 1) \times (t - 1)$ d.f. remaining after the main factors have been taken into account are in fact attributable to a combination of interaction and error, but without the possibility to untangle them.

This being clear, how could one conceive a plan with two crossed factors, no replication but with a possibility of testing the interaction? The answer is: in limited cases, when at least one of the two factors can be coded with less than $(s - 1)$ or $(t - 1)$ d.f., and/or if the interaction can be coded with less than $(s - 1) \times (t - 1)$ d.f. The situation addressed here is the one of community surveys made at a given set of locations and repeated over time, for instance for long-term monitoring purposes such as those performed by governmental agencies in many countries around the world. This type of sampling produces an unreplicated repeated-measures design, which is a type of two-way factorial design. Sites and survey times cannot be replicated; at best, an adjacent site would be a pseudo-replicate. The same applies to survey times.

In such data, there is no replication at the level of individual sites: each site is sampled only once in a given survey. However, interaction is an important indicator in such situations: if present, it can mean either that the spatial structure of the community has changed over time, or, conversely, that the temporal evolution of the community is not the same at all sites. In fact, interaction is so important for such repeated surveys that it should be "*the first statistical indication ecologists should look for*" (Legendre et al. 2010). As in any ANOVA, the presence of an interaction means that one should run separate analyses of the time factor at each site and separate analyses of the spatial structure at each point in time.

Let us consider the ANOVA as a multiple regression using a design matrix of explanatory variables, i.e., a matrix of dummy variables coding the levels of the factors. We saw in Sect. 6.3.2.9 how to code a factor into a set of orthogonal Helmert contrasts. In the present case, to code for the spatial and temporal factors with s sites and t times, we need $(s - 1)$ and $(t - 1)$ contrasts, which correspond to the number of d.f. of the two factors. If the interaction could be tested, its design matrix would be created by multiplying each Helmert contrast in the space matrix by each contrast of the temporal matrix; the result would be a set of $(s - 1) \times (t - 1)$ new contrasts, orthogonal to the contrasts representing the main factors. Now, the proposition of the space-time interaction analysis is to replace either the spatial or the temporal contrasts, or the interaction terms, or all of these, by a subset of the dbMEM variables than can be computed to represent the spatial or temporal design. The subset is chosen to capture all variation with positive spatial or temporal correlation. Since the subset contains less than $(s-1)$ or $(t-1)$ variables (space or time are said to be *under-fitted*), or since the interaction subset contains less than $(s-1) \times (t-1)$ variables, one is saving some degrees of freedom that can be used to estimate the residual error, and thus to test the interaction term.

On this basis, Legendre et al. (2010) presented the following ANOVA models with their partitioning of variance and degrees of freedom:

- Model 1: standard, two-way crossed ANOVA design for data with replication. $(s-1)$, $(t-1)$ and $(s-1) \times (t-1)$ Helmert contrasts form the design matrices representing space, time and interaction. $st(r-1)$ degrees of freedom remain for the error term.
- Model 2: standard two-way crossed ANOVA design without replication. $(s-1)$ and $(t-1)$ Helmert contrasts form the design matrices representing space and time. The remaining $(s-1) \times (t-1)$ d.f. are used to estimate the error variance. Interaction may be present but cannot be tested for.
- Model 3a: two-way crossed ANOVA design without replication, but with space under-fitted: u dbMEM spatial variables replace the $(s-1)$ spatial Helmert contrasts; $u < (s-1)$; an interaction design matrix can be constructed, containing $u \times (t-1)$ terms. $(s-u-1) \times$ t d.f. remain to estimate the error variance. A test of the interaction is thus possible.
- Model 3b: two-way crossed ANOVA design without replication, but with time under-fitted: v dbMEM spatial variables replace the $(t-1)$ temporal Helmert contrasts; $v < (t-1)$; an interaction design matrix can be constructed, containing $v \times (s-1)$ terms. $(t-v-1) \times$ s d.f. remain to estimate the error variance. A test of the interaction is thus possible.
- Model 4: two-way crossed ANOVA design without replication, but with space and time under-fitted: u dbMEM spatial variables replace the $(s-1)$ spatial Helmert contrasts and v dbMEM variables replace the $(t-1)$ temporal contrasts; $u < (s-1)$ and $v < (t-1)$; an interaction design matrix can be constructed, containing $u \times v$ terms. $[s \times t-(u+v+u \times v)-1]$ d.f. remain to estimate the error variance. A test of the interaction is thus possible.
- Model 5: two-way crossed ANOVA design without replication, but with interaction under-fitted: the main factors are represented by their $(s-1)$ and $(t-1)$ Helmert contrasts, but the interaction variables are created using the u and v dbMEM variables used in Model 4. The d.f. of the interaction term are $u \times v$ and $[((s-1) \times (t-1)-u \times v]$ d.f. remain for he residual error.
- Model 6 is a special case where one wants to test the significance of a spatial effect, or of a temporal effect, in the presence of space-time interaction. One approach, called Model 6a in the Legendre et al. (2010) paper (but not in the **R** function; see below), tests for the presence of spatial structure in turn for each of

the *t* times, correcting for multiple testing. The spatial variables used are the dbMEM constructed for Models 3a, 4 and 5. This variant is not implemented in the **stimodels()** function. Another approach, called Model 6b in the paper (and used in the **R** function), involves a simultaneous test for spatial structure in all t times, using a staggered matrix of spatial dbMEM variables. The same approaches can be applied to test for temporal structure in the presence of space-time interaction, by constructing a staggered matrix of temporal dbMEM for each site. Readers are referred to the Appendix C of the Legendre et al. (2010) paper for more details. Beware: in the **stimodels()** function presented below, the models called 6a and 6b are both of the staggered type, 6a for testing space and 6b for testing time.

Models 3, 4 and 5 allow the testing of the interaction, but the drawback is that there is some lack-of-fit in the terms coded with dbMEM variables and in the interaction (which contains dbMEM in the three models). Since this lack-of-fit is part of the residual error, some power is lost in the tests. Also, the permutation tests of the interaction and main factors can handle the lack-of-fit of the various terms differently depending on the model. In all, Legendre et al. recommend Model 5 to test the interaction, mainly because its permutation test has a correct type I error and this Model provides the highest power to detect an interaction. The authors' recommendations are as follows: "*if one has replicates, Model 1 is the correct choice. If not, then space-time interaction can safely (in terms of Type I error rate) be tested using Model 5. If interaction turns out to be nonsignificant, one can test for the main effects using Model 2.* [...] *In contrast, if the interaction is significant, one should perform separate analyses for the different sampling campaigns and/or separate time series analyses for the different spatial units*". Another argument that strongly speaks in favour of Model 5 for testing interaction is that this model leaves the smallest number of d.f. in the residuals, which increases the power of the test. These analyses of the main factors in the presence of an interaction can then be carried out under Models 6a or 6b of function **stimodels()** or directly with function **quicksti()** (see below).

7.6.2 Testing the Space-Time Interaction with the sti Functions

As a Supplement to their paper, Legendre et al. (2010) distributed a package called **STI** that has never been submitted to CRAN. Recently, the functions of this package, now called **stimodels()** and **quicksti()**, have been integrated in the package **adespatial**. Beware: there is an **STI** (uppercase) package available on the CRAN web site, that has nothing to do with the one addressed here and carries out quite different analyses.

The two data sets used in most of this book do not lend themselves to a space-time interaction analysis. Therefore, we will use a dataset that is provided in the **adespatial** package, and is also used as an example in the Legendre et al. (2010) paper. It consists in the counts over 10 periods of 10 days of the abundances of 56 adult Trichoptera (Insecta) species emerging in 22 emergence traps laid along the outflow stream of Lac Cromwell on the territory of the Station de Biologie des Laurentides (Université de Montréal). We have thus a linear spatial layout of 22 points and 10 time points, yielding 220 observations.

Among the two functions to perform space-time interaction analysis, **stimodels()** offers more options, and **quicksti()** provides a quick way to run an sti analysis under Model 5 followed by tests of the main factors. The latter are run under Models 6a and 6b if the interaction is significant, and under Model 5 (or 2: standard test of the main factors using their Helmert contrasts) otherwise. Let us compute an sti analysis with the function **stimodels()**.

Beware: the spatial and temporal eigenfunctions constructed by the original functions of the STI package (distributed as a Supplement to the paper) are actually first-generation PCNMs. By default the functions retained half of the PCNM variables, which, at least in case of regular sampling designs, correspond to positive autocorrelation. The **stimodels()** and **quicksti()** functions incorporated in package **adespatial** compute dbMEM functions, retaining those with positive spatial autocorrelation.

```
# Load the Trichoptera data
data(trichoptera)
names(trichoptera) # Species codes
```

The first two columns contain the site and date numbers, respectively.

```
# Hellinger transformation
tricho.hel <- decostand(trichoptera[ ,-c(1,2)], "hel")
# sti analysis
stimodels(tricho.hel, S = 22, Ti = 10, model = "5")
```

It is useless to store the results into an object unless one wants to use the function in a workflow involving the retrieval of specific sti results; the output object only contains the numerical results of the tests in a raw form. The analysis, run as described above, produces the following on-screen display:

```
=======================================================
       Space-time ANOVA without replicates
Pierre Legendre, Miquel De Caceres, Daniel Borcard
=======================================================

 Number of space points (s) = 22
 Number of time points (tt) = 10
 Number of observations (n = s*tt) = 220
 Number of response variables (p) = 56

 Computing dbMEMs to code for space
 Truncation level for space dbMEMs = 1
 Computing dbMEMs to code for time
 Truncation level for time dbMEMs = 1

 Number of space coding functions = 10
 Number of time coding functions = 4

 MODEL V: HELMERT CONTRAST FOR TESTING MAIN FACTORS.
          SPACE AND TIME dbMEMs FOR TESTING INTERACTION.
   Number of space variables = 21
   Number of time variables = 9
   Number of interaction variables = 40
   Number of residual degrees of freedom = 149

 Interaction test:   R2 = 0.1835   F = 2.313   P( 999 perm) = 0.001
 Space test:         R2 = 0.2522   F = 6.0569   P( 999 perm) = 0.001
 Time test:          R2 = 0.2688   F = 15.0608   P( 999 perm) = 0.001
-------------------------------------------------------
     Time for this computation = 2.899000  sec
=======================================================
```

The output first presents some global statistics about the data. After having displayed the number of space and time dbMEMs computed, it presents the final number of space (21 Helmert contrasts), time (9 Helmert contrasts) and interaction variables. In this example, 10 spatial and 4 temporal dbMEMs produce $10 \times 4 = 40$ interaction variables. There are 220 observations; hence, the number of residual d.f. is equal to $220 - 21 - 9 - 40 - 1 = 149$.

The result shows that the interaction is highly significant ($F = 2.313$, $p = 0.001$). This means that the pattern of distribution of the trichopteran communities along the stream changes over time or, conversely, that the temporal changes in community structure do not follow the same course at the different sampling locations (emergence traps). In this situation, the tests of the main effects cannot be readily interpreted. One must resort to Model 6 tests for that. We could run these tests using the **`stimodels()`** function twice with argument `model = "6a"` and `model = "6b"` respectively. Since function **`quicksti()`** does this automatically, let us run it for demonstration purpose.

```
# Quick and easy sti analysis using function quicksti()
quicksti(tricho.hel, S = 22, Ti = 10)
```

The result of the interaction test is of course the same as above; the tests of spatial and temporal structures using staggered matrices give the following results:

```
---------------------------------------------------------------------
 Testing for the existence of separate spatial structures (model 6a)
---------------------------------------------------------------------

   Number of space variables = 100
   Number of time variables = 9
   Number of residual degrees of freedom = 110

 Space test:  R2 = 0.4739   F = 2.0256   P( 999 perm) = 0.001

-----------------------------------------------------
 Testing for separate temporal structures (model 6b)
-----------------------------------------------------

   Number of space variables = 21
   Number of time variables = 88
   Number of residual degrees of freedom = 110

Time test:    R2 = 0.4981   F = 2.4936   P( 999 perm) = 0.001
```

These tests are valid when an interaction is present. They show that there is both a significant spatial structure and a significant temporal change in the trichopteran community. More precisely, they mean that there is at least one site showing significant temporal change and at least one time point where significant spatial structure is present.

7.7 Conclusion

Spatial analysis of ecological data has undergone huge developments during the last decades. The paradigm shift announced by Legendre (1993) has been accompanied by an increasing awareness, not only of the importance of spatial structures *per se*, but also of the need for refined modelling tools to identify, represent and explain the complex structures by which ecological interactions manifest themselves in living communities. While an entire family of techniques aimed at prediction and mapping has been developed in the field of geostatistics and some of them can be applied to ecological problems, the specific questions and data in the field of ecology demanded other approaches more directly related to the multivariate and multiscale structure of communities and their relationship to the environment. We have presented the most important among them in this chapter, encouraging the readers to apply them to their own data in a creative way.

Chapter 8
Community Diversity

8.1 Objectives

Today's concerns about worldwide environmental degradation and the issues of ecological assessment, monitoring, conservation and restoration, have increased the interest in measuring the diversity of organisms or other constituents of ecological communities (cf. Loreau 2010). Its simplest component, species richness, is a single number that can be easily used outside the scientific community to illustrate ecological issues to the media and decision makers. However, species richness is only a small element of the story.

In many people's minds, the words "diversity" or "biodiversity" simply refer to the number of species in a given area, but actually there are far more dimensions to the concept of biodiversity. Species diversity itself can be defined and measured in a variety of ways (cf. Magurran 2004; de Bello et al. 2010; Legendre and Legendre 2012 Sect. 6.5). Other types of diversity exist at various levels of organization of the living world, ranging from genome to landscape. At the community level, functional diversity, i.e., the diversity of functional traits, has received much attention in recent years, as well as phylogenetic diversity. As we did in previous chapters of this book, we will mostly focus here on species and communities and explore first various facets of taxonomic diversity.

In this chapter you will:

- get an overview of the concept of diversity in ecology;
- compute various measures of alpha species diversity;
- explore the concept of beta diversity;
- partition beta diversity into its local and species contributions;
- partition beta diversity into replacement, richness difference and nestedness;
- get a brief introduction to the concept of functional diversity

D. Borcard et al., *Numerical Ecology with R*, Use R!,
https://doi.org/10.1007/978-3-319-71404-2_8

8.2 The Multiple Facets of Diversity

8.2.1 *Introduction*

Life on Earth is diversified at multiple spatial scales, ranging from the molecular level (genes) to continent-encompassing biomes. Therefore, diversity expresses itself at all these scales, and no single number or method can account for all this complexity. Every level of organization has its own rules and structures and its diversity must be addressed accordingly.

Genetic diversity is a rapidly expanding topic, where huge amounts of data call for computer-efficient methods of data reduction that fall outside the scope of this book. Readers are referred to manuals addressing these topics in a way related to this book, e.g. Paradis (2012), Cadotte and Davies (2016).

Species diversity is at the core of the community-level approach that underlies this book. For example, the objective of many of the methods presented in Chap. 7 is to test hypotheses about the processes that generate spatial variation in community composition, or beta diversity, in ecosystems. In the present chapter, we will explore some important components of the study of taxonomic (community) diversity.

Switching from species to ecological traits is a mean of generalizing ecological models, so that they can be applied to ecologically similar habitats irrespective of the identity of the species found in them. This is the goal of *functional ecology*, based on the proposition by Southwood (1977), who stated: "*habitat provides the templet on which evolution forges characteristic life-history strategies*". Functional ecology is in rapid development. A small subset of its questions are addressed in Sect. 6.11 (fourth-corner problem). However, as of this writing, many different avenues are being explored, and no unifying framework has been proposed yet. Consequently, we will refrain from delving into this matter, except in the form of a very short section (Sect. 8.5).

8.2.2 *Species Diversity Measured by a Single Number*

8.2.2.1 Species Richness and Rarefaction

The simplest measure of species diversity is q, the number of species or species richness. Although it looks straightforward, there is a problem with its estimation. Indeed, what we must rely upon is a sample of the area (or volume in aquatic environments) of interest. Consequently, the true total number of species in that area or volume is out of reach in practice. Every sampling unit contains a certain number of individuals belonging to a certain number of species, and, given the fact that some species are more rare than others and therefore less likely to be detected, the total number of species of a sampling unit or a set of sampling units increases with the sampled area or volume and the number of detected individuals. Consequently, the comparison of the species richness of two sampling units and/or two groups of individuals, which are estimates of the true numbers of species, is biased.

Magurran (2004) points out the distinction between species density, defined as the number of species per specified collection area (Hurlbert 1971) and numerical species richness , which is the number of species per specified number of individuals or biomass unit (Kempton 1979). To ensure comparability between two sites, Sanders (1968) proposed a rarefaction method, which estimates the number of species in sampling units containing the same number of individuals; it is therefore based on the concept of numerical species richness. An important point is that rarefaction can only be computed on true (untransformed) counts of individuals. Sanders' formula has been corrected by Hurlbert (1971). It estimates the number q' of species in a standardized sampling unit of n' individuals based on a real sampling unit containing q species, n individuals and n_i individuals belonging to species i. Hurlbert's equation is the following (Legendre and Legendre 2012 Sect. 6.5):

$$E\left(q'\right) = \sum_{i=1}^{q} \left[1 - \frac{\binom{n - n_i}{n'}}{\binom{n}{n'}}\right] \tag{8.1}$$

where $n' \leq (n - n_1)$, n_1 is the number of individuals in the most abundant species, and the terms in parentheses are combinations. For example:

$$\binom{n}{n'} = \frac{n!}{n'!(n - n')!}$$

8.2.2.2 Species Abundance Diversity Components: Richness and Evenness

A vector of species abundances can be seen as a qualitative variable where each species is a state, and the abundance profile is the frequency distribution of the observations. Under this logic, the *dispersion* of this qualitative variable can be computed on the basis of the relative frequencies p_i of the q states (species) using the well-known Shannon equation (Shannon 1948):

$$H = -\sum_{i=1}^{q} p_i \log p_i \tag{8.2}$$

Shannon's index increases when the number of species increases, but another factor is also at play. Actually, the index takes two components into account: (i) the number of species (species richness) and (ii) the *evenness* or *equitability* of the species frequency distribution. For any number of individuals, H takes its maximum when all species are represented by equal abundances:

$$H_{\max} = -\sum\nolimits_{i=1}^{q} \frac{1}{q} \log \frac{1}{q} = \log q \tag{8.3}$$

Therefore, the Pielou evenness J can be defined as follows (Pielou 1966):

$$J = H/H_{\max} \tag{8.4}$$

Note that this ratio is independent of the base of the logarithms used for the calculation. Curiously, despite its poor performance, due to its long recognized strong dependence on species richness (Sheldon 1969, Hurlbert 1971), Pielou's evenness is still the most widely used evenness index in the ecological literature.

Evenness can be related to the shape of *species abundance models*, i.e., functions describing the shape of rank/abundance plots where the abscissa ranks the species in order of decreasing abundances and the ordinate represents the log-transformed abundances. The four main models are the geometric, log and lognormal series, and the broken stick model. In that order, they represent a progression ranging from the geometric series where a few species are very dominant and the others quite rare, to the broken stick model where species share the abundances most evenly, but not to the point of having equal abundances, a situation that never occurs in the real world. Thus, evenness is increasing from one model to the next in this sequence. Rank/abundance plots can be drawn using the **`radfit()`** function of the **`vegan`** package, which allows one to fit various species abundance models.

The Code It Yourself corner #4

Write a function to compute the Shannon-Weaver entropy for a site vector containing species abundances. The formula is:

$$H' = -\sum [p_i \times log(p_i)]$$

where $p_i = n_i/N$ and n_i = abundance of species i and N = total abundance of all species.

After that, display the code of **`vegan`** *'s function* **`diversity()`** *to see how it has been coded among other indices by Jari Oksanen and Bob O'Hara. Nice and compact, isn't it?*

Other measures of diversity have been proposed. One often used in ecology is Simpson's (1949) concentration index that gives the probability that two randomly chosen organisms belong to the same species:

$$\lambda = \sum\nolimits_{i=1}^{q} \frac{n_i(n_{i-1})}{n(n-1)} = \frac{\sum_{i=1}^{q} n_i(n_i - 1)}{n(n-1)} \tag{8.5}$$

where q is the number of species. When n is large, n_i becomes close to $(n - 1)$ and the equation simplifies to:

$$\lambda = \sum_{i=1}^{q} \left(\frac{n_i}{n}\right)^2 = \sum_{i=1}^{q} p_i^2 \tag{8.6}$$

Actually this quantity increases when the probability that two organisms are conspecific is large (i.e., when species richness is low), so that it is generally transformed to a diversity form either as D = 1 – λ (Gini-Simpson index; Greenberg 1956) or D = 1/λ (inverse Simpson index; Hill 1973). The latter version makes the index less sensitive to changes in the abundances of the (usually few) very abundant species. The Gini-Simpson index can be converted to D = (1 – λ)/λ. This index is the ratio between the total possible interspecific interactions and the possible intraspecific interactions (Margalef and Gutiérrez 1983).

Species richness, Shannon's entropy and Simpson's diversity are actually special cases of Rényi's *generalized entropy* formula (Rényi 1961), as noted by Hill (1973) and Pielou (1975):

$$H_a = \frac{1}{1-a} \log \sum_{i=1}^{q} p_i^a \tag{8.7}$$

where a is the order of the entropy measure ($a = 0, 1, 2\ldots$). Hill (1973) proposed to use the corresponding *diversity numbers*:

$$N_a = e^{H_a} \tag{8.8}$$

Following Hill (1973), Rényi's first three entropies H_a (with $a = 0$, 1, and 2) and the corresponding diversity numbers N_a are listed in Table 8.1. The parameter a quantifies the importance of the species abundances, and thus of evenness: when $a = 0$ diversity is simply the number of species (i.e., presence-absence); when a increases more and more importance is given to the most abundant species. a can be generalized and take values above 2 (see Sect. 8.4.1, Fig. 8.3).

With this notation, and following Pielou (1975), Shannon's evenness becomes H_1/H_0. Hill (1973) proposed to apply the following ratios: $E_1 = N_1/N_0$ (his version of Shannon's evenness) and $E_2 = N_2/N_0$ (Simpson's evenness). Many community ecologists argue nowadays for the use of Hill's numbers for taxonomic diversity and Hill's ratios for evenness, instead of Shannon's entropy and Pielou's evenness, respectively (e.g. Jost 2006) because these numbers, sometimes called "numbers equivalents", are more easily interpretable: they represent "*the number of equally*

Table 8.1 Rényi's first three entropies H_a and the corresponding Hill's diversity numbers N_a

Entropy number	Diversity
$H_0 = \log q$	$\mathbf{N_0} = q$ (q = number of species)
$\boldsymbol{H_1} = -\sum p_i \log p_i = H$	$N_1 = \exp(H)$
$H_2 = -\log \sum p_i^2$	$\mathbf{N_2} = 1/\lambda$

The three most widely used quantities are in boldface

likely elements (individuals, species, etc.) needed to produce the observed value of the diversity index" (Ellison 2010; sentence modified from Jost 2007). Diversity numbers are also preferable for interpretation through linear models because they are more likely to be in linear relationship with environmental variables. These true diversity indices (*sensu* Jost 2006) are part of a unified framework for community diversity, including taxonomic, functional and phylogenetic facets and its partitioning into alpha, beta and gamma components (de Bello et al. 2010). This topic is addressed in Sect. 8.5.2.

8.2.3 Taxonomic Diversity Indices in Practice

8.2.3.1 Species Diversity Indices of the Fish Communities of the Doubs River

Let's start by loading the packages and the species datasets.

```
# Load the required packages
library(ade4)
library(adegraphics)
library(adespatial)
library(vegan)
library(vegetarian)
library(ggplot2)
library(FD)
library(taxize)

# Source additional functions that will be used later in this
# Chapter. Our scripts assume that files to be read are in
# the working directory.
source("panelutils.R")
source("Rao.R")

# Load the Doubs data. The file Doubs.Rdata is assumed to be in
# the working directory
load("Doubs.RData")
# Remove empty site 8
spe <- spe[-8, ]
env <- env[-8, ]
spa <- spa[-8, ]

# Load the oribatid mite data. The file mite.Rdata is assumed
# to be in the working directory.
load("mite.RData")
```

One can easily compute classical alpha taxonomic diversity indices for the fish data. Let us do it with the help of function **diversity()** of the **vegan** package for some indices.

```
# Get help on the diversity() function
?diversity

# Compute alpha diversity indices of the fish communities
N0 <- rowSums(spe > 0)              # Species richness
N0 <- specnumber(spe)               # Species richness (alternate)
H <- diversity(spe)                 # Shannon entropy (base e)
Hb2 <- diversity(spe, base = 2) # Shannon entropy (base 2)
N1 <- exp(H)                        # Shannon diversity (base e)
                                    # (number of abundant species)
N1b2 <- 2^Hb2                       # Shannon diversity (base 2)
N2 <- diversity(spe, "inv")         # Simpson diversity
                                    # (number of dominant species)
J <- H / log(N0)                    # Pielou evenness
E10 <- N1 / N0                      # Shannon evenness (Hill's ratio)
E20 <- N2 / N0                      # Simpson evenness (Hill's ratio)
(div <- data.frame(N0, H, Hb2, N1, N1b2, N2, E10, E20, J))
```

Hint *Note the special use of function* **rowSums()** *for the computation of species richness N0. Normally,* **rowSums(array)** *computes the* ***sums*** *of the rows in that array. Here, argument* `spe > 0` *calls for the sum of the* ***cases*** *where the value is greater than 0.*

Hill numbers (N), which are all expressed in the same units (species number equivalent), and Hill ratios (E) derived from these numbers, should be used to compute diversity indices instead of the popular formulae for Shannon entropy (H) and Pielou evenness (J).

Contrary to Shannon entropy, Shannon diversity number N_1 is independent of the choice of the logarithm base (2, e or 10). Because Shannon entropy is zero when species richness is one, Pielou evenness cannot be calculated for site #1. Moreover, it has been proved that, contrary to Hill ratios, Pielou evenness is biased because it is *systematically* positively correlated with species richness, as shown in the correlation matrix (Fig. 8.1):

```
# Correlations among diversity indices
cor(div)
pairs(div[-1, ],
      lower.panel = panel.smooth,
      upper.panel = panel.cor,
      diag.panel = panel.hist,
      main = "Pearson Correlation Matrix"
)
```

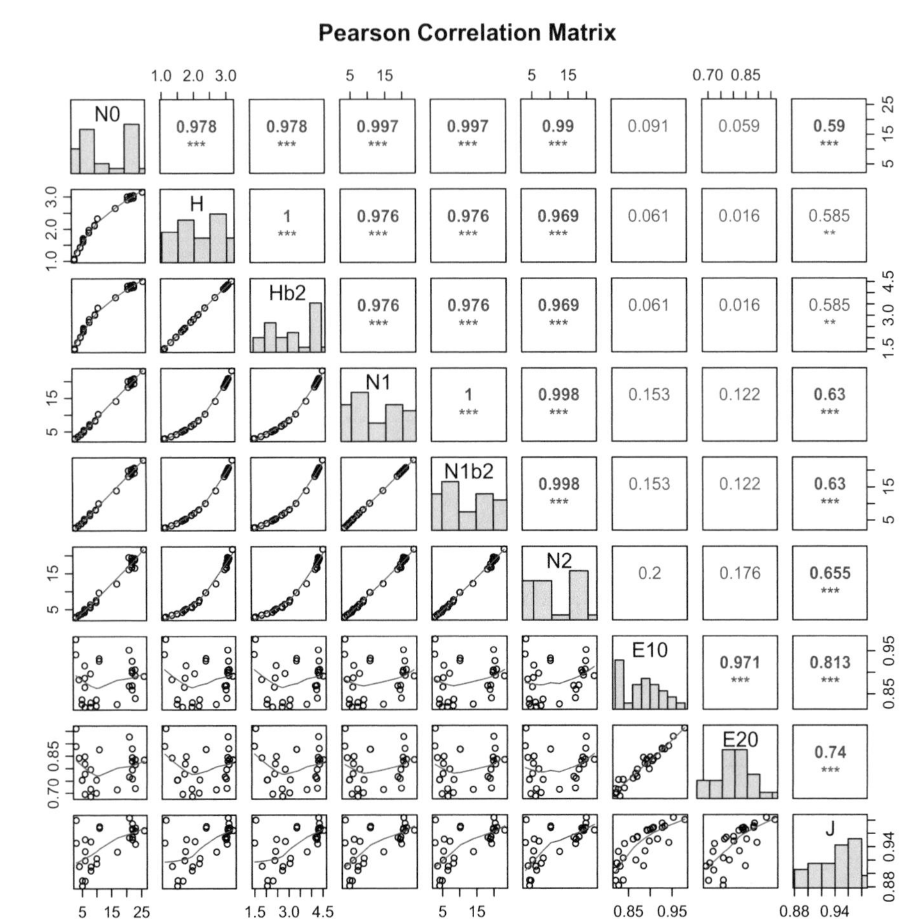

Fig. 8.1 Scatter plots and correlation matrix of pairs of alpha taxonomic diversity indices obtained from the fish abundance data

8.2.3.2 Rarefaction Analysis of the Mite Communities

Hurlbert's (1971) rarefaction can be computed by means of the function **`rarefy()`** of package **`vegan`**. The Doubs fish species data cannot be used in this example because the abundances are measured on a 0–5 scale. Let us compute a rarefaction analysis of the mite data.

Generally, one computes the number of expected species (one "rarefies") to a "sample size" (remember that here this refers to the number of individuals) equal to the number of individuals of the sampling unit where the abundance is the smallest. However, in this data set, most units (soil cores) contain 80 or more individuals, and only a handful contain fewer, with the minimum being 8. Therefore, let us rarefy the data set to a sample size of 80 individuals (and receive a warning because of that).

Prior to that, we compute some statistics about the observed species richness and number of individuals.

```
# Number of species in the 70 moss or soil cores
(mite.nbsp <- specnumber(mite))
# Cores with minimum and maximum observed species richness
mite.nbsp[mite.nbsp == min(mite.nbsp)]
mite.nbsp[mite.nbsp == max(mite.nbsp)]
range(mite.nbsp)
# Total abundance in the 70 cores
(mite.abund <- rowSums(mite))
range(mite.abund)
# Abundance in the cores with smallest number of species
mite.abund[mite.nbsp == min(mite.nbsp)]
# Abundance in the core with largest number of species
mite.abund[mite.nbsp == max(mite.nbsp)]
# Number of species in the core with smallest abundance
mite.nbsp[mite.abund == min(mite.abund)]
# Number of species in the core with largest abundance
mite.nbsp[mite.abund == max(mite.abund)]

# Rarefaction to 80 individuals
mite.rare80 <- rarefy(mite, sample = 80)
# Compare ranking of observed and rarefied cores
sort(mite.nbsp)
sort(round(mite.rare80))
# Cores with minimum and maximum estimated species richness
mite.rare80[mite.rare80 == min(mite.rare80)]
mite.rare80[mite.rare80 == max(mite.rare80)]
# Observed core with smallest predicted species richness
mite[which(mite.rare80 == min(mite.rare80)),]
# Observed core with largest predicted species richness
mite[which(mite.rare80 == max(mite.rare80)),]
```

As revealed by the first lines of code, the actual numbers of species in the data set range from 5 to 25. Also, observe that one of the two cores with the smallest observed number of individuals (core #57) is also the one with the smallest number of species; however, the core with the largest abundance (core #67) also contains very few species (only 6). The core with the highest observed species richness is core #11 (25 species).

Now, these comparisons are made on unstandardized samples. The rarefaction to 80 individuals modifies the ranking of the core. The core with the smallest estimated number of species is core #67. As seen above, it is the core with the largest observed abundance. A look at the observed abundance profile explains this feature: of the 781 individuals observed in that core, 723 belong to a single species (*Limncfci*). With 6 observed species for a total of 783 individuals, the estimation for 80 individuals drops to 3.8 species. At the other extreme, the core with the highest estimated species richness is core #30 with 19.1 species for 80 individuals. The actual numbers

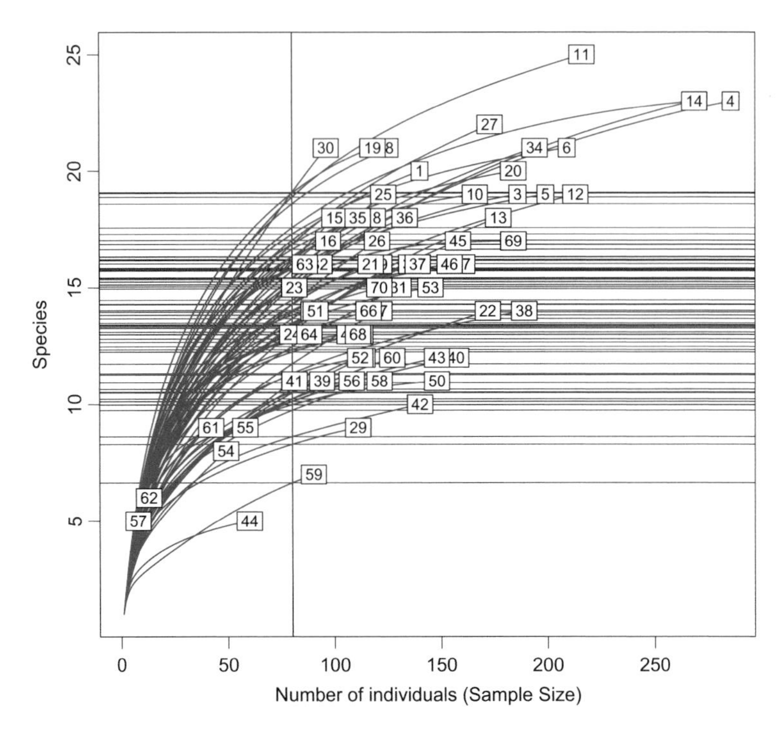

Fig. 8.2 Rarefaction curves for 69 of the 70 cores of the mite data. Core #67 was removed because it compressed all the other cores to the left of the graph. Labels: core numbers. The vertical line corresponds to 80 individuals. The horizontal lines give the number of expected species for each core and 80 individuals

of species and individuals are 21 species and 96 individuals. Therefore, the estimated species richness for 80 individuals is close.

Another use of rarefaction consists in plotting the expected number of species as a function of the number of individuals. This rarefaction curve can be produced for each site by using function **`rarecurve()`**. It is displayed in Fig. 8.2.

```
rarecurve(
  mite[-67,],
  step = 1,
  sample = 80,
  xlab = "Number of individuals (Sample Size)",
  ylab = "Species",
  label = TRUE,
  col = "blue"
)
```

8.3 When Space Matters: Alpha, Beta and Gamma Diversities

Species diversity varies through space. Trivial as it may seem, this observation had to wait two seminal papers by Whittaker (1960, 1972) to receive appropriate attention. In Whittaker's conception, alpha (α) is local diversity, beta (β) is the spatial variation of diversity, and gamma (γ) is regional diversity. These concepts have since gained huge interest among ecologists.

Alpha diversity has been presented in Sect. 8.2.2. In various ways it measures the variation in species identity of individuals observed at a given site (Legendre and Legendre 2012 Sect. 6.5).

Gamma diversity is the species diversity of the whole region concerned by a study. Since the region is generally not known in its entirety, one estimates gamma diversity by pooling the sampling units available and computing the same indices as for alpha diversity, under the assumption that the sampling units are representative of the species composition of the region.

Beta diversity is another matter. It can be construed as the variation of species composition among sites in the area of interest (Legendre et al. 2005, Anderson et al. 2006, Anderson et al. 2011) and can be studied in many different ways. It deserves a separate section.

8.4 Beta Diversity

Beta diversity can be estimated on the basis of either presence-absence or quantitative data. Different approaches have been proposed to study beta diversity. Some of them attempt to summarize it by a single number. Others take advantage of the *variation* component of beta diversity and study this variation for itself using methods rooted in the concepts presented in Chaps. 3, 5, 6 and 7 of this book. We will present some of these methods below (Sect. 8.4.2 and 8.4.3), but let us first briefly summarize the simplest ways of measuring beta diversity.

8.4.1 Beta Diversity Measured by a Single Number

Whittaker proposed a first measure of beta diversity for species presence-absence data: $\beta = S/\overline{\alpha}$, where S is the number of species in the pooled community composition vector (i.e., gamma diversity) and $\overline{\alpha}$ is the mean number of species at the n sampling sites. β tells us how many more species are present in the whole area than in an average individual site. This index is said to be *multiplicative* since it is a ratio.

Other multiplicative indices have been proposed. See Legendre and Legendre (2012, p. 260) and references therein.

An alternative, additive, ANOVA-type approach also exists. The principle, exposed in detail by Lande (1996), is to compute a quantity $D_T = D_{among} - \bar{D}_{within}$ where D_T is gamma diversity and D_{within} is an average measure of diversity at the level of the sites (alpha diversity). Thus, D_{among} is the variation of diversity among sites, i.e., beta diversity. The diversity components D can be based on species richness (N_0), Shannon information H_1 or Simpson diversity $1 - \lambda$ (but not $N_2 = 1/\lambda$). See Lande (1996) and the *Forum* section in *Ecology* (2010: 1962–1992) for details.

Another approach to multiplicative or additive partitioning of community diversity has been recently advocated, compatible with the general framework based on Hill numbers (Jost 2006, Jost 2007, de Bello et al. 2010).

Gamma species richness, i.e. the number of species observed in a collection of communities sampled within a given regional range, is simple to obtain from the species data frame, since it is the number of columns of the data frame. Different algorithms are available to estimate the *species pool*, including species that could have been missed by the sampling. The **`specpool()`** function of package **`vegan`** does this job. Predictions are based on the number of the less frequent species in the dataset (species observed in only one or two sites).

```
# Gamma richness and expected species pool
?specpool
(gobs <- ncol(spe))
(gthe <- specpool(spe))
```

> *Compare the observed gamma species richness with the expected species pool predicted by four models (Chao, first and second order jackknife, bootstrap). How many unobserved species may contribute to the "dark diversity"?*

Using the **`d()`** function of package **`vegetarian`** is the easiest way of computing mean alpha, multiplicative beta and gamma diversities based on Hill numbers. The argument `lev` is used to select the component and the argument `q` the order of the diversity index (any null or positive numeric value, integer or real; the higher the order, the higher the importance given to the abundant species; argument `q` corresponds to the subscript a in Table 8.1). Thanks to this framework, it is possible to plot multiplicative beta diversity as a function of the order of the diversity index.

```
# Multiplicative partitioning of Hill numbers (Jost 2006, 2007)
?d

# Mean alpha species richness
d(spe, lev = "alpha", q = 0)
# Mean alpha Shannon diversity
d(spe, lev = "alpha", q = 1)
# Mean alpha Simpson diversity
d(spe, lev = "alpha", q = 2, boot = TRUE)

# Multiplicative beta species richness
d(spe, lev = "beta", q = 0)
# Multiplicative beta Shannon diversity
d(spe, lev = "beta", q = 1)
# Multiplicative beta Simpson diversity
d(spe, lev = "beta", q = 2, boot = TRUE)

# Gamma species richness
d(spe, lev = "gamma", q = 0)
# Gamma Shannon diversity
d(spe, lev = "gamma", q = 1)
# Gamma Simpson diversity
d(spe, lev = "gamma", q = 2, boot = TRUE)

# Plot multiplicative beta diversity vs order
mbeta <- data.frame(order = 0:20, beta = NA, se = NA)
for (i in 1:nrow(mbeta)) {
  out <- d(spe, lev = "beta", q = mbeta$order[i], boot = TRUE)
  mbeta$beta[i] <- out$D.Value
  mbeta$se[i] <- out$StdErr
}
mbeta
ggplot(mbeta, aes(order, beta)) +
  geom_point() +
  geom_line() +
  geom_errorbar(aes(order, beta, ymin = beta - se,
                    ymax = beta + se), width = 0.2) +
  labs(y = "Multiplicative beta diversity",
       x = "Order of the diversity measure")
```

Hint *Note the non-conventional syntax of function* **ggplot()** *of the package* **ggplot2** *for the line plot of Fig. 8.3.*

It turns out that multiplicative beta diversity increases from about 2 to 7 when increasing the order, i.e. when giving more and more importance to the evenness component against the richness component of species diversity (Fig. 8.3).

Another useful function to achieve additive partitioning of taxonomic, but also functional and phylogenetic diversities in a unified framework is **Rao()**; it is not

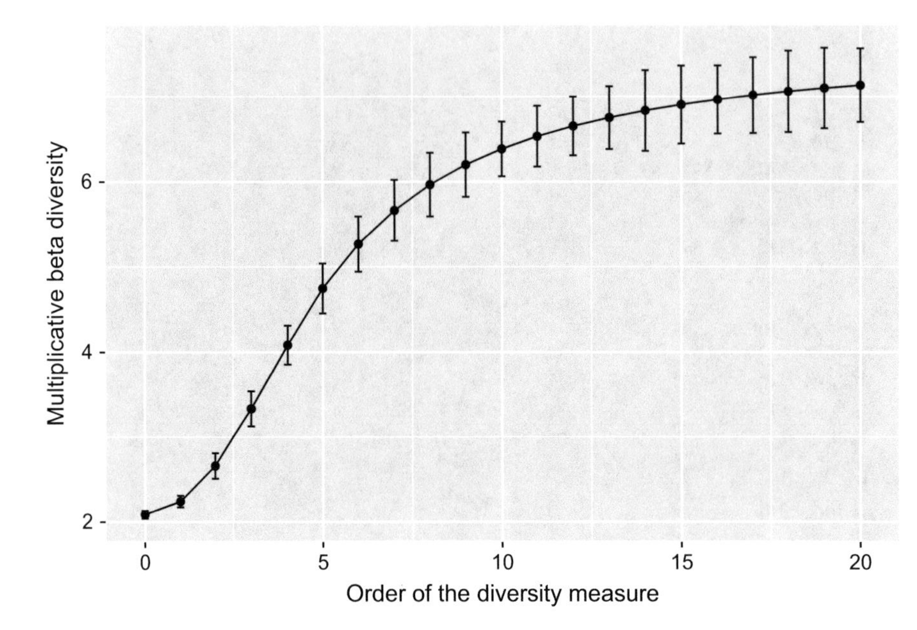

Fig. 8.3 Multiplicative beta diversity of the fish communities as a function of the order of the diversity index. Error bars represent standard errors computed by bootstrap

included in any **R** package but it is provided as an online appendix of a synthesis paper by de Bello et al. (2010). All indices computed by this function are based on the Rao quadratic entropy, corresponding in the taxonomic case to (inverse) Simpson diversity. We shall present an application of this function to taxonomic, phylogenetic and functional diversity of the fish communities of the Doubs River in Sect. 8.5.

Now, a number is still. . . only a number. Beta diversity expresses itself across the area under study by the variation in species composition and abundance among all sites. Therefore, one can study beta diversity in much more detail by analysing the data matrix itself, using different approaches. The following sections are devoted to this family of approaches.

8.4.2 Beta Diversity as the Variance of the Community Composition Table: SCBD and LCBD Indices

8.4.2.1 Introduction

Beta diversity reflects the processes that generate and maintain the variety of living communities in a region. Therefore, it is of great importance to ecologists, and a large corpus of literature is devoted to its definition and measurement. This section proposes one approach to the concept, linked to the variance of the site-by-species table and to many of the analyses described in this book, which analyse or partition that variance in different manners.

Beta diversity has been defined in various ways. Whittaker's (1960) definition is the variation in species composition among sites within a geographic area of interest. Different equations exist to measure that variation. From the research on beta diversity two approaches emerge: (1) Beta diversity can be construed as turnover, i.e., the directional change in community composition along a predefined spatial, temporal or environmental gradient. (2) Beta diversity can also be defined as the variation in community composition among sampling units, without reference to an explicit gradient. Both concepts are within the scope of Whittaker's definition.

Many papers have proposed indices of diversity at the three levels. Most of the proposed measurements of beta diversity are linked to alpha and gamma. In the multiplicative and additive approaches mentioned in Sect. 8.4.1, beta diversity is a derived quantity, i.e. it is linked to, and can be obtained only from the measurement of alpha and gamma. In his introduction to the *Forum* section in *Ecology* (2010: 1962–1992), Ellison (2010) called for a measure of beta diversity that would be computationally independent from the two other levels. Legendre and De Cáceres' (2013) method fulfils this wish. It uses the total variance of the site-by-species Table **Y**, Var(**Y**), as an estimate of beta diversity, in the line of Pélissier et al. (2003), Legendre et al. (2005) and Anderson et al. (2006). Var(**Y**) is calculated independently from alpha and gamma diversity. Note that the simple ordination methods presented in Chap. 5 and the constrained ordination methods of Chap. 6 can be applied to community composition data and be used to analyse and interpret the patterns of variation of the community composition among sites. Likewise, the methods of spatial eigenfunction analysis described in Chap. 7 have been developed to decompose the spatial variation of community data among spatial scales. Therefore, all these methods can be seen as methods of analysis of beta diversity.

8.4.2.2 Computing Beta Diversity as Var(Y)

The total variance Var(**Y**) is obtained in three steps. First, compute a matrix **S** of squared deviations $[s_{ij}]$ of the y_{ij} abundance values from the corresponding column means $\bar{y}_j$:

$$s_{ij} = \left(y_{ij} - \bar{y}_j\right)^2 \tag{8.9}$$

The total sum-of-squares of **Y** is the sum of the squared values in matrix **S**:

$$\mathrm{SS}_{\mathrm{Total}} = \sum_{i=1}^{n} \sum_{j=1}^{p} s_{ij} \tag{8.10}$$

The total variance of **Y**, Var(**Y**), is the total sum-of-squares divided by (n – 1):

$$\mathrm{BD_{Total}} = \mathrm{Var}(\mathbf{Y}) = \mathrm{SS_{Total}}/(n-1) \tag{8.11}$$

We call this variance $\mathrm{BD_{Total}}$ since it is our measure of total beta diversity.

Remember that if you want to compute $\mathrm{BD_{Total}}$ in this way, you must pre-transform the species abundances, for instance using the Hellinger or chord pre-transformation (Sect. 3.5). Otherwise, the underlying measure would be the Euclidean distance, which is almost always inappropriate for community composition data.

Var(**Y**) can also be obtained directly from a dissimilarity matrix D (proof in Legendre and Fortin 2010, Appendix 1). The dissimilarity measure can be any one deemed appropriate for the data at hand (but see Sect. 8.4.2.3 below). $\mathrm{SS_{Total}}$ is obtained by summing the dissimilarities D_{hi} in the upper triangular portion of the dissimilarity matrix and dividing by n:

$$\mathrm{SS_{Total}} = \mathrm{SS}(\mathbf{Y}) = \frac{1}{n}\sum_{h=1}^{n-1}\sum_{i=h+1}^{n} D_{hi}^2 \tag{8.12}$$

Note that n is the number of sites, not the number of dissimilarities.

For non-Euclidean dissimilarity matrices D where $\left[\sqrt{D_{hi}}\right]$ is Euclidean (e.g., Jaccard, Sørensen, percentage difference), compute:

$$\mathrm{SS_{Total}} = \mathrm{SS}(\mathbf{Y}) = \frac{1}{n}\sum_{h=1}^{n-1}\sum_{i=h+1}^{n} D_{hi} \tag{8.13}$$

$\mathrm{BD_{Total}}$ is then computed with Eq. 8.11.

The second method allows one to use many dissimilarity measures, but Legendre and De Cáceres (2013) showed that not all are appropriate for estimating beta diversity.

A major strength of the Var(**Y**) approach is that beta diversity can be decomposed into local contributions of the sites to beta diversity (LCBD indices) and also, when Var(**Y**) is computed from the species data table and not from a dissimilarity matrix, to species contributions to beta diversity (SCBD indices).

8.4.2.3 Properties of Dissimilarity Measures with Respect to Beta Diversity

Legendre and De Cáceres (2013) analysed 16 quantitative dissimilarity measures with respect to 14 properties. Among these and beyond some trivial ones, they pointed out six properties that they deemed indispensable for an appropriate assessment of beta diversity (see their Appendix S3 for details): double-zero asymmetry (P4); sites without species in common have the largest dissimilarity (P5); dissimilarity does not decrease in series of nested assemblages; e.g. the dissimilarity should not decrease when the number of unique species in one or both sites increases (P6);

species replication invariance: a community composition table with the columns in two or several copies should produce the same dissimilarities among sites as the original data Table (P7); invariance to the measurement units; e.g. the dissimilarity between two sites should be the same if biomass is measured in g or in mg (P8); existence of a fixed upper bound (P9). Properties P1 to P3 were trivial and shared by all indices. The authors compared the 16 dissimilarity measures on the basis of the remaining 11 properties by means of a PCA and identified 5 types of coefficients. Among the indices already described in this book that are appropriate for beta diversity measurement, let us mention the Hellinger and chord distances as well as the percentage difference (aka Bray-Curtis) dissimilarity. On the contrary, the Euclidean and chi-square distances failed in one or several properties. All appropriate indices have values between 0 and 1, or between 0 and $\sqrt{2}$, the maximum value being obtained when two sites have entirely different species compositions. The Ružička dissimilarity, not included in that study, was later shown by Legendre (2014) to be also appropriate. Furthermore, the Jaccard, Sørensen and Ochiai coefficients for presence-absence data, which are the binary forms of quantitative indices in the previous list, are also appropriate. Correspondences between quantitative and binary indices are presented in Legendre and De Cáceres (2013, Table 1).

For [0, 1] dissimilarity indices, the maximum possible value of $\mathrm{BD}_{\mathrm{Total}}$ is 0.5. For [0, $\sqrt{2}$] indices, the maximum value of $\mathrm{BD}_{\mathrm{Total}}$ is 1. That maximum is reached when all sites have entirely different species compositions when compared to one another (Legendre and De Cáceres 2013).

8.4.2.4 Local Contributions to Beta Diversity (LCBD)

The contribution of site i to the overall beta diversity is the sum (SS_i) of the centred and squared values for site (or row) i in matrix **S**:

$$SS_i = \sum_{j=1}^{p} s_{ij} \tag{8.14}$$

The **relative** contribution of site i to beta diversity, called the **local contribution to beta diversity (LCBD)**, is:

$$\mathrm{LCBD}_i = \mathrm{SS}_i/\mathrm{SS}_{\mathrm{Total}} \tag{8.15}$$

where $\mathrm{SS}_{\mathrm{Total}} = \sum_{i=1}^{n} ss_i$ (Eq. 8.10).

LCBD indices can be mapped on the territory under study. They represent, in Legendre and De Cáceres' words, "*the degree of uniqueness of the sampling units in terms of community composition*". They can be tested for significance by random, independent permutations within the columns of matrix **Y**. The null hypothesis of the test is that the species are distributed at random among the sites, independently from one another; the species abundance distributions of the data are preserved in the permutations, but the association of the species to the site ecological conditions and their spatial distributions are destroyed. Alpha diversity is destroyed as well in the permutations.

Legendre and De Cáceres (2013) also showed how to calculate LCBD indices from a dissimilarity matrix. The computation involves the same transformation of the dissimilarity matrix as the one used in PCoA.

8.4.2.5 Species Contributions to Beta Diversity (SCBD)

Decomposition of beta diversity into species contributions can only be computed from a site-by-species data table, not from a dissimilarity matrix since the species abundances at the sites have been lost in the calculation of the dissimilarities.

The contribution of species j to the overall beta diversity is the sum (SS_j) of the centred and squared values for species (or column) j in matrix **S**:

$$SS_j = \sum_{i=1}^{n} s_{ij} \tag{8.16}$$

Again, $\mathrm{SS}_{\mathrm{Total}} = \sum_{i=1}^{n} ss_i$ (Eq. 8.10).

The **relative** contribution of species j to beta diversity, called the **species contribution to beta diversity (SCBD)**, is:

$$\mathrm{SCBD}_j = \mathrm{SS}_j / \mathrm{SS}_{\mathrm{Total}} \tag{8.17}$$

8.4.2.6 Computation of LCBD and SCBD using `beta.div()` of Package `adespatial`

Legendre and De Cáceres (2013) wrote function **`beta.div()`** to compute beta diversity as Var(**Y**) and its decomposition into LCBD and SCBD. This function is now part of the **`adespatial`** package. Let us apply it to the Doubs fish data, as the authors did in their paper, but here we will use the Hellinger instead of the chord transformation, which is also appropriate. Legendre and De Cáceres (2013) have shown that the LCBD indices computed from the 11 dissimilarity coefficients suitable for beta diversity assessment were highly concordant.

```
# Computation using beta.div {adespatial} on
# Hellinger-transformed species data
spe.beta <- beta.div(spe, method = "hellinger", nperm = 9999)
summary(spe.beta)
spe.beta$beta   # SSTotal and BDTotal

# Which species have a SCBD larger than the mean SCBD?
spe.beta$SCBD[spe.beta$SCBD >= mean(spe.beta$SCBD)]
```

You could plot the species with the largest SCBD along the river, as in Fig. 2.3.

```
# LCBD values
spe.beta$LCBD
# p-values
spe.beta$p.LCBD
# Holm correction
p.adjust(spe.beta$p.LCBD, "holm")
# Sites with significant Holm-corrected LCBD value
row.names(spe[which(p.adjust(spe.beta$p.LCBD, "holm") <= 0.05),])

# Plot the LCBD values on the river map
plot(spa,
     asp = 1,
     cex.axis = 0.8,
     pch = 21,
     col = "white",
     bg = "brown",
     cex = spe.beta$LCBD * 70,
     main = "LCBD values",
     xlab = "x coordinate (km)",
     ylab = "y coordinate (km)"
)
lines(spa, col = "light blue")
text(85, 11, "***", cex = 1.2, col = "red")
text(80, 92, "***", cex = 1.2, col = "red")
```

We obtain $SS_{Total} = 14.07$ and $BD_{Total} = 0.5025$. These values differ slightly from those of Legendre and De Cáceres (2013) because we used a different distance coefficient. Five species have an SCBD higher than the mean SCBD: the brown trout (`Satr`), Eurasian minnow (`Phph`), bleak (`Alal`), stone loach (`Babl`) and (to a lesser extent) roach (`Ruru`). The (Hellinger-transformed) abundances of these species vary the most among sites, which makes them interesting as ecological indicators.

The largest LCBD values are concentrated in three zones around sites 1, 13 and 23 (Fig. 8.4). The permutation tests show that the LCBD values of sites 1 and 23 are significant after a Holm correction for 29 simultaneous tests. Both sites stand out because they harbour very few species, which makes them different from most of the other sites. This point allows us to emphasize that sites having high LCBD values are not automatically "special" in a good sense, e.g. by harbouring rare species or by being exceptionally rich. Any departure from the overall species abundance pattern increases the LCBD value. In the present example, site 1 was a pristine site at the head of the river, harbouring a single species, the brown trout (`Satr`), whereas sites 23–25 suffered from urban pollution and were in need of rehabilitation.

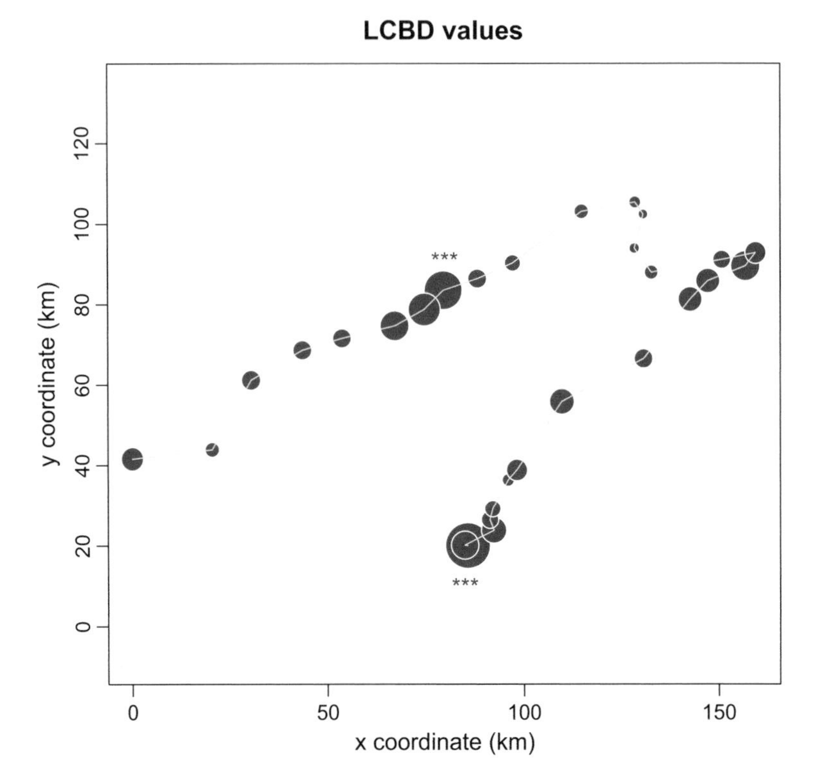

Fig. 8.4 LCBD indices of the 29 sites of the Doubs River, based on a computation on Hellinger-transformed species abundances. Bubble sizes are proportional to the LCBD values; so the largest bubbles correspond to sites that contribute most to the overall beta diversity of the data set. The *** indicate the two significant LCBD indices ($p \leq 0.05$) after Holm correction

8.4.3 Partitioning Beta Diversity into Replacement, Richness Difference and Nestedness Components

Harrison et al. (1992), Williams (1996) and Lennon et al. (2001) proposed that the differences among communities (beta diversity) result from two processes: species replacement (sometimes called turnover) and richness difference (species gain and loss; or its special case, nestedness).

Species replacement is what is observed along ecological gradients where some species are found on a limited range of the gradient according to their ecological optima and tolerances. Some species are observed at one end of the gradient, then disappear and are replaced by others, and so on. The causes of these patterns include environmental forcing, but also, potentially, competition and historical events (i.e., disturbances and other processes that have occurred in the past and left their marks in the communities).

Richness difference may be caused by local species disappearances, differences in the local abiotic conditions leading to different numbers of ecological niches, or other ecological processes leading to communities with higher or lower numbers of species. *Nestedness* is a special case of richness difference where the species at poorer sites are a strict subset of the species present at richer sites.

This new vision of beta diversity led to the design of several methods to partition a dissimilarity matrix into these two components. In Appendix S1 of his paper, Legendre (2014) reviewed the development of these methods, and the main article proposes a unifying algebraic framework to compare the published formulae. Legendre distinguished two main families of indices, that he dubbed the Podani and the Baselga families respectively, after the names of the authors that led two groups of researchers who developed these indices. References can be found in the paper. The presentation and tables below are inspired from the Legendre (2014) paper.

The common quantities of the two families for comparing community composition at two sites, 1 and 2, are:

- the common species (a: number of common species; A: sum of the abundances, over all species, that sites 1 and 2 have in common);
- the species unique to site 1 (b: number of species unique to site 1; B: sum of the abundances, over all species, that are unique to site 1);
- the species unique to site 2 (c: number of species unique to site 2; C: sum of the abundances, over all species, that are unique to site 2).

Important: for quantitative data, this decomposition of dissimilarities is not possible with relative counts (e.g. percentage cover in vegetation analysis). Indeed, the proportional share of a given species is conditioned by that of all the others, which makes it impossible to compare species proportional abundances between sites.

The two tables below (Tables 8.2 and 8.3) illustrate these quantities by means of fictitious pairs of sites.

The quantities a, b and c are the components used to compute two dissimilarity measures that we saw in Chap. 3: the Jaccard (S_7) and Sørensen (S_8) indices. The quantitative elements A, B and C are used to build the corresponding quantitative dissimilarity coefficients, i.e., the Ružička index (quantitative equivalent of Jaccard)

Table 8.2 A fictitious pair of sites with presence-absence data, exemplifying the quantities a, b and c used to compute species replacement and richness difference coefficients

	Sp01	Sp02	Sp03	Sp04	Sp05	Sp06	Sp07	Sp08	Sp09	Sp10	Sp11	Sp12
Site 1	1	1	1	1	1	1	1	1	1	0	0	0
Site 2	1	1	1	1	0	0	0	0	0	1	1	1
		a					***b***				***c***	

Here, $a = 4$, $b = 5$ and $c = 3$

Table 8.3 A fictitious pair of sites with abundance data, exemplifying the quantities A, B and C used to compute abundance replacement and abundance difference coefficients

	Sp01	Sp02	Sp03	Sp04	Sp05	Sp06	Sp07	Sp08	Sp09	Sum
Site 1	15	8	5	10	19	7	0	0	14	78
Site 2	10	2	20	30	0	0	25	11	14	112
Min(1,2)	10	2	5	10	0	0	0	0	14	41 = A
Unique 1	5	6	0	0	19	7	0	0	0	37 = B
Unique 2	0	0	15	20	0	0	25	11	0	71 = C

Min(1,2): smallest abundance in sites 1 and 2; Unique 1 and Unique 2: abundance unique to site 1 or 2. The sum of Min(1,2) is equal to A, the sum of Unique 1 is equal to B and the sum of Unique 2 is equal to C. Here, $A = 41$, $B = 37$ and $C = 71$

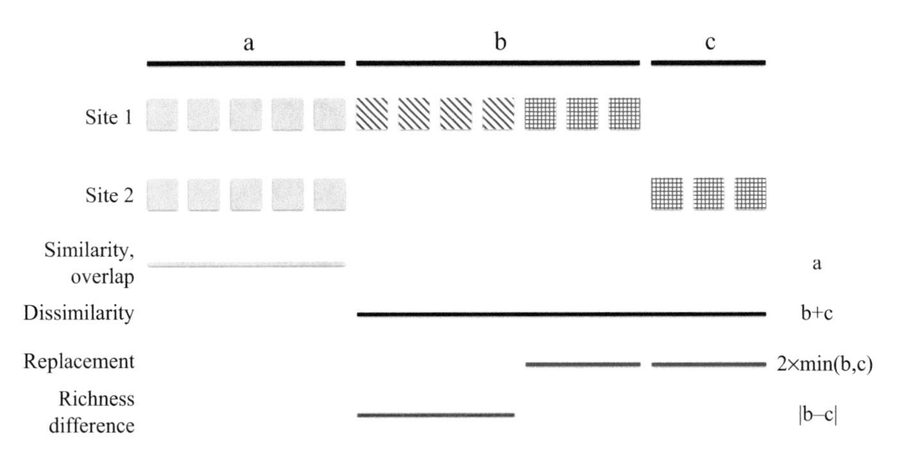

Fig. 8.5 Two fictitious sites with 12 (site 1) and 8 (site 2) species, with the quantities a, b and c used to construct indices of dissimilarity, replacement, richness difference and nestedness for presence-absence data. After Legendre (2014)

and the percentage difference D_{14} (aka Bray-Curtis, quantitative equivalent of Sørensen).

In the papers reviewed by Legendre (2014), the quantities estimating species replacement, used as numerators in the various indices for presence-absence data, are either min(b, c) or 2 × min(b, c). Indeed, among the b and c unique species, the smaller number is considered as the number of species replacing the equivalent number at the other site. For quantitative data, the corresponding quantities min(B, C) or 2 × min(B, C) are used. For richness difference, the statistic of interest is $|b–c|$; for instance in Table 8.2, $b = 5$ and $c = 3$, so 3 species at site 1 are considered to replace 3 species at site 2, and the remaining 2 species represent the richness difference between the two sites. The corresponding quantity for differences in abundance is $|B – C|$, which is also the difference between the total abundances of the two sites, i.e. $|71–37| = 34$ in Table 8.3. Fig. 8.5 illustrates the roles played by a, b and c to construct the various indices based on presence-absence data.

Table 8.4 Formulae of the Podani group of indices for presence-absence and quantitative data, with examples based on the fictitious values of Tables 8.2 and 8.3

	Presence-absence	Abundance
Numerator		
Replacement Examples	$2 \times \min(b, c)$ $2 \times 3 = 6$	$2 \times \min(B, C)$ $2 \times 37 = 74$
Richn. or abund. Diff. Examples	$\lvert b - c\rvert$ $\lvert 5–3\rvert = 2$	$\lvert B - C\rvert$ $\lvert 37–71\rvert = 34$
Dissimilarity Examples	$(b + c)$ $5 + 3 = 8$	$(B + C)$ $37 + 71 = 108$
Jaccard – Ružička group		
Denominator Examples	$(a + b + c)$ $4 + 5 + 3 = 12$	$(A + B + C)$ $41 + 37 + 71 = 149$
Dissimilarity Examples	$D_J = (b + c)/(a + b + c)$ $D_J = 8 / 12 = 0.667$	$D_{Ru} = (B + C)/(A + B + C)$ $D_{Ru} = 108 / 149 = 0.725$
Replacement Examples	$Repl_J = 2 \times \min(b,c)/(a + b + c)$ $Repl_J = 6 / 12 = 0.500$	$Repl_{Ru} = 2 \times \min(B,C)/(A + B + C)$ $Repl_{Ru} = 74 / 149 = 0.497$
Richn. or abund. Diff. Examples	$RichDiff_J = \lvert b - c\rvert/(a + b + c)$ $RichDiff_J = 2 / 12 = 0.167$	$RichDiff_{Ru} = \lvert B - C\rvert/(A + B + C)$ $RichDiff_{Ru} = 34 / 149 = 0.228$
Sørensen – Percentage difference group		
Denominator Examples	$(2a + b + c)$ $2 \times 4 + 5 + 3 = 16$	$(2A + B + C)$ $2 \times 41 + 37 + 71 = 190$
Dissimilarity Examples	$D_S = (b + c)/(2a + b + c)$ $D_S = 8 / 16 = 0.500$	$D_{14} = (B + C)/(2A + B + C)$ $D_{14} = 108 / 190 = 0.568$
Replacement Examples	$Repl_S =$ $2 \times \min(b, c)/(2a + b + c)$ $Repl_S = 6 / 16 = 0.375$	$Repl_{D14} =$ $2 \times \min(B,C)/(2A + B + C)$ $Repl_{D14} = 74 / 190 = 0.389$
Richn. or abund. Diff. Examples	$RichDiff_S = \lvert b–c\rvert/(2a + b + c)$ $RichDiff_S = 2 / 16 = 0.125$	$RichDiff_{D14} = \lvert B–C\rvert/(2A + B + C)$ $RichDiff_{D14} = 34 / 190 = 0.179$

$D_J = 1 - S_7$; $D_S = 1 - S_8$

8.4.3.1 The Podani Family of Indices

The indices of replacement and richness (or abundance) difference described by the Podani group of authors are presented in Table 8.4 with numerical examples based on Tables 8.2 and 8.3. The indices are built with the numerators described above, scaled by denominators that depend on the group of indices.

Observe that in all cases the replacement and richness (or abundance) difference indices add up to the corresponding dissimilarity index.

Podani and Schmera (2011) also proposed an *index of nestedness* defined as $N = a + \lvert b - c\rvert$ if $a > 0$ and $N = 0$ if $a = 0$. N can be relativized (scaled between 0 and 1) by dividing it by $(a + b + c)$. The authors' argument for this index is that richness difference only represents a part of nestedness, and a correct assessment of nestedness should include the common species (a). See Sect. 8.4.3.3 for a comparison of nestedness indices.

8.4.3.2 The Baselga Family of Indices

This family contains indices of replacement (sometimes called *turnover*) and *nestedness*. The replacement indices can be easily compared to those of the Podani family, but the nestedness indices are different in their purpose and formulation from the richness and abundance difference indices of the Podani family. They also differ from the Podani and Schmera (2011) index of nestedness.

Baselga (2010) described a replacement index (called *turnover* in his paper) based on the Sørensen dissimilarity D_S, and proposed to subtract it from the Sørensen dissimilarity to obtain the nestedness component. These two indices are called $Repl_{BS}$ and Nes_{BS} in Table 8.5. The same line of reasoning led to indices summing up to Jaccard's dissimilarity D_J. For quantitative data, the quantities *a*, *b* and *c* are replaced by *A*, *B* and *C* respectively. These indices are presented (with numerical examples) in Table 8.5.

Table 8.5 Replacement (turnover) and nestedness indices of the Baselga family, with numerical examples based on Tables 8.2 and 8.3

	Presence-absence	Abundance
Jaccard – Ružička group		
Dissimilarity Examples	$D_J = (b + c)/(a + b + c)$ 8 / 12 = 0.667	$D_{Ru} = (B + C)/(A + B + C)$ 108 / 149 = 0.725
Replacement (turnover) Examples	$Repl_{BJ} = \frac{2\min(b,c)}{a+2\min(b,c)}$ 6 / [4 + (2 × 3)] = 0.600	$Repl_{BR} = \frac{2\min(B,C)}{A+2\min(B,C)}$ (2 × 37)/[41 + (2 × 37)] = 0.643
Nestedness Examples	$Nes_{BJ} = \frac{\lvert b-c\rvert}{a+b+c} \times \frac{a}{a+2\min(b,c)}$ (2/12) × [4/(4 + 2 × 3)] = 0.067	$Nes_{BR} = \frac{\lvert B-C\rvert}{A+B+C} \times \frac{A}{A+2\min(B,C)}$ Nes_{BR} = (34/149) × [41/(41 + (2 × 37))] = 0.081
Sørensen – % difference group		
Dissimilarity Examples	$D_S = (b + c)/(2a + b + c)$ 8 / 16 = 0.500	$D_{14} = (B + C)/(2A + B + C)$ 108 / 190 = 0.568
Replacement (turnover) Examples	$Repl_{BS} = \frac{\min(b,c)}{a+\min(b,c)}$ 3 / (4 + 3) = 0.429	$Repl_{B.D14} = \frac{\min(B,C)}{A+\min(B,C)}$ (37)/[41 + 37] = 0.474
Nestedness Examples	$Nes_{BS} = \frac{\lvert b-c\rvert}{2a+b+c} \times \frac{a}{a+\min(b,c)}$ (2/16) × [4/(4 + 3)] = 0.071	$Nes_{BS} = \frac{\lvert B-C\rvert}{2A+B+C} \times \frac{A}{A+\min(B,C)}$ Nes_{BR} = (34/190) × [41/(41 + 37)] = 0.094

One can easily get confused by all these indices. In his 2014 paper, Legendre put things in order and shed some light into the resemblances and differences among these coefficients. The conclusions of this work can be summarized as follows.

1. It is the numerators of the proposed indices that estimate (1) replacement and (2) either richness difference or nestedness. The denominators proposed by the various authors can all be applied to scale the indices. However, the choice of a denominator influences the positioning of the sites in an ordination based on these quantities, when compared to the use of the numerators only.
2. The indices of the Podani family measure replacement and richness (or abundance) difference. Those in the Baselga family are replacement (sometimes called turnover) and nestedness indices. Richness difference is not equal to nestedness. See Sect. 8.4.3.3 below.
3. In the two families, replacement and richness difference (Podani) or replacement and nestedness (Baselga) sum to four dissimilarity measures that are appropriate for beta diversity assessment, following the criteria of Legendre and De Cáceres (2013) (see Sect. 8.4.2.3).
4. The replacement and richness difference indices of the Podani family are easy to interpret in ecological terms, as well as the replacement indices in the Baselga family. Baselga's nestedness indices are less obvious, but logical nevertheless.
5. Matrices of indices can be used to produce ordinations of the sites through principal coordinate analysis (PCoA, Sect. 5.5). In this context, the Podani indices of richness or abundance difference in the Sørensen group are particularly well adapted since they are Euclidean. See Legendre (2014, Table S1.4).
6. There have been some discussions about the overestimation of species replacement by Baselga's index $Repl_{BS}$ (Carvalho et al. 2012) and, conversely, of underestimation by Podani's $Repl_J$ (Baselga 2012). These properties result from the denominators used to scale the indices. In the opinion of Legendre (2014), "*one can* [...] *scale the indices* [...] *with denominators of one's choice, depending on the purpose of the study*".

8.4.3.3 The Flavours of Nestedness

While the concepts of species (or abundance) replacement and richness (or abundance) difference are easy to understand, nestedness is more elusive and its definition and measurement have caused some controversies in the literature; see Legendre (2014) for details. Here we show several examples with various numbers of common and unique species to highlight and compare the behaviour of the Podani and Baselga nestedness indices.

The examples, which are constructed on the indices derived from the Jaccard dissimilarity, are presented in Table 8.6. They range from the case of complete dissimilarity (no species in common) to complete similarity (all species in common), going through different proportions of unique species. The formulae used to compute the nestedness values below are the ones presented above:

$$\text{Podani: } N = \frac{a + |b - c|}{a + b + c} \text{ if } a > 0 \text{ and } N = 0 \text{ if } a = 0$$

$$\text{Baselga: } Nes_{\mathrm{BJ}} = \frac{|b - c|}{a + b + c} \times \frac{a}{a + 2\min(b, c)}$$

On the basis of Table 8.6, let us examine the behaviour of the Podani and Baselga indices of nestedness from selected points of view.

Minimum value – Both indices equal 0 when $a = 0$ (case 1 in Table 8.6). This is a logical basic property since a group of species cannot be a subset of another if the sites are fully different in their species contents. Note that the Baselga formula incorporates this property (by means of the second term with a as a numerator) while the Podani index includes it as an exception to the equation. Baselga's nestedness Nes_{BJ} is also equal to 0 when $b = c$ (cases 2 and 9), which means that in this author's definition, nestedness can only exist when there is a difference in richness between the sites. Podani's nestedness index, on the other hand, produces a value equal to the Jaccard similarity S_7 for two sites that have equal richness. $Repl_{\mathrm{J}}$ and Podani's nestedness sum to 1 when $b = c$ (case 2).

Maximum value – The scaled Podani's index can reach a value of 1; Baselga's index culminates at Jaccard's dissimilarity D_{J} when b or c is 0 (case 8). For both indices the maximum value is attained when there are common species ($a > 0$) and either b or c is zero, and this irrespective of the number of unique species. An important difference is that when complete similarity is attained (case 9, where $b = c = 0$) Podani's nestedness index also has its maximum value, while Baselga's index is equal to 0. This highlights a major difference between the two definitions of nestedness: Podani and Schmera (2011) propose that a is a major component of nestedness, to the point of considering two fully similar sites as completely nested. Baselga, on the contrary, puts more emphasis on the richness difference, i.e., $|b - c|$, with only a modest contribution of a. Moreover, for fixed b and c, Podani's nestedness increases monotonically with a, whereas Baselga's nestedness increases to some maximum and then decreases (cases 3–4-5).

Table 8.6 Example cases for the comparison of the Podani and Schmera (2011) nestedness index N and Baselga's Nes_{BJ} index

Case	a	b	c	Podani nestedness	Baselga nestedness	Jaccard dissimilarity
1	0	5	4	0.000	0.000	1.000
2	1	4	4	0.111	0.000	0.889
3	2	3	2	0.429	0.048	0.714
4	4	3	2	0.556	0.056	0.556
5	8	3	2	0.692	0.051	0.385
6	8	5	4	0.529	0.029	0.529
7	2	100	1	0.981	0.481	0.981
8	4	5	0	1.000	0.556	0.556
9	9	0	0	1.000	0.000	0.000

Note also that Podani's index's fixed upper bound (1) is easily attained, while Baselga's upper bound $D_J = (1 - S_7)$ can never reach 1. Indeed, a Jaccard dissimilarity of 1 corresponds to the case where no common species are found, and this, by definition, leads to a Baselga nestedness $Nes_{BJ} = 0$.

Nestedness = Jaccard dissimilarity – As shown above, Baselga's nestedness is equal to the Jaccard dissimilarity D_J when $a > 0$ and either b or $c = 0$ (case 8). Podani's index is equal to the Jaccard dissimilarity when $a = b + c - |b - c|$ (cases 4, 6 and 7), which means that the larger the difference between b and c is, the smaller a has to be for the nestedness to be equal to $D_J = (1 - S_7)$ (case 7). In other words, Podani's nestedness can be very high between two very different sites, not only when a is large.

The following features emerge from these comparisons: (1) Podani and Baselga agree on the fact that nestedness is possible only when there are species in common ($a > 0$). (2) In both indices the maximum value is attained when either b or c is zero. (3) In both indices nestedness increases when $|b - c|$ increases. (4) Podani's index considers the common species a as direct contributors to nestedness, contrary to Baselga's Nes_{BJ}. Consequently, when $b = c = 0$, Podani's index reaches the maximum value of 1, and Baselga's Nes_{BJ} equals zero. (5) The contribution of a to nestedness is clear in Podani's index, but intricate and non monotonic in Baselga's Nes_{BJ}. (6) Podani's nestedness has a fixed and attainable upper bound of 1; the maximum value of Baselga's Nes_{BJ} changes for every pair of sites, being equal to the corresponding $D_J = (1 - S_7)$ dissimilarity except for its maximum value 1.

8.4.3.4 Computing Replacement, Richness Difference and Nestedness Using `beta.div.comp()` of package `adespatial`

Our practical application is based on the Doubs fish river data, like the one presented in Legendre (2014). The first step is to compute the dissimilarity, replacement (or turnover) and richness or abundance difference (or nestedness) matrices. We will do this for Podani's Jaccard-based indices, and then we will extract and plot some results for further examination.

```
# Jaccard-based Podani indices (presence-absence data)
fish.pod.j <- beta.div.comp(spe, coef = "J", quant = FALSE)
# What is in the output object?
summary(fish.pod.j)
# Display summary statistics:
fish.pod.j$part
```

The output object produced by **`beta.div.comp()`** is a list containing three matrices of class "dist": the replacement (`$repl`), the richness/abundance difference (or nestedness in the case of Baselga indices) (`$rich`) and the chosen dissimilarity matrix (Jaccard in the example above) (`$D`). Furthermore, the output object contains a vector (`$part`) with the following global results: (1) BD_{Total},

(2) total replacement diversity, (3) total richness diversity (or nestedness), (4) total replacement diversity/ BD_{Total}, (5) total richness diversity (or nestedness)/ BD_{Total}. A final item (`$note`) gives the name of the dissimilarity coefficient.

In addition to **`beta.div.comp()`**, function **`LCBD.comp()`** found in **`adespatial`** allows users to compute the total variance (*BDtotal*) associated with any dissimilarity matrix with class "dist", or any similar matrix representing an additive decomposition of *BDtotal* into *Repl*, *RichDiff*, *Nes* or other components, computed by other **R** functions. The function also computes the LCBD indices derived from that matrix.

The diversities associated to the Jaccard *D* matrix and its $Repl_J$ and $RichDiff_J$ components are computed on square-rooted values because the Jaccard dissimilarity is not Euclidean but its square-root is. Calculation of total diversities is done by computing SS_{Total} and BD_{Total} from a *D* matrix, as in Eqs. 8.13 and 8.11.

In our example, the vector of global results gives the following numbers:

```
[1] 0.3258676 0.0925460 0.2333216 0.2839988 0.7160012
```

It is easy to verify that the sum of the second and third values = (rounded) 0.09 + 0.23 = 0.32, i.e., that total replacement diversity and total richness diversity sum to BD_{Total}. In this example, the total richness diversity accounts for the larger proportion (71.6%) of BD_{Total}.

In Legendre's case study, the last site (site 30) acts as a reference for all plots because the fish necessarily colonized the river from its lower part. We will study Podani's Jaccard-based richness difference ($RichDiff_J$ in Table 8.4) on that basis. Site 30 is rich in species (see Sect. 8.2.3). Therefore, if the various species did reach different points along the river, richness difference is expected to increase upstream from site 30. Is this increase constant or did some events or special conditions along the stream produce another patterns?

To produce the plot, we must first extract the appropriate values from the richness difference matrix and create a vector of site numbers. The result is presented in Fig. 8.6.

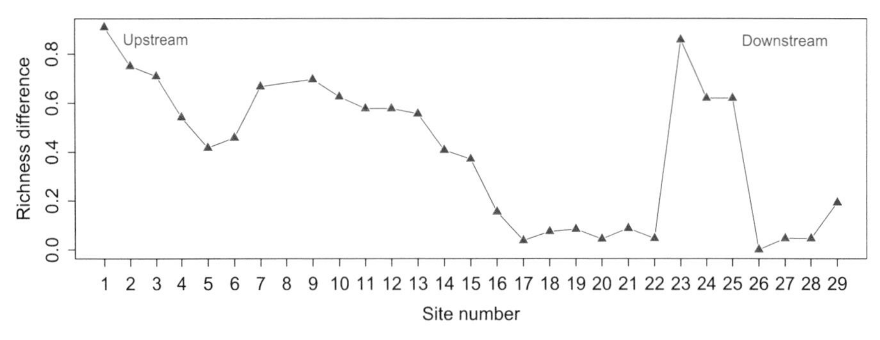

Fig. 8.6 Doubs fish data: richness difference of sites 1–29 (excluding 8) with respect to site 30. Jaccard-based Podani index $RichDiff_J$

```
# Extraction of the richness difference matrix
fish.rich <- as.matrix(fish.pod.j$rich)
# Plot of the richness difference with respect to site 30
fish.rich.30 <- fish.rich[29, ][-29]
site.names <- seq(1, 29)[-8]
plot(site.names,
  fish.rich.30,
  type = "n",
  xaxp = c(1, 29, 28),
  main = "Doubs fish data: richness difference with respect
          to site 30",
  xlab = "Site number",
  ylab = "Richness difference"
)
lines(site.names, fish.rich.30, pch = 24, col="red")
points(
  site.names,
  fish.rich.30,
  pch = 24,
  cex = 1.5,
  col = "white",
  bg = "red"
)
text(3, 0.85, "Upstream", cex = 1, col = "red")
text(27, 0.85, "Downstream", cex = 1, col = "red")
```

Hint *Observe how the richness difference between sites 1 to 29 and site 30 has been obtained: the richness difference object of class "dist" is first transformed into a square matrix (*`fish.rich`*) and its last row is extracted, excluding the last value (richness difference of site 30 with itself; remember that site 8, where no fish had been captured, has been removed from the data set).*

Figure 8.6 shows that upstream from site 30 there is little richness difference up to site 17, with the notable exception of sites 23, 24 and 25. In the mid twentieth century, these three sites were heavily polluted (urban pollution). Since more species are found upstream from these sites, one can deduce that the species pool has been depleted in these three sites due to major anthropic perturbation. Upstream from site 17, richness difference steadily increases to site 9, then drops at sites 6, 5 and 4, then increases again up to site 1, where it is the highest. For some reasons (probably pollution downstream of the city of Pontarlier at the time of sampling in the 1960's), species richness is quite low in sites 7–13 when compared to the other sites in the upper course of the river.

We will now plot together Podani's Jaccard-based replacement ($Repl_J$) and richness difference ($RichDiff_J$) indices as well as the Jaccard dissimilarity. Contrary to Legendre's example, however, we will plot the values between neighbouring sites, not with respect to site 30 (Fig. 8.7), to emphasize local (pairwise) features.

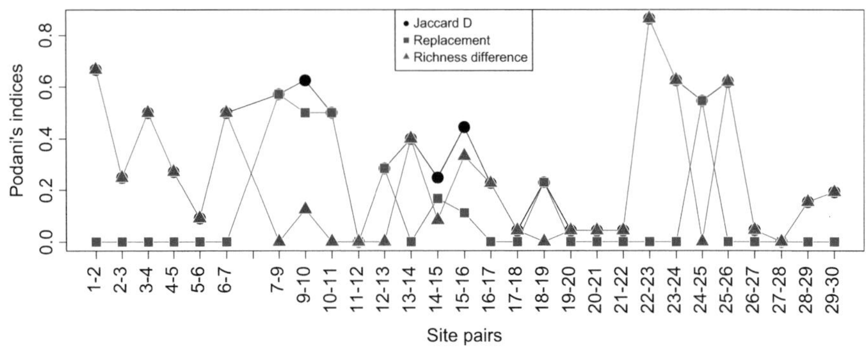

Fig. 8.7 Doubs fish data: Podani's replacement $Repl_J$ and richness difference $RichDiff_J$ indices, and Jaccard dissimilarity $D_J = (1 - S_7)$, computed between neighbouring sites. At each pair of sites $Repl_J + RichDiff_J = D_J$. The Jaccard D values symbols are often hidden by the points representing one of the components

```
# Extraction of the replacement matrix
fish.repl <- as.matrix(fish.pod.j$repl)
# Extraction of the Jaccard dissimilarity DJ matrix
fish.jac <- as.matrix(fish.pod.j$D)
# Plot of the Jaccard, replacement and richness difference indices
# between nearest neighbours
# First, extract the subdiagonals of the square dissimilarity
# matrices
fish.repl.neigh <- diag(fish.repl[-1, ]) # Replacement
fish.rich.neigh <- diag(fish.rich[-1, ]) # Richness difference
fish.jac.neigh <- diag(fish.jac[-1, ]) # Jaccard DJ index
label.pairs = c("1-2", "2-3", "3-4", "4-5", "5-6", "6-7", " ",
  "7-9", "9-10", "10-11", "11-12", "12-13", "13-14", "14-15",
  "15-16", "16-17", "17-18", "18-19", "19-20", "20-21", "21-22",
  "22-23", "23-24", "24-25", "25-26", "26-27", "27-28", "28-29",
  "29-30")
plot(
  absc,
  fish.jac.neigh,
  type = "n",
  xaxt = "n",
  main = "Replacement - Richness difference - Jaccard -
          nearest neighbours",
  xlab = "Site pairs",
  ylab = "Podani's indices"
)
axis(side = 1, 2:30, labels = label.pairs, las = 2, cex.axis = 0.9)
lines(absc, fish.jac.neigh, col = "black")
```

```
points(
  absc,
  fish.jac.neigh,
  pch = 21,
  cex = 2,
  col = "black",
  bg = "black"
)
lines(absc, fish.repl.neigh, col = "blue")
points(
  absc,
  fish.repl.neigh,
  pch = 22,
  cex = 2,
  col = "white",
  bg = "blue"
)
lines(absc, fish.rich.neigh, col = "red")
points(
  absc,
  fish.rich.neigh,
  pch = 24,
  cex = 2,
  col = "white",
  bg = "red"
)
legend(
  "top",
  c("Jaccard D", "Replacement", "Richness difference"),
  pch = c(16, 15, 17),
  col = c("black", "blue", "red")
)
```

Hint *Observe how the replacement, richness and Jaccard dissimilarity indices between* neighbouring *sites have been obtained: the object of class "dist" is first transformed into a square matrix (e.g.* `fish.repl`*), its first row is removed and the diagonal of the remaining matrix (i.e., the relevant part of the subdiagonal of the original matrix) is extracted.*

Of course, this figure is completely different from Fig. 2a of Legendre (2014). Here we compare each site with the previous one, which gives very different information. One can observe that from site 1 to 7, the dissimilarities are due exclusively to richness differences. Species replacement occurs mainly between neighbouring sites in the intermediate section of the river. Note also the large richness difference between sites 22-23, 23-24 and 25-26, due to the drop in species

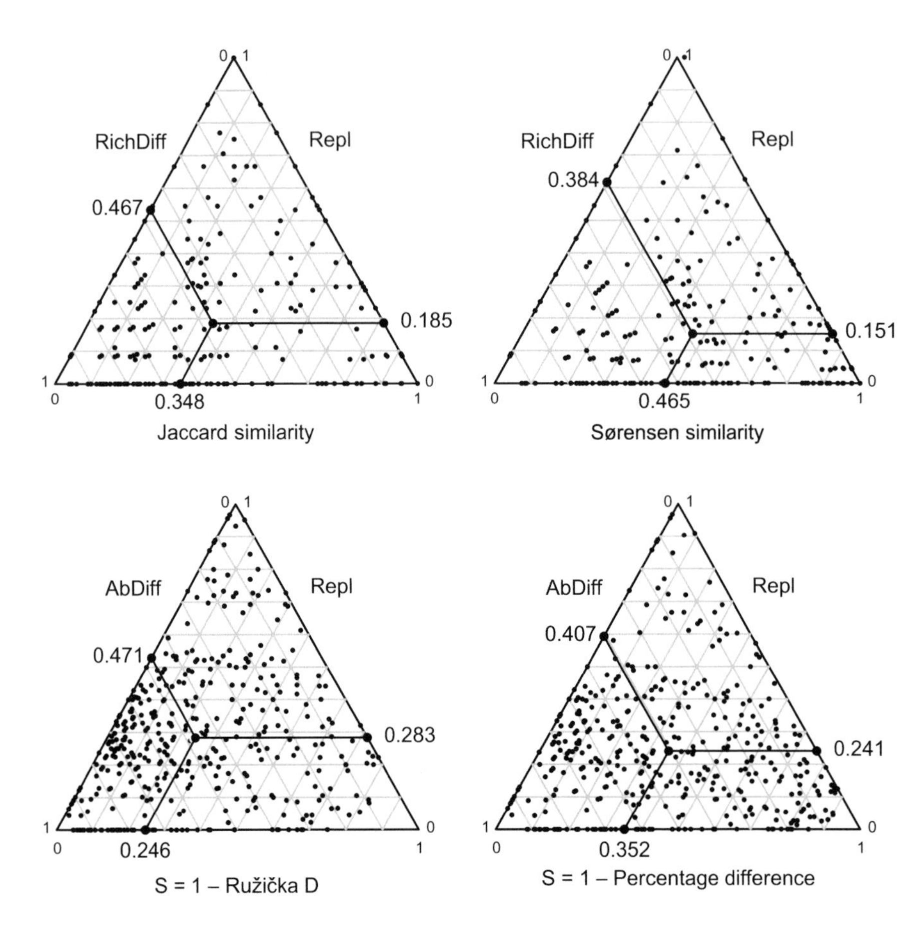

Fig. 8.8 Triangle plots of all pairs of sites in the Doubs fish data set, based on the Podani indices of richness or abundance difference, replacement, and corresponding *similarities*. In each plot, the mean values of the indices, as well as the position of a "mean pair of sites", are represented by larger black dots

richness in sites 23, 24 and 25. Interestingly, in all but three pairs of sites (9-10, 14-15 and 15-16), either richness difference or replacement accounts for the whole dissimilarity value.

Another interesting way of representing the richness difference and replacement values along with the corresponding *similarity* index is a triangular plot or simplex (Legendre 2014, Fig. 4), using the function **`triangle.plot()`** of package **`ade4`**. In a triangle plot, each point is a pair of sites represented by a triplet of values: $S = (1 - D)$, *Repl,* and *RichDiff* or *AbDiff*. In Fig. 8.8, we have added the means of these three quantities onto each plot. The four plots represent the Podani indices and corresponding similarities for presence-absence and quantitative data.

```
# Jaccard
fish.pod.J <- beta.div.comp(spe, coef = "J", quant = FALSE)
# Sorensen
fish.pod.S <- beta.div.comp(spe, coef = "S", quant = FALSE)
# Ruzicka
fish.pod.qJ <- beta.div.comp(spe, coef = "J", quant = TRUE)
# Percentage difference
fish.pod.qS <- beta.div.comp(spe, coef = "S", quant = TRUE)
# Data frames for the triangular plots
fish.pod.J.3 <- cbind((1-fish.pod.J$D),
                      fish.pod.J$repl,
                      fish.pod.J$rich)
colnames(fish.pod.J.3) <- c("Similarity", "Repl", "RichDiff")
fish.pod.S.3 <- cbind((1-fish.pod.S$D),
                      fish.pod.S$repl,
                      fish.pod.S$rich)
colnames(fish.pod.S.3) <- c("Similarity", "Repl", "RichDiff")
fish.pod.qJ.3 <- cbind((1-fish.pod.qJ$D),
                      fish.pod.qJ$repl,
                      fish.pod.qJ$rich)
colnames(fish.pod.qJ.3) <- c("Similarity", "Repl", "AbDiff")
fish.pod.qS.3 <- cbind((1-fish.pod.qS$D),
                      fish.pod.qS$repl,
                      fish.pod.qS$rich)
colnames(fish.pod.qS.3) <- c("Similarity", "Repl", "AbDiff")

par(mfrow = c(2, 2))
triangle.plot(as.data.frame(fish.pod.J.3[, c(3, 1, 2)]),
  show = FALSE,
  labeltriangle = FALSE,
  addmean = TRUE
)
text(-0.45, 0.5, "RichDiff", cex = 1.5)
text(0.4, 0.5, "Repl", cex = 1.5)
text(0, -0.6, "Jaccard similarity", cex = 1.5)
triangle.plot(as.data.frame(fish.pod.S.3[, c(3 ,1 ,2)]),
  show = FALSE,
  labeltriangle = FALSE,
  addmean = TRUE
)
text(-0.45, 0.5, "RichDiff", cex = 1.5)
text(0.4, 0.5, "Repl", cex = 1.5)
text(0, -0.6, "Sørensen similarity", cex = 1.5)
```

```
triangle.plot(as.data.frame(fish.pod.qJ.3[, c(3, 1, 2)]),
  show = FALSE,
  labeltriangle = FALSE,
  addmean = TRUE
)
text(-0.45, 0.5, "AbDiff", cex = 1.5)
text(0.4, 0.5, "Repl", cex = 1.5)
text(0, -0.6, "S = 1 - Ružička D", cex = 1.5)
triangle.plot(as.data.frame(fish.pod.qS.3[, c(3, 1, 2)]),
  show = FALSE,
  labeltriangle = FALSE,
  addmean = TRUE
)
text(-0.45, 0.5, "AbDiff", cex = 1.5)
text(0.4, 0.5, "Repl", cex = 1.5)
text(0, -0.6, "S = 1 - Percentage difference", cex = 1.5)
# Display values of the mean points in the triangular plots
colMeans(fish.pod.J.3[, c(3, 1, 2)])
colMeans(fish.pod.S.3[, c(3, 1, 2)])
colMeans(fish.pod.qJ.3[, c(3, 1, 2)])
colMeans(fish.pod.qS.3[, c(3, 1, 2)])
```

8.4.3.5 Explaining Replacement and Richness Difference

Is there a link between replacement or richness difference and the environmental variables? These questions can be addressed by means of canonical ordination. Indeed, replacement and richness difference matrices can be submitted, together with the environmental variables, to distance-based RDA (db-RDA, Sect. 6.3.3). Let us apply this method by using function **dbrda()** of package **vegan**.

```
# Replacement
repl.dbrda <- dbrda(fish.repl ~ ., data = env , add = "cailliez")
anova(repl.dbrda)
RsquareAdj(repl.dbrda)

# Richness difference
rich.dbrda <- dbrda(fish.rich ~ ., data = env , add = "cailliez")
anova(rich.dbrda)
RsquareAdj(rich.dbrda)
plot(
  rich.dbrda,
  scaling = 1,
  display = c("lc", "cn"),
  main = "Richness difference explained by environmental variables"
)
```

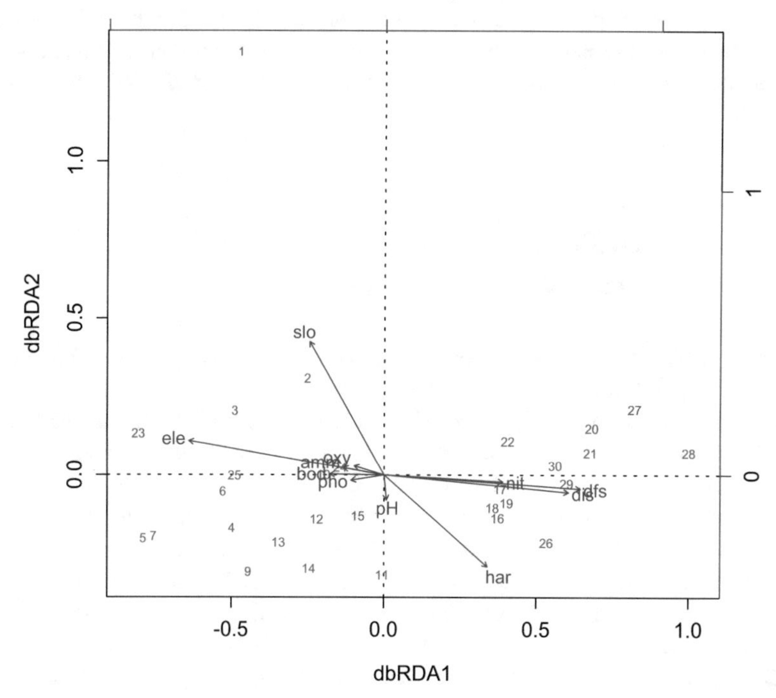

Fig. 8.9 Distance-based redundancy analyses of the matrix of richness difference $RichDiff_J$ explained by the environmental variables. Scaling 1

The first db-RDA result (replacement) shows a very weak relationship of the replacement matrix with the environmental variables ($R^2_{adj} = 0.038$, $p = 0.007$). It is not advisable to apply this analysis to strongly non-Euclidean matrices, as it is the case for most *Repl* matrices. We will refrain from drawing a biplot, which might give a false impression of containing useful information. On the other hand, richness difference is well linked to our environmental variables ($R^2_{adj} = 0.168$ $p = 0.001$). A biplot of the two first axes (Fig. 8.9) confirms that large richness differences are mainly found among the higher elevation sites, and small richness difference among the sites in the lower part of the river, which are those where the discharge is the largest and the nitrogen content is also large[1].

[1]Note that a matrix of richness difference in the Podani family and the Sørensen group would have been fully Euclidean (Legendre 2014, Appendix S1, Table S1.4). Readers could apply this analysis to such a matrix and discover if the R^2_{adj} is larger. Legendre (2014, Fig. S6.2b) shows an example of PCoA ordination of a $RichDiff_S$ matrix.

The analyses presented in Sects. 8.4.3.4 and 8.4.3.5 have all been conducted with the Jaccard-based Podani replacement and richness difference indices. We suggest readers to repeat the exercise with the corresponding quantitative Ružička-based indices, with the Sørensen-percentage difference indices, and also with the Baselga family of indices.

8.5 Functional Diversity, Functional Composition and Phylogenetic Diversity of Communities

Although we focus in this chapter on taxonomic (species) diversity, we cannot end it without briefly introducing what has become a popular research field in community ecology and functional ecology. *Functional* diversity is, with *taxonomic* and *phylogenetic* diversities, one of the three facets of biodiversity at the community level.

Here, we shall restrict ourselves to the principles and applications of two recent approaches that are widely used nowadays in functional community ecology. The first approach is a flexible distance-based framework to compute alpha functional diversity indices from multiple traits (Villéger et al. 2008, Laliberté and Legendre 2010). The second one is part of the unified framework that we mentioned in Sect. 8.4.1 about the partitioning of diversity into alpha, beta and gamma components (de Bello et al. 2010), and is based on Rao's quadratic entropy (Botta-Dukát 2005).

8.5.1 Alpha Functional Diversity

Using a distance-based approach, Villéger et al. (2008) proposed to distinguish three independent components in the functional diversity of a given community: *functional richness* (FRic) represents the amount of functional space filled by a community, i.e. the volume of the minimum convex hull that includes all species present in the multidimensional space of functional traits; *functional evenness* (FEve) measures the regularity of the abundance distribution of the species along the minimum spanning tree that links the species points in multidimensional functional space; *functional divergence* (FDiv) relates to how species abundances are distributed within the functional trait space. FEve and FDiv are constrained between 0 and 1. The functional trait space may be either the raw species trait matrix if all traits are quantitative and continuous, or the space obtained from a PCoA of the Gower dissimilarity matrix computed on the mixed trait matrix.

Another distance-based multi-trait diversity index, called *functional dispersion* (FDis), has been added to the framework by Laliberté and Legendre (2010). It is the mean distance of individual species to the centroid of all species in the multidimensional trait space, taking into account the species relative abundances. These authors implemented all these functional indices in the **R** package **FD**, together with Rao quadratic entropy, community-weighted means of trait values (CWMs) and functional group richness (FGR).

The Rao quadratic entropy has been advocated as a simple and universal diversity index (see Sect. 8.5.2) and it is strongly related to functional dispersion. CWMs are not strictly diversity indices but represent altogether the *functional composition* of the community[2]. FGR is the number of functional groups in a given community, these groups being defined from a clustering of species based on their functional traits.

We cannot develop these theoretical explanations here and we invite the interested reader to refer to the cited articles. Let us now apply this distance-based framework to the fish data. For this, we need two data frames, one containing species abundances at each site (`spe`) and the other the functional traits of each species (`tra`).

```
summary(fishtraits)
rownames(fishtraits)
names(spe)
names(fishtraits)
tra <- fishtraits [ , 6:15]
tra
```

The trait data frame contains four quantitative variables and six binary variables describing the diet. Values are taken from various sources, mainly fishbase.org (Froese and Pauly 2017). Quantitative functional traits are: `BodyLength` (the average total body length of the adult fish), `BodyLengthMax` (the maximum total body length of the adult fish), `ShapeFactor` (the ratio of total body length to maximum body height) and `TrophicLevel` (the relative positions of the species in the trophic chains).

The **dbFD()** function of the **FD** package will be used to compute all distance-based functional diversity indices, as well as CWMs and FGR. For FGR, we must specify the number of data-defined functional groups (or the fusion level) after looking at the dendrogram (select at least 5 groups to improve the resulting pattern):

[2]Relating standard CWM matrices based on quantitative traits to environmental variables representing gradients through linear models (correlation, regression or RDA) has been strongly criticized by Peres-Neto et al. (2017). In particular, when only the traits or only the environmental variables are important in structuring the species distributions, tests of correlations based on this approach have strongly inflated type I error. The authors recommend the fourth-corner approach (Sect. 6.11) instead.

```
# Distance-based functional diversity indices
?dbFD
res <-
  dbFD(
    tra,
    as.matrix(spe),
    asym.bin = 5:10,
    stand.FRic = TRUE,
    clust.type = "ward.D",
    CWM.type = "all",
    calc.FGR = TRUE
  )
# g  # cut the dendrogram using the number of groups as criterion
# 10 # choose the number of functional groups
res

# Add these indices to the div data frame
div$FRic <- res$FRic
div$FEve <- res$FEve
div$FDiv <- res$FDiv
div$FDis <- res$FDis
div$RaoQ <- res$RaoQ
div$FGR  <- res$FGR
div
```

Hint *The* `stand.FRic` *argument allows us to standardize FRic by the global FRic that includes all species. As a result, FRic is constrained between 0 and 1.*

To visualize the output, let us plot the diversity indices on the map of the Doubs River (Fig. 8.10; see the code in the accompanying **R** script).

Compare the maps. How do you interpret these patterns?

CWMs were computed by the **dbFD()** function, and they can also be obtained directly using the **functcomp()** function:

```
# Community-weighted means of trait values (CWMs)
functcomp(tra, as.matrix(spe), CWM.type = "all")
```

Hint *Note the argument* `CWM.type`*, which applies to binary or qualitative traits, for which the user can choose between simply returning the dominant class when* `CWM.type` *is* `"dom"` *(e.g. 0 or 1 for diet binary variables) or computing the sum of relative abundances for each individual class when* `CWM.type` *is* `"all"`.

Can you detect any trend in the variation of CWMs along the Doubs River? Plot these mean traits on the map of the river.

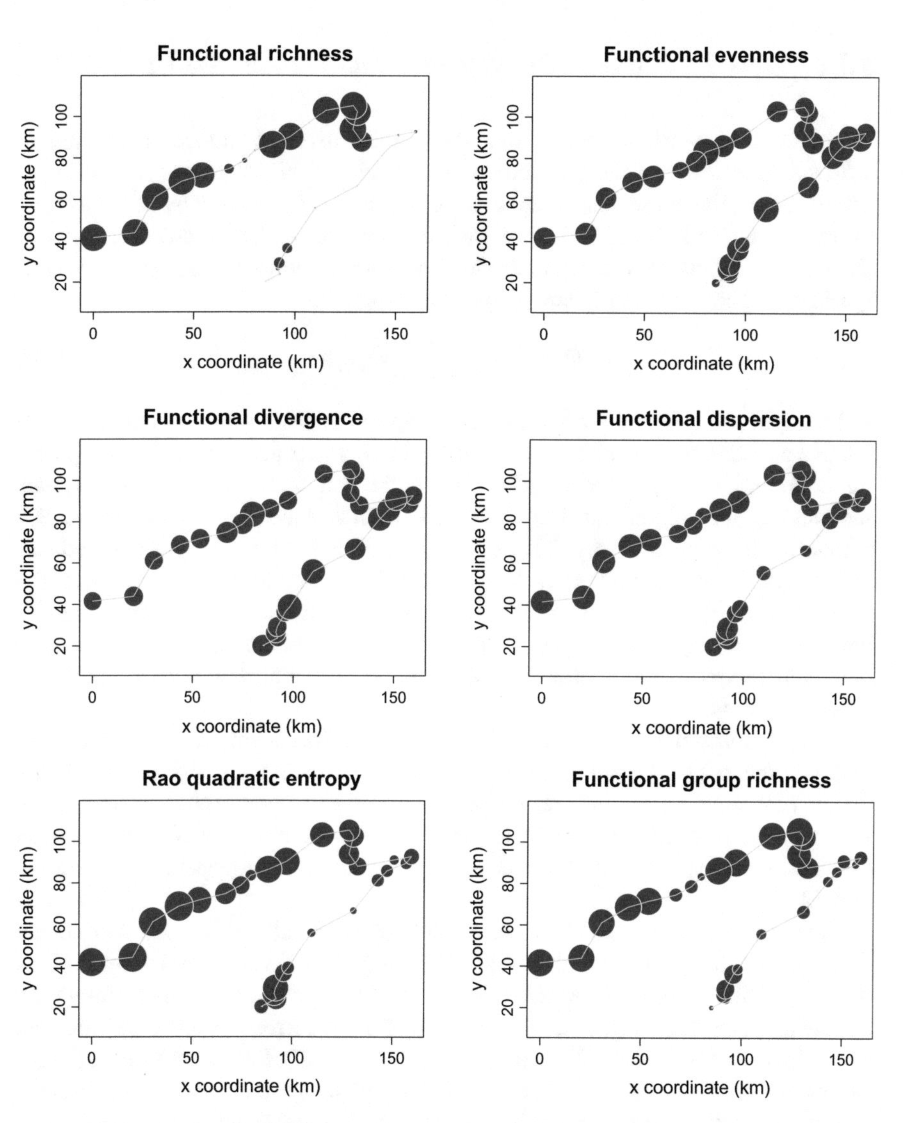

Fig. 8.10 Variation in functional richness (*top left*), evenness (*top right*), divergence (*middle left*), dispersion (*middle right*), Rao quadratic entropy (*bottom left*) and functional group richness (*bottom right*) of the fish communities along the Doubs River

8.5.2 Beta Taxonomic, Phylogenetic and Functional Diversities

A unified framework for computing community diversity indices according to different facets (taxonomic [generally at the species level], phylogenetic and functional) and spatial components (alpha, beta, gamma) was proposed by de Bello et al. (2010). It is based on the "Rao quadratic entropy" or *quadratic diversity index Q* (Rao 1982), computed as the sum of the dissimilarities between all possible pairs of species, weighted by the product of species proportions:

$$Q = \sum_{i=1}^{q} \sum_{j=1}^{q} d_{ij} p_{ic} p_{jc} \quad (8.18)$$

where d_{ij} is the difference (dissimilarity) between species i and species j, p_{ic} the relative abundance of species i in community c, p_{jc} the relative abundance of species j in community c, q the number of species in the community (or in the whole data set), and Q the (alpha) quadratic diversity or Rao index of community c. For taxonomic (species) diversity (TD), $d_{ij} = 1$ for all $i \neq j$ or 0 for $i = j$, and Q reduces to the inverse Simpson diversity N_2 (see Sect. 8.2.2.2). For phylogenetic diversity (PD), differences are pairwise distances between species measured along the branches of a phylogenetic tree (or of a hierarchical classification based on this phylogenetic tree). For functional diversity (FD), differences are dissimilarities in functional traits.

As d_{ij} is constrained to vary from 0 (species i and j are identical) to 1, the inverse Simpson diversity (TD) represents the potential maximum value that the Rao index (PD or FD) could reach if all species were completely different (phylogenetically or functionally, respectively), whereas the difference or ratio between the inverse Simpson diversity and the Rao index is a measure of phylogenetic or functional redundancy.

Spatial components (i.e. partitioning into alpha and beta) of TD, PD and FD are derived from the application of the Rao index to gamma diversity and unbiased beta diversity (either additive or multiplicative). Beta diversity is made unbiased, i.e. independent from alpha diversity, by applying a correction based on numbers equivalents (see Sect. 8.2.2.2), as proposed by Jost (2007). See de Bello et al. (2010) for details. An example is presented to help readers better understand this approach.

Let us apply this framework to the fish data. At first, we might derive a cophenetic distance matrix from a simplified phylogenetic tree of the fish species. The state-of-the-art method would be to reconstruct a phylogenetic tree from appropriate nucleotide sequences, but this would lead us far beyond the scope of this book. For the sake of simplicity, we shall directly retrieve the hierarchical classification (currently accepted taxonomic hierarchy) from the species list using the **`classification()`** function of the **`taxize`** package, and use the topology of this tree to compute the simplified phylogenetic dissimilarity matrix:

```
# Distance matrix based on a simplified phylogenetic classification
# Retrieve hierarchical classification from species list
splist <- as.character(fishtraits$LatinName)
spcla <- classification(splist, db = "gbif")
# Compute the distance matrix and the phylogenetic tree
tr <- class2tree(spcla)
tr$classification
tr$distmat
# Convert the tree to a cophenetic matrix
# constrained between 0 and 1
phylo.d <- cophenetic(tr$phylo) / 100
# Replace full species names by name codes
rownames(phylo.d) <- names(spe)
colnames(phylo.d) <- names(spe)
```

Hint *We used the argument* `db = "gbif"` *to retrieve the current taxonomic classification from the GBIF database (Global Biodiversity Information Facility, www.gbif.org). Other databases are available, such as* `"ncbi"`, *related to the Genbank database.*

We also need a functional dissimilarity matrix of the fish species based on their functional traits. Since the trait matrix contains both quantitative and binary variables, we compute Gower dissimilarities with the **`gowdis()`** function. We can then plot the tree of the simplified phylogenetic classification and the dendrogram of a Ward clustering of the functional dissimilarity matrix (Fig. 8.11).

```
# Functional dissimilarity matrix (Gower dissimilarity)
trait.d <- gowdis(tra, asym.bin = 4:9)

# Plot the tree and the dendrogram
trait.gw <- hclust(trait.d, "ward.D2")
par(mfrow = c(1, 2))
plot(tr)
plot(trait.gw, hang = -1, main = "")
```

We are now ready to run the **`Rao()`** function to compute species (i.e., taxonomic without reference to phylogeny), phylogenetic and functional diversity indices and their decomposition into alpha, beta and gamma components.

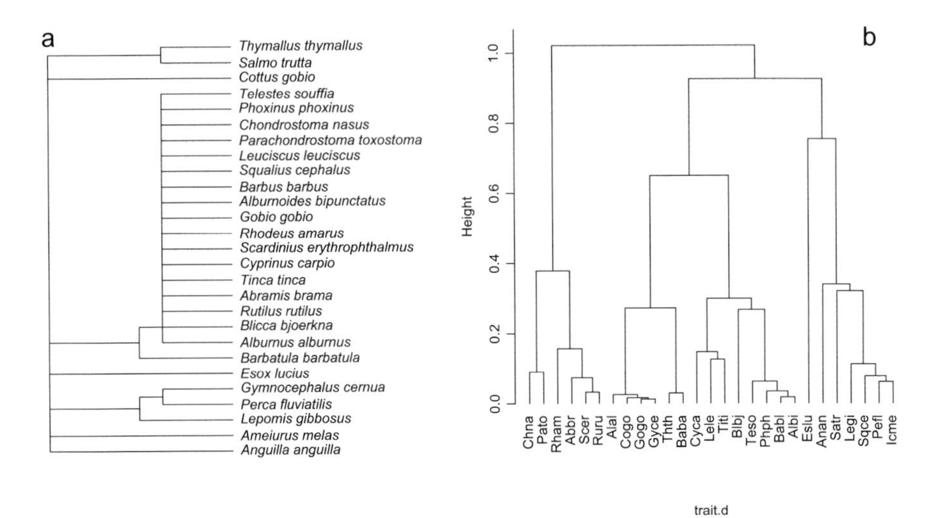

Fig. 8.11 Simplified phylogenetic tree of the fish species (**a**) and Ward hierarchical clustering of the species based on their functional traits (**b**)

```
# Additive partitioning of TD, FD and PD (de Bello et al. 2010)
spe.rao <- Rao(
  sample = t(spe),
  dfunc = trait.d, # optional functional dissimilarity matrix
  dphyl = phylo.d, # optional phylogenetic distance matrix
  weight = FALSE,
  Jost = TRUE,
  structure = NULL
)
names(spe.rao)

# Species diversity (Simpson)
names(spe.rao$TD)
# Mean alpha Simpson diversity
spe.rao$TD$Mean_Alpha
# Gamma Simpson diversity
spe.rao$TD$Gamma
# Additive beta Simpson diversity
spe.rao$TD$Beta_add
spe.rao$TD$Gamma - spe.rao$TD$Mean_Alpha
# Beta diversity expressed as percentage of gamma
spe.rao$TD$Beta_prop
spe.rao$TD$Beta_add / spe.rao$TD$Gamma
# Multiplicative beta Simpson diversity
spe.rao$TD$Gamma / spe.rao$TD$Mean_Alpha
```

```
# Phylogenetic diversity (Rao)
names(spe.rao$PD)
# Mean alpha Rao phylogenetic diversity
spe.rao$PD$Mean_Alpha
# Gamma Rao phylogenetic diversity
spe.rao$PD$Gamma
# Additive beta Rao phylogenetic diversity
spe.rao$PD$Beta_add
spe.rao$PD$Gamma - spe.rao$PD$Mean_Alpha
# Beta phylogenetic diversity expressed as percentage of gamma
spe.rao$PD$Beta_prop
spe.rao$PD$Beta_add / spe.rao$PD$Gamma
# Multiplicative beta Rao phylogenetic diversity
spe.rao$PD$Gamma / spe.rao$PD$Mean_Alpha

# Functional diversity (Rao)
names(spe.rao$FD)
# Mean alpha Rao functional diversity
spe.rao$FD$Mean_Alpha
# Gamma Rao functional diversity
spe.rao$FD$Gamma
# Additive beta Rao functional diversity
spe.rao$FD$Beta_add
spe.rao$FD$Gamma - spe.rao$FD$Mean_Alpha
# Beta functional diversity expressed as percentage of gamma
spe.rao$FD$Beta_prop
spe.rao$FD$Beta_add / spe.rao$FD$Gamma
# Multiplicative beta Rao functional diversity
spe.rao$FD$Gamma / spe.rao$FD$Mean_Alpha

# Variation of alpha TD, FD and PD along the Doubs river
spe.rao$TD$Alpha
spe.rao$PD$Alpha
spe.rao$FD$Alpha

# Add Rao-based diversity indices to the div data frame
div$alphaTD <- spe.rao$TD$Alpha
div$alphaPD <- spe.rao$PD$Alpha
div$alphaFD <- spe.rao$FD$Alpha

# Save the div data frame as a CSV file
write.csv(div, file = "diversity.csv", quote = FALSE)
```

In addition, we can plot the alpha diversity indices of each site on the map of the Doubs River (Fig. 8.12; see the code in the accompanying **R** script):

Compare these spatial patterns with those of Figure 8.10. How do you interpret the difference between taxonomic and phylogenetic diversity patterns?

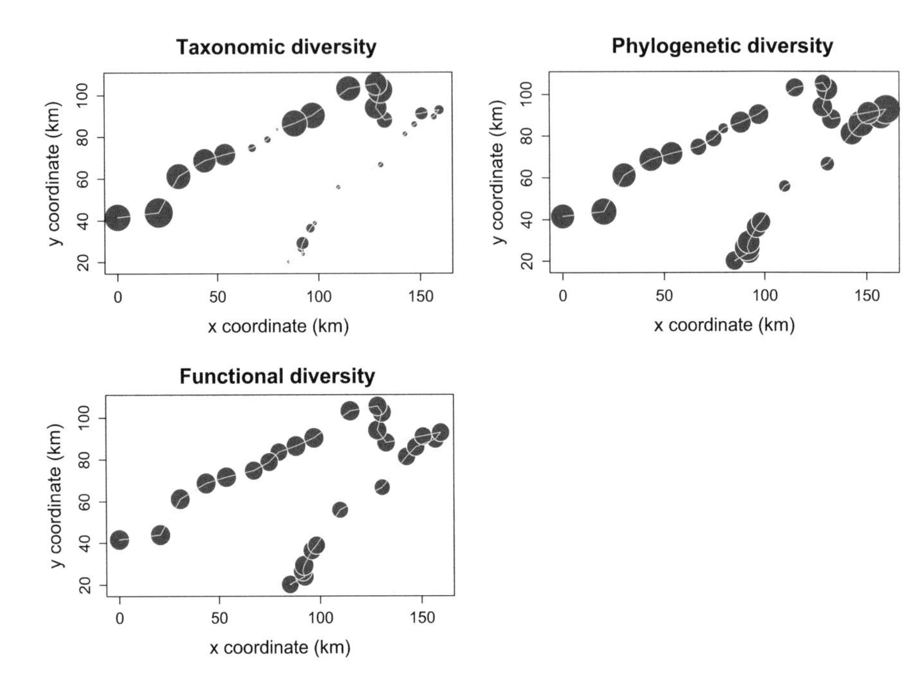

Fig. 8.12 Taxonomic, phylogenetic and functional Rao-based alpha diversity indices of the fish communities along the Doubs River

8.6 Conclusion

The study of ecological diversity in its diverse aspects is one of the most exciting challenges of today's quantitative ecology, especially now that we understand that any method of data analysis that decomposes the total variance of a site-by-species data table is an analysis of beta diversity. The statistical methods studied in Chaps. 5, 6, and 7 of this book decompose that variance into simple or canonical ordination axes, among experimental factors, among matrices of explanatory variables, or among spatial scales.

Numerous questions remain to be resolved about the many approaches to the measurement of diversity (Chap. 8). But the most fruitful outcomes will likely come when we will be able to fully put in relationship the multiscale complexity of genetic, specific, ecological and landscape diversity.

The multifaceted nature of ecological problems can nowadays be addressed in a much deeper way than before, and the authors of the methods are themselves constantly surprised at the range of applications ecologists make of their statistical offspring. Many more developments will certainly be made in the forthcoming years, and we wish to conclude by inviting readers to participate in this effort, both by asking new and challenging ecological questions and by devoting themselves to the exciting task of methodological development.

Bibliography

Abdi, H., Williams, L.J., Valentin, D.: Multiple factor analysis: principal component analysis for multitable and multiblock data sets. WIREs Comput Stat. **5**, 149–179 (2013)

Anderson, M.J.: Distinguishing direct from indirect effects of grazers in intertidal estuarine assemblages. J. Exp. Mar. Biol. Ecol. **234**, 199–218 (1999)

Anderson, M.J.: Distance-based tests for homogeneity of multivariate dispersions. Biometrics. **62**, 245–253 (2006)

Anderson, M.J., Ellingsen, K.E., McArdle, B.H.: Multivariate dispersion as a measure of beta diversity. Ecol. Lett. **9**, 683–693 (2006)

Anderson, M.J., Crist, T.O., Chase, J.M., Vellend, M., Inouye, B.D., Freestone, A.L., Sanders, N.J., Cornell, H.V., Comita, L.S., Davies, K.F., Harrison, S.P., Kraft, N.J.B., Stegen, J.C., Swenson, N.G.: Navigating the multiple meanings of β diversity: a roadmap for the practicing ecologist. Ecol. Lett. **14**, 19–28 (2011)

Baselga, A.: Partitioning the turnover and nestedness components of beta diversity. Glob. Ecol. Biogeogr. **19**, 134–143 (2010)

Baselga, A.: The relationship between species replacement, dissimilarity derived from nestedness, and nestedness. Glob. Ecol. Biogeogr. **21**, 1223–1232 (2012)

Beamud, S.G., Diaz, M.M., Baccala, N.B., Pedrozo, F.L.: Analysis of patterns of vertical and temporal distribution of phytoplankton using multifactorial analysis: acidic Lake Caviahue, Patagonia, Argentina. Limnologica. **40**, 140–147 (2010)

Benjamini, Y., Hochberg, Y.: Controlling the false discovery rate: a practical and powerful approach to multiple testing. J R Stat Soc B. **57**, 289–300 (1995)

Bernier, N., Gillet, F.: Structural relationships among vegetation, soil fauna and humus form in a subalpine forest ecosystem: a Hierarchical Multiple Factor Analysis (HMFA). Pedobiologia. **55**, 321–334 (2012)

Bivand, R.S., Pebesma, E.J., Gomez-Rubio, V.: Applied spatial data analysis with R. In: Use R Series, 2nd edn. Springer, New York (2013)

Blanchet, F.G., Legendre, P., Borcard, D.: Forward selection of explanatory variables. Ecology. **89**, 2623–2632 (2008a)

Blanchet, F.G., Legendre, P., Borcard, D.: Modelling directional spatial processes in ecological data. Ecol. Model. **215**, 325–336 (2008b)

Borcard, D., Legendre, P.: Environmental control and spatial structure in ecological communities: an example using Oribatid mites (Acari, Oribatei). Environ. Ecol. Stat. **1**, 37–61 (1994)

Blanchet, F. G., Legendre, P, Maranger, R., Monti, D., Pepin, P.: Modelling the effect of directional spatial ecological processes at different scales. Oecologia. **166**, 357–368 (2011)

D. Borcard et al., *Numerical Ecology with R*, Use R!,
https://doi.org/10.1007/978-3-319-71404-2

Borcard, D., Legendre, P.: All-scale spatial analysis of ecological data by means of principal coordinates of neighbour matrices. Ecol. Model. **153**, 51–68 (2002)

Borcard, D., Legendre, P.: Is the Mantel correlogram powerful enough to be useful in ecological analysis? A simulation study. Ecology. **93**, 1473–1481 (2012)

Borcard, D., Legendre, P., Drapeau, P.: Partialling out the spatial component of ecological variation. Ecology. **73**, 1045–1055 (1992)

Borcard, D., Legendre, P., Avois-Jacquet, C., Tuomisto, H.: Dissecting the spatial structure of ecological data at multiple scales. Ecology. **85**, 1826–1832 (2004)

Borcard, D., Gillet, F., Legendre, P.: Numerical Ecology with R. UseR! Series. Springer, New York (2011)

Borthagaray, A.I., Arim, M., Marquet, P.A.: Inferring species roles in metacommunity structure from species co-occurrence networks. Proc. R. Soc. B. **281**, 20141425 (2014)

Botta-Dukát, Z.: Rao's quadratic entropy as a measure of functional diversity based on multiple traits. J. Veg. Sci. **16**, 533–540 (2005)

Breiman, L., Friedman, J.H., Olshen, R.A., Stone, C.G.: Classification and Regression Trees. Wadsworth International Group, Belmont (1984)

Cadotte, M.W., Davies, T.J.: Phylogenies in Ecology: A Guide to Concepts and Methods. Princeton University Press, Princeton (2016)

Carlson, M.L., Flagstad, L.A., Gillet, F., Mitchell, E.A.D.: Community development along a proglacial chronosequence: are above-ground and below-ground community structure controlled more by biotic than abiotic factors? J. Ecol. **98**, 1084–1095 (2010)

Carvalho, J.C., Cardoso, P., Gomes, P.: Determining the relative roles of species replacement and species richness differences in generating beta-diversity patterns. Glob. Ecol. Biogeogr. **21**, 760–771 (2012)

Chessel, D., Lebreton, J.D., Yoccoz, N.: Propriétés de l'analyse canonique des correspondances; une illustration en hydrobiologie. Revue de Statistique Appliquée. **35**, 55–72 (1987)

Choler, P.: Consistent shifts in Alpine plant traits along a mesotopographical gradient. Arct. Antarct. Alp. Res. **37**, 444–453 (2005)

Chytrý, M., Tichy, L., Holt, J., Botta-Duka, Z.: Determination of diagnostic species with statistical fidelity measures. J. Veg. Sci. **13**, 79–90 (2002)

Clua, E., Buray, N., Legendre, P., Mourier, J., Planes, S.: Behavioural response of sicklefin lemon sharks *Negaprion acutidens* to underwater feeding for ecotourism purposes. Mar. Ecol. Prog. Ser. **414**, 257–266 (2010)

Davé, R.N., Krishnapuram, R.: Robust clustering methods: a unified view. IEEE Trans. Fuzzy Syst. **5**, 270–293 (1997)

de Bello, F., Lavergne, S., Meynard, C., Lepš, J., Thuiller, W.: The partitioning of diversity: showing Theseus a way out of the labyrinth. J. Veg. Sci. **21**, 992–1000 (2010)

De Cáceres, M., Legendre, P.: Associations between species and groups of sites: indices and statistical inference. Ecology. **90**, 3566–3574 (2009)

De Cáceres, M., Font, X., Oliva, F.: The management of numerical vegetation classifications with fuzzy clustering methods. J. Veg. Sci. **21**, 1138–1151 (2010)

De'ath, G.: Multivariate regression trees: a new technique for modeling species-environment relationships. Ecology. **83**, 1105–1117 (2002)

Declerck, S.A.J., Coronel, J.S., Legendre, P., Brendonck, L.: Scale dependency of processes structuring metacommunities of cladocerans in temporary pools of High-Andes wetlands. Ecography. **34**, 296–305 (2011)

Dolédec, S., Chessel, D.: Co-inertia analysis: an alternative method to study species-environment relationships. Freshw. Biol. **31**, 277–294 (1994)

Doledec, S., Chessel, D., ter Braak, C.J.F., Champely, S.: Matching species traits to environmental variables: a new three-table ordination method. Environ. Ecol. Stat. **3**, 143–166 (1996)

Dray, S., Legendre, P.: Testing the species traits-environment relationships: the fourth-corner problem revisited. Ecology. **89**, 3400–3412 (2008)

Dray, S., Pettorelli, N., Chessel, D.: Matching data sets from two different spatial samplings. J. Veg. Sci. **13**, 867–874 (2002)

Dray, S., Chessel, D., Thioulouse, J.: Co-inertia analysis and the linking of ecological data tables. Ecology. **84**, 3078–3089 (2003)
Dray, S., Legendre, P., Peres-Neto, P.R.: Spatial modelling: a comprehensive framework for principal coordinate analysis of neighbour matrices (PCNM). Ecol. Model. **196**, 483–493 (2006)
Dray, S., Choler, P., Doledec, S., Peres-Neto, P.R., Thuiller, W., Pavoine, S., ter Braak, C.J.F.: Combining the fourth-corner and the RLQ methods for assessing trait responses to environmental variation. Ecology. **95**, 14–21 (2014)
Dufrêne, M., Legendre, P.: Species assemblages and indicator species: the need for a flexible asymmetrical approach. Ecol. Monogr. **67**, 345–366 (1997)
Dungan, J.L., Perry, J.N., Dale, M.R.T., Legendre, P., Citron-Pousty, S., Fortin, M.-J., Jakomulska, A., Miriti, M., Rosenberg, M.S.: A balanced view of scaling in spatial statistical analysis. Ecography. **25**, 626–640 (2002)
Efron, B.: Bootstrap methods: another look at the jackknife. Ann. Stat. **7**, 1–26 (1979)
Efron, B., Halloran, E., Holmes, S.: Bootstrap confidence levels for phylogenetic trees. Proc Nat Acad Sci USA. **93**, 13429–13434 (1996)
Ellison, A.M.: Partitioning diversity. Ecology. **91**, 1962–1963 (2010)
Escofier, B., Pagès, J.: Multiple factor analysis (AFMULT package). Comput Stat Data Anal. **18**, 121–140 (1994)
Escoufier, Y.: The duality diagram: a means of better practical applications. In: Legendre, P., Legendre, L. (eds.) Developments in Numerical Ecology, NATO ASI Series Series, Series G: Ecological Sciences, vol. 14, pp. 139–156. Springer, Berlin (1987)
Ezekiel, M.: Methods of Correlational Analysis. Wiley, New York (1930)
Faith, D.P., Minchin, P.R., Belbin, L.: Compositional dissimilarity as a robust measure of ecological distance. Vegetatio. **69**, 57–68 (1987)
Felsenstein, J.: Confidence limits on phylogenies: an approach using the bootstrap. Evolution. **39**, 783–791 (1985)
Froese, R., Pauly, D. (Eds): FishBase. World Wide Web electronic publication. www.fishbase.org (2017)
Geary, R.C.: The contiguity ratio and statistical mapping. Inc Stat. **5**, 115–145 (1954)
Geffen, E., Anderson, M.J., Wayne, R.K.: Climate and habitat barriers to dispersal in the highly mobile gray wolf. Mol. Ecol. **13**, 2481–2490 (2004)
Gordon, A.D.: Classification in the presence of constraints. Biometrics. **29**, 821–827 (1973)
Gordon, A.D., Birks, H.J.B.: Numerical methods in quaternary palaeoecology. I. Zonation of pollen diagrams. New Phytol. **71**, 961–979 (1972)
Gordon, A.D., Birks, H.J.B.: Numerical methods in quaternary palaeoecology. II. Comparison of pollen diagrams. New Phytol. **73**, 221–249 (1974)
Gower, J.C.: Some distance properties of latent root and vector methods used in multivariate analysis. Biometrika. **53**, 325–338 (1966)
Gower, J.C.: Comparing classifications. In: Felsenstein, J. (ed.) Numerical Taxonomy. NATO ASI Series, vol. G-1, pp. 137–155. Springer, Berlin (1983)
Gower, J.C., Legendre, P.: Metric and Euclidean properties of dissimilarity coefficients. J. Classif. **3**, 5–48 (1986)
Gray, D.K., Arnott, S.E.: Does dispersal limitation impact the recovery of zooplankton communities damaged by a regional stressor? Ecol. Appl. **21**, 1241–1256 (2011)
Greenacre, M., Primicerio, R.: Multivariate Analysis of Ecological Data. Fundación BBVA, Bilbao (2013)
Greenberg, J.H.: The measurement of linguistic diversity. Language. **32**, 109–115 (1956)
Griffith, D.A., Peres-Neto, P.R.: Spatial modeling in ecology: the flexibility of eigenfunction spatial analyses. Ecology. **87**, 2603–2613 (2006)
Grimm, E.C.: CONISS: A FORTRAN 77 program for stratigraphically constrained cluster analysis by the method of incremental sum of squares. Comput. Geosci. **13**, 13–35 (1987)
Guénard, G., Legendre, P., Boisclair, D., Bilodeau, M.: Assessment of scale-dependent correlations between variables. Ecology. **91**, 2952–2964 (2010)

Hardy, O.J.: Testing the spatial phylogenetic structure of local communities: statistical performances of different null models and test statistics on a locally neutral community. J. Ecol. **96**, 914–926 (2008)

Harrison, S., Ross, S.J., Lawton, J.H.: Beta-diversity on geographic gradients in Britain. J. Anim. Ecol. **61**, 151–158 (1992)

Hill, M.O.: Diversity and evenness: a unifying notation and its consequences. Ecology. **54**, 427–432 (1973)

Hill, M.O., Smith, A.J.E.: Principal component analysis of taxonomic data with multi-state discrete characters. Taxon. **25**, 249–255 (1976)

Holm, S.: A simple sequentially rejective multiple test procedure. Scand. J. Stat. **6**, 65–70 (1979)

Hurlbert, S.H.: The non-concept of species diversity: a critique and alternative parameters. Ecology. **52**, 577–586 (1971)

Isaaks, E.H., Srivastava, R.M.: An Introduction to Applied Geostatistics. Oxford University Press, New York (1989)

Jaccard, P.: Étude comparative de la distribution florale dans une portion des Alpes et du Jura. Bulletin de la Société Vaudoise des Sciences Naturelles. **37**, 547–579 (1901)

Jongman, R.H.G., ter Braak, C.J.F., van Tongeren, O.F.R.: Data Analysis in Community and Landscape Ecology. Cambridge University Press, Cambridge (1995)

Josse, J., Husson, F.: Handling missing values in exploratory multivariate data analysis methods. Journal de la Société Française de Statistique. **153**, 79–99 (2012)

Josse, J., Pagès, J., Husson, F.: Testing the significance of the RV coefficient. Comput Stat Data Anal. **53**, 82–91 (2008)

Jost, L.: Entropy and diversity. Oikos. **113**, 363–375 (2006)

Jost, L.: Partitioning diversity into independent alpha and beta components. Ecology. **88**, 2427–2439 (2007)

Kaufman, L., Rousseeuw, P.J.: Finding Groups in Data: An Introduction to Cluster Analysis. Wiley, New York (2005)

Kempton, R.A.: Structure of species abundance and measurement of diversity. Biometrics. **35**, 307–322 (1979)

Laliberté, E., Legendre, P.: A distance-based framework for measuring functional diversity from multiple traits. Ecology. **91**, 299–305 (2010)

Laliberté, E., Paquette, A., Legendre, P., Bouchard, A.: Assessing the scale-specific importance of niches and other spatial processes on beta diversity: a case study from a temperate forest. Oecologia. **159**, 377–388 (2009)

Lamentowicz, M., Lamentowicz, L., van der Knaap, W.O., Gabka, M., Mitchell, E.A.D.: Contrasting species-environment relationships in communities of testate amoebae, bryophytes and vascular plants along the fen-bog gradient. Microb. Ecol. **59**, 499–510 (2010)

Lance, G.N., Williams, W.T.: A generalized sorting strategy for computer classifications. Nature. **212**, 218 (1966)

Lance, G.N., Williams, W.T.: A general theory of classificatory sorting strategies. I. Hierarchical systems. Comput. J. **9**, 373–380 (1967)

Lande, R.: Statistics and partitioning of species diversity, and similarity among multiple communities. Oikos. **76**, 5–13 (1996)

Le Dien, S., Pagès, J.: Analyse factorielle multiple hiérarchique. Revue de statistique appliquée. **51**, 47–73 (2003)

Lear, G., Anderson, M.J., Smith, J.P., Boxen, K., Lewis, G.D.: Spatial and temporal heterogeneity of the bacterial communities in stream epilithic biofilms. FEMS Microbiol. Ecol. **65**, 463–473 (2008)

Legendre, P.: Quantitative methods and biogeographic analysis. In: Garbary, D.J., South, R.R. (eds.) Evolutionary Biology of the Marine Algae of the North Atlantic. NATO ASI Series, vol. G22, pp. 9–34. Springer, Berlin (1990)

Legendre, P.: Spatial autocorrelation: trouble or new paradigm? Ecology. **74**, 1659–1673 (1993)

Legendre, P.: Species associations: the Kendall coefficient of concordance revisited. J. Agric. Biol. Environ. Stat. **10**, 226–245 (2005)

Legendre, P.: Coefficient of concordance. In: Salking, N.J. (ed.) Encyclopedia of Research Design, vol. 1, pp. 164–169. SAGE Publications, Los Angeles (2010)
Legendre, P.: Interpreting the replacement and richness difference components of beta diversity. Glob. Ecol. Biogeogr. **23**, 1324–1334 (2014)
Legendre, P., Anderson, M.J.: Distance-based redundancy analysis: testing multi-species responses in multi-factorial ecological experiments. Ecol. Monogr. **69**, 1–24 (1999)
Legendre, P., Borcard, D.: Box-Cox-chord transformations for community composition data prior to beta diversity analysis. Ecography. **41**, 1–5 (2018)
Legendre, P., De Cáceres, M.: Beta diversity as the variance of community data: dissimilarity coefficients and partitioning. Ecol. Lett. **16**, 951–963 (2013)
Legendre, P., Fortin, M.J.: Comparison of the Mantel test and alternative approaches for detecting complex multivariate relationships in the spatial analysis of genetic data. Mol. Ecol. Resour. **10**, 831–844 (2010)
Legendre, P., Gallagher, E.D.: Ecologically meaningful transformations for ordination of species data. Oecologia. **129**, 271–280 (2001)
Legendre, L., Legendre, P.: Écologie Numérique. Masson, Paris *and* Les Presses de l'Université du Québec, Québec (1979)
Legendre, L., Legendre, P.: Numerical Ecology. Elsevier, Amsterdam (1983)
Legendre, P., Legendre, L.: Numerical Ecology, 3rd English edn. Elsevier, Amsterdam (2012)
Legendre, P., Rogers, D.J.: Characters and clustering in taxonomy: a synthesis of two taximetric procedures. Taxon. **21**, 567–606 (1972)
Legendre, P., Dallot, S., Legendre, L.: Succession of species within a community: chronological clustering with applications to marine and freshwater zooplankton. Am. Nat. **125**, 257–288 (1985)
Legendre, P., Oden, N.L., Sokal, R.R., Vaudor, A., Kim, J.: Approximate analysis of variance of spatially autocorrelated regional data. J. Classif. **7**, 53–75 (1990)
Legendre, P., Galzin, R., Harmelin-Vivien, M.L.: Relating behavior to habitat: solutions to the fourth-corner problem. Ecology. **78**, 547–562 (1997)
Legendre, P., Dale, M.R.T., Fortin, M.-J., Gurevitch, J., Hohn, M., Myers, D.: The consequences of spatial structure for the design and analysis of ecological field surveys. Ecography. **25**, 601–615 (2002)
Legendre, P., Borcard, D., Peres-Neto, P.R.: Analyzing beta diversity: partitioning the spatial variation of community composition data. Ecol. Monogr. **75**, 435–450 (2005)
Legendre, P., De Cáceres, M., Borcard, D.: Community surveys through space and time: testing the space-time interaction in the absence of replication. Ecology. **91**, 262–272 (2010)
Legendre, P., Oksanen, J., ter Braak, C.J.F.: Testing the significance of canonical axes in redundancy analysis. Methods Ecol. Evol. **2**, 269–277 (2011)
Lennon, J.J., Koleff, P., Greenwood, J.J.D., Gaston, K.J.: The geographical structure of British bird distributions: diversity, spatial turnover and scale. J. Anim. Ecol. **70**, 966–979 (2001)
Loreau, M.: The challenges of biodiversity science. In: Kinne, O. (ed.) Excellence in Ecology, 17. International Ecology Institute, Oldendorf/Luhe (2010)
Margalef, R. and Gutiérrez, E.: How to introduce connectance in the frame of an expression for diversity. American Naturalist. **121**, 601–607 (1983)
Magurran, A.E.: Measuring Biological Diversity. Wiley-Blackwell, London (2004)
McArdle, B.H., Anderson, M.J.: Fitting multivariate models to community data: a comment on distance-based redundancy analysis. Ecology. **82**, 290–297 (2001)
McCune, B.: Influence of noisy environmental data on canonical correspondence analysis. Ecology. **78**, 2617–2623 (1997)
McCune, B., Grace, J.B.: Analysis of Ecological Communities. MjM Software Design, Gleneden Beach (2002)
McGarigal, K., Cushman, S., Stafford, S.: Multivariate Statistics for Wildlife and Ecology Research. Springer, New York (2000)
Méot, A., Legendre, P., Borcard, D.: Partialling out the spatial component of ecological variation: questions and propositions in the linear modelling framework. Environ. Ecol. Stat. **5**, 1–27 (1998)

Miller, J.K.: The sampling distribution and a test for the significance of the bimultivariate redundancy statistic: a Monte Carlo study. Multivar. Behav. Res. **10**, 233–244 (1975)
Milligan, G.W., Cooper, M.C.: An examination of procedures for determining the number of clusters in a data set. Psychometrika. **50**, 159–179 (1985)
Moran, P.A.P.: Notes on continuous stochastic phenomena. Biometrika. **37**, 17–23 (1950)
Murtagh, F., Legendre, P.: Ward's hierarchical agglomerative clustering method: which algorithms implement Ward's criterion? J. Classif. **31**, 274–295 (2014)
Oden, N.L., Sokal, R.R.: Directional autocorrelation: an extension of spatial correlograms to two dimensions. Syst. Zool. **35**, 608–617 (1986)
Oksanen, J., Blanchet, F. G., Friendly, M., Kindt, R., Legendre, P., McGlinn, D., Minchin, P. R., O'Hara, R. B., Simpson, G. L., Solymos, P., Stevens, M. H. H., Szoecs, E., Wagner, H. vegan: Community Ecology Package. R package version 2.5-0. (2017)
Olesen, J.M., Bascompte, J., Dupont, Y.L., Jordano, P.: The modularity of pollination networks. Proc. Natl. Acad. Sci. **104**, 19891–19896 (2007)
Orlóci, L., Kenkel, N.C.: Introduction to Data Analysis. International Co-operative Publishing House, Burtonsville (1985)
Paradis, E.: Analysis of Phylogenetics and Evolution with R Use R! Series, 2nd edn. Springer, New York (2012)
Pélissier, R., Couteron, P., Dray, S., Sabatier, D.: Consistency between ordination techniques and diversity measurements: two strategies for species occurrence data. Ecology. **84**, 242–251 (2003)
Peres-Neto, P.R., Legendre, P.: Estimating and controlling for spatial structure in the study of ecological communities. Glob. Ecol. Biogeogr. **19**, 174–184 (2010)
Peres-Neto, P.R., Legendre, P., Dray, S., Borcard, D.: Variation partitioning of species data matrices: estimation and comparison of fractions. Ecology. **87**, 2614–2625 (2006)
Peres-Neto, P.R., Dray, S., ter Braak, C.J.F.: Linking trait variation to the environment: critical issues with community-weighted mean correlation resolved by the fourth-corner approach. Ecography. **40**, 806–816 (2017)
Pielou, E.C.: The measurement of diversity in different types of biological collections. Theor. Biol. **13**, 131–144 (1966)
Pielou, E.C.: Ecological Diversity. Wiley, New York (1975)
Pillai, K.C.S., Hsu, Y.S.: Exact robustness studies of the test of independence based on four multivariate criteria and their distribution problems under violations. Ann. Inst. Stat. Math. **31**, 85–101 (1979)
Podani, J.: Extending Gower's general coefficient of similarity to ordinal characters. Taxon. **48**, 331–340 (1999)
Podani, J., Schmera, D.: A new conceptual and methodological framework for exploring and explaining pattern in presence-absence data. Oikos. **120**, 1625–1638 (2011)
Rao, C.R.: Diversity and dissimilarity coefficients: a unified approach. Theor. Popul. Biol. **21**, 24–43 (1982)
Raup, D.M., Crick, R.E.: Measurement of faunal similarity in paleontology. J. Paleontol. **53**, 1213–1227 (1979)
Rényi, A.: On measures of entropy and information. In J. Neyman (ed) Proceedings of the fourth Berkeley Symposium on mathematical statistics and probability. Univerity of California Press, Berkeley (1961)
Robert, P., Escoufier, Y.: A unifying tool for linear multivariate statistical methods: the RV-coefficient. Appl. Stat. **25**, 257–265 (1976)
Sanders, H.L.: Marine benthic diversity: a comparative study. American Naturalist. 102, 243–282 (1968)
Shannon, C.E.: A mathematical theory of communications. Bell Syst. Tech. J. **27**, 379–423 (1948)
Sharma, S., Legendre, P., De Cáceres, M., Boisclair, D.: The role of environmental and spatial processes in structuring native and non-native fish communities across thousands of lakes. Ecography. **34**, 762–771 (2011)
Sheldon. A. L.: Equitability indices: dependence on the species count. Ecology. **50**, 466–467 (1969)

Shimodaira, H.: An approximately unbiased test of phylogenetic tree selection. Syst. Biol. **51**, 492–508 (2002)
Shimodaira, H.: Approximately unbiased tests of regions using multistep- multiscale bootstrap resampling. Ann. Stat. **32**, 2616–2641 (2004)
Sidák, Z.: Rectangular confidence regions for the means of multivariate normal distributions. J. Am. Stat. Assoc. **62**, 626–633 (1967)
Simpson, E.H.: Measurement of diversity. Nature (Lond.) **163**, 688 (1949)
Sokal, R.R.: Spatial data analysis and historical processes. In: Diday, E., et al. (eds.) Data Analysis and Informatics IV, pp. 29–43. North-Holland, Amsterdam (1986)
Southwood, T.R.E.: Habitat, the templet for ecological strategies? J. Anim. Ecol. **46**, 337–365 (1977)
Suzuki, R., Shimodaira, H.: Pvclust: an R package for assessing the uncertainty in hierarchical clustering. Bioinformatics. **22**, 1540–1542 (2006)
ter Braak, C.J.F.: Canonical correspondence analysis: a new eigenvector technique for multivariate direct gradient analysis. Ecology. **67**, 1167–1179 (1986)
ter Braak, C.J.F.: The analysis of vegetation-environment relationships by canonical correspondence analysis. Vegetatio. **69**, 69–77 (1987)
ter Braak, C.J.F.: Partial canonical correspondence analysis. In: Bock, H.H. (ed.) Classification and Related Methods of Data Analysis, pp. 551–558. North-Holland, Amsterdam (1988)
ter Braak, C.J.F.: Fourth-corner correlation is a score test statistic in a log-linear trait–environment model that is useful in permutation testing. Environ. Ecol. Stat. **24**, 219–242 (2017)
ter Braak, C.J.F., Schaffers, A.P.: Co-correspondence analysis: a new ordination method to relate two community compositions. Ecology. **85**, 834–846 (2004)
ter Braak, C.J.F., Šmilauer, P.: CANOCO Reference Manual and CanoDraw for Windows user's Guide: Software for Canonical Community Ordination (ver. 4.5). Microcomputer Power, New York (2002)
ter Braak, C., Cormont, A., Dray, S.: Improved testing of species traits–environment relationships in the fourth corner problem. Ecology. **93**, 1525–1526 (2012)
van den Brink, P.J., ter Braak, C.J.F.: Multivariate analysis of stress in experimental ecosystems by Principal Response Curves and similarity analysis. Aquat. Ecol. **32**, 163–178 (1998)
van den Brink, P.J., ter Braak, C.J.F.: Principal response curves: analysis of time-dependent multivariate responses of biological community to stress. Environ. Toxicol. Chem. **18**, 138–148 (1999)
Verneaux, J.: Cours d'eau de Franche-Comté (Massif du Jura). Recherches écologiques sur le réseau hydrographique du Doubs. Essai de biotypologie. Thèse d'état, Besançon, France (1973)
Verneaux, J., Schmitt, A., Verneaux, V., Prouteau, C.: Benthic insects and fish of the Doubs River system: typological traits and the development of a species continuum in a theoretically extrapolated watercourse. Hydrobiologia. **490**, 63–74 (2003)
Villéger, S., Mason, N.W.H., Mouillot, D.: New multidimensional functional diversity indices for a multifaceted framework in functional ecology. Ecology. **89**, 2290–2301 (2008)
Wackernagel, H.: Multivariate Geostatistics, 3rd edn. Springer, Berlin (2003)
Wagner, H.H.: Spatial covariance in plant communities: integrating ordination, variogram modeling, and variance testing. Ecology. **84**, 1045–1057 (2003)
Wagner, H.H.: Direct multi-scale ordination with canonical correspondence analysis. Ecology. **85**, 342–351 (2004)
Ward, J.H.: Hierarchical grouping to optimize an objective function. J. Am. Stat. Assoc. **58**, 236–244 (1963)
Whittaker, R.H.: Vegetation of the Siskiyou mountains, Oregon and California. Ecol. Monogr. **30**, 279–338 (1960)
Whittaker, R.H.: Evolution and measurement of species diversity. Taxon. **21**, 213–251 (1972)
Wiens, J.A.: Spatial scaling in ecology. Funct. Ecol. **3**, 385–397 (1989)
Wildi, O.: Data Analysis in Vegetation Ecology, 2nd edn. Wiley-Blackwell, Chichester (2013)
Williams, P.H.: Mapping variations in the strength and breadth of biogeographic transition zones using species turnover. Proc. R. Soc. B Biol. Sci. **263**, 579–588 (1996)

Williams, W.T., Lambert, J.M.: Multivariate methods in plant ecology. I. Association-analysis in plant communities. J. Ecol. **47**, 83–101 (1959)
Wright, S.P.: Adjusted P-values for simultaneous inference. Biometrics. **48**, 1005–1013 (1992)
Zuur, A.F., Ieno, E.N., Smith, G.M.: Analysing Ecological Data. Springer, New York (2007)

References – R packages (in alphabetical order)

The list below provides references pertaining to the packages used or cited in the book. The references are those provided by the CRAN web site or the author(s) of the packages in the documentation. Some refer directly to R, others are bibliographical references. For additional information, type https://cran.r-project.org/web/packages/

ade4 – Version used: 1.7–8

Chessel, D., Dufour, A.B., Thioulouse, J.: The ade4 package-I- one-table methods. R News. **4**, 5–10 (2004)
Dray, S., Dufour, A.B.: The ade4 package: implementing the duality diagram for ecologists. J. Stat. Softw. **22**, 1–20 (2007)
Dray, S., Dufour, A.B., Chessel, D.: The ade4 package-II: two-table and K-table methods. R News. **7**. 47–52 (2007). R package version 1.7-8 (2017)

adegraphics – Version used: 1.0-8

Dray, S., Siberchicot, A. and with contributions from Thioulouse, J. Based on earlier work by Julien-Laferrière, A.: adegraphics: An S4 Lattice-Based Package for the Representation of Multivariate Data. R package version 1.0-8 (2017)

adespatial – Version used: 0.0-9

Dray, S., Blanchet, F. G., Borcard, D., Clappe, S., Guénard, G., Jombart, T., Larocque, G., Legendre, P., Madi, N., Wagner, H. adespatial: Multivariate Multiscale Spatial Analysis. R package version 0.0-9 (2017)

agricolae – Version used: 1.2-4

Felipe de Mendiburu: agricolae: Statistical Procedures for Agricultural Research. R package version 1.2-4 (2016)

ape – Version used: 4.1

Paradis, E., Claude, J., Strimmer, K.: APE: analyses of phylogenetics and evolution in R language. Bioinformatics. **20**, 289–290 (2004). R package version 4.1 (2017)

base and stats – Version used: 3.4.1

R Development Core Team: R: A language and environment for statistical computing. R Foundation for Statistical Computing, Vienna, Austria. URL http://www.R-project.org (2017)

cluster – Version used: 2.0.6

Maechler, M., Rousseeuw, P., Struyf, A., Hubert, M., Hornik, K.: cluster: Cluster Analysis Basics and Extensions. R package version 2.0.6 (2017)

cocorresp – Version used: 0.3-99

Simpson, G.L.: cocorresp: Co-correspondence analysis ordination methods (2009). R package version 0.3-99 (2016)

colorspace – Version used: 1.3-2

Ihaka, R., Murrell, P., Hornik, K., Fisher, J.C., Zeileis, A.: colorspace: Color Space Manipulation. R package version 1.3-2 (2016)

dendextend – Version used: 1.5.2

Galili, T.: dendextend: an R package for visualizing, adjusting, and comparing trees of hierarchical clustering. Bioinformatics. https://doi.org/10.1093/bioinformatics/btv428 (2015). R package version 1.5.2 (2017)

Ellipse – Version used: 0.3-8

Murdoch, D., Chow, E.D.: ellipse: Functions for drawing ellipses and ellipse-like confidence regions. R package version 0.3-8 (2013)

FactoMineR – Version used: 1.36

Le, S., Josse, J., Husson, F.: FactoMineR: an R package for multivariate Analysis. J Stat Soft. **25**, 1–18 (2008).

FD – Version used: 1.0-12

Laliberté, E., Legendre, P., Shipley, B.: FD: measuring functional diversity from multiple traits, and other tools for functional ecology. R package version 1.0-12 (2014)

gclus – Version used: 1.3.1

Hurley, C.: gclus: Clustering Graphics. R package version 1.3.1 (2012)

ggplot2 – Version used: 2.2.1

Wickham, H.: ggplot2: Elegant Graphics for Data Analysis. Springer-Verlag New York (2009) R package version 2.2.1 (2016)

googleVis – Version used: 0-6-2

Gesmann, M., de Castillo, D.: Using the Google visualisation API with R. R J. **3**, 40–44 (2011). R package version 0-6-2 (2017)

igraph – Version used: 1.1.2

Csardi G., Nepusz T: The igraph software package for complex network research, InterJournal, Complex Systems 1695 (2006). R package version 1.1.2 (2017)

indicspecies – Version used: 1.7.6

De Cáceres, M., Legendre, P.: Associations between species and groups of sites: indices and statistical inference. Ecology. **90**, 3566–3574 (2009). R package version 1.7.6 (2016)

labdsv – Version used: 1.8-0

Roberts, D.W. : labdsv: Ordination and Multivariate Analysis for Ecology. R package version 1.8-0 (2016)

MASS – Version used: 7.3-47

Venables, W.N., Ripley, B.D.: Modern Applied Statistics with S, 4th edn. Springer, New York (2002). ISBN 0-387-95457-0. R package version 7.3-47 (2017)

missMDA – Version used: 1.11

Josse, J., Husson, F.: missMDA: A Package for Handling Missing Values in Multivariate Data Analysis. J. Stat. Softw. **70**, 1–31 (2016). R package version 1.11 (2017)

mvpart – Version used: 1.6-2

rpart by Therneau, T. M. and Atkinson, B. R port of rpart by Ripley, B. Some routines from vegan -- Oksanen, J. Extensions and adaptations of rpart to mvpart by De'ath, G.: mvpart: Multivariate partitioning. R package version 1.6-2 (2014)

MVPARTwrap – Version used: 0.1-9.2

Ouellette, M. H. and with contributions from Legendre, P.: MVPARTwrap: Additional features for package mvpart. R package version 0.1-9.2 (2013)

picante – Version used: 1.6-2

Kembel, S.W., Cowan, P.D., Helmus, M.R., Cornwell, W.K., Morlon, H., Ackerly, D.D., Blomberg, S.P., Webb, C.O.: Picante: R tools for integrating phylogenies and ecology. Bioinformatics. **26**, 1463–1464 (2010). R package version 1.6-2 (2014)

pvclust – Version used: 2.0-0

Suzuki, R., Shimodaira, H.: pvclust: Hierarchical Clustering with P-Values via Multiscale Bootstrap Resampling. R package version 2.0-0 (2015)

RColorBrewer – Version used: 1.1-2

Neuwirth, E.: RColorBrewer: ColorBrewer Palettes. R package version 1.1-2 (2014)

rgexf – Version used: 0.15.3

Vega Yon, G., Fábrega Lacoa, J., Kunst, J.B. : rgexf: Build, Import and Export GEXF Graph Files. R package version 0.15.3 (2015)

RgoogleMaps – Version used: 1.4.1

Loecher, M., Ropkins, K.: RgoogleMaps and loa: Unleashing R Graphics Power on Map Tiles. J. Stat. Softw. **63**, 1–18 (2015). R package version 1.4.1 (2016)

rioja – Version used: 0.9-15

Juggins, S.: rioja: Analysis of Quaternary Science Data. R package version 0.9-15 (2017)

rrcov – Version used: 1.4-3

Todorov, V., Filzmoser, P.: An object-oriented framework for robust multivariate analysis. J. Stat. Softw. **32**, 1–47 (2009). R package version 1.4-3 (2016)

SoDA – Version installed: 1.0-6

John M. Chambers: SoDA: Functions and Examples for "Software for Data Analysis". R package version 1.0-6 (2013)

spdep – Version used: 0.6-15

Bivand, R., Piras, G.: Comparing implementations of estimation methods for spatial econometrics. J. Stat. Softw. **63**, 1–36 (2015)

Bivand, R. with many contributors: spdep: Spatial dependence: weighting schemes, statistics and models. R package version 0.6-15 (2017)

taxize – Version used: 0.8.9

Chamberlain, S., Szocs, E., Boettiger, C., Ram, K., Bartomeus, I., Baumgartner, J., Foster, Z., O'Donnell, J: taxize: Taxonomic information from around the web. R package version 0.8.9 (2017)

vegan – Version used: 2.4-4

Oksanen, J., Blanchet, F. G., Friendly, M., Kindt, R., Legendre, P., McGlinn, D., Minchin, P. R., O'Hara, R. B., Simpson, G. L., Solymos, P., Stevens, M. H. H., Szoecs, E., Wagner, H.: vegan: Community Ecology Package. R package version 2.4-4 (2017)

vegan3d – Version used: 1.1-0

Oksanen, J., Kindt, R., Simpson, G. L.: vegan3d: Static and Dynamic 3D Plots for the 'vegan' Package. R package version 1.1-0 (2017)

vegclust – Version used: 1.7.1

De Cáceres, M., Font, X. and Oliva, F.: The management of vegetation classifications with fuzzy clustering. J. Veg. Sci. 21, 1138–1151 (2010). R package version 1.7.1 (2017)

vegetarian – Version used: 1.2

Charney, N., Record, S.: vegetarian: Jost Diversity Measures for Community Data. R package version 1.2 (2012)

Index

D. Borcard et al., *Numerical Ecology with R*, Use R!,
https://doi.org/10.1007/978-3-319-71404-2

D

V

W

Printed in the United States
By Bookmasters